Soil Quality, Sustainable Agriculture and Environmental Security in Central and Eastern Europe

NATO Science Series

A Series presenting the results of activities sponsored by the NATO Science Committee. The Series is published by IOS Press and Kluwer Academic Publishers, in conjunction with the NATO Scientific Affairs Division.

A. **Life Sciences**	IOS Press
B. **Physics**	Kluwer Academic Publishers
C. **Mathematical and Physical Sciences**	Kluwer Academic Publishers
D. **Behavioural and Social Sciences**	Kluwer Academic Publishers
E. **Applied Sciences**	Kluwer Academic Publishers
F. **Computer and Systems Sciences**	IOS Press

1. **Disarmament Technologies**	Kluwer Academic Publishers
2. **Environmental Security**	Kluwer Academic Publishers
3. **High Technology**	Kluwer Academic Publishers
4. **Science and Technology Policy**	IOS Press
5. **Computer Networking**	IOS Press

NATO-PCO-DATA BASE

The NATO Science Series continues the series of books published formerly in the NATO ASI Series. An electronic index to the NATO ASI Series provides full bibliographical references (with keywords and/or abstracts) to more than 50000 contributions from international scientists published in all sections of the NATO ASI Series.
Access to the NATO-PCO-DATA BASE is possible via CD-ROM "NATO-PCO-DATA BASE" with user-friendly retrieval software in English, French and German (WTV GmbH and DATAWARE Technologies Inc. 1989).

The CD-ROM of the NATO ASI Series can be ordered from: PCO, Overijse, Belgium

Series 2. Environment Security – Vol. 69

Soil Quality, Sustainable Agriculture and Environmental Security in Central and Eastern Europe

edited by

M. J. Wilson

Macaulay Land Use Research Institute,
Aberdeen, U.K.

and

B. Maliszewska-Kordybach

Institute of Soil Science and Plant Cultivation,
Pulawy, Poland

Kluwer Academic Publishers

Dordrecht / Boston / London

Published in cooperation with NATO Scientific Affairs Division

Proceedings of the NATO Advanced Research Workshop on
Soil Quality in Relation tot Sustainable Development of Agriculture and Environmental Security in Central and Eastern Europe
Pulawy, Poland
13–17 October 1997

A C.I.P. Catalogue record for this book is available from the Library of Congress.

ISBN 0-7923-6377-9

Published by Kluwer Academic Publishers,
P.O. Box 17, 3300 AA Dordrecht, The Netherlands.

Sold and distributed in North, Central and South America
by Kluwer Academic Publishers,
101 Philip Drive, Norwell, MA 02061, U.S.A.

In all other countries, sold and distributed
by Kluwer Academic Publishers,
P.O. Box 322, 3300 AH Dordrecht, The Netherlands.

Printed on acid-free paper

Printed in the Netherlands.

CONTENTS

PREFACE

To varying extents the countries of Central and Eastern Europe (CEE) are all in a state of transition towards a market-oriented economy whereby trade policy is being liberalized and the private sector encouraged. Agriculture is a crucial part of the economies of many of these countries and all face a variety of environmental and socio-economic problems connected with the transformations that are currently affecting the agricultural sector. The concept of sustainable development provides a convenient over arching framework for the formulation of government environmental policy for the CEE countries, particularly for those that aspire to join the EU and which would be required to meet EU environmental standards. Sustainable development in the context of agriculture inevitably involves consideration of appropriate strategies for balancing crop and animal production with protection of the quality of national soil and water resources, as well as the consideration of the socio-economic consequences of adopting such strategies. As part of this process there is an urgent need to compile, exchange and evaluate current information concerning the quality of soils in CEE countries as well as to assess the potential impact of new management practices on the soil and on the wider environment. This will help in the development of agricultural strategies that may be accepted at a regional level, particularly if the socio-economic element can also be taken into account.

It was in this context that a NATO Advanced Research Workshop was convened to meet at the Institute of Soil Science and Plant Cultivation in Pulawy, Poland towards the end of 1997. The overall objective of the Workshop was to evaluate the current and future impact of pollution and changing management practices upon the quality and sustainable use of soils for productive agriculture in CEE countries. In addition the Workshop attempted to review existing knowledge and data and to look at strategies for implementing western agro-environmental standards in CEE countries and conditions. In particular, it was decided to focus upon:

(a) Evaluation of soil quality in CEE countries, particularly as this quality is affected by contamination with various pollutants, acidification, erosion etc.
(b) Assessment of the effects of current and future management practices, such as increased fertilisation, minimum tillage, greater use of pesticides, application of sewage sludges,etc, upon soil quality and the wider environment.
(c) Utilisation of relevant experimental data from western countries so as to clarify the interaction between agricultural production and consequences for soil quality with changing management and
(d) Evaluation of socio-economic problems related to new market-orientated agricultural strategies in the CEEs.

The Workshop was based upon the contributions of well-known scientists drawn from both CEE and western countries and the programme was built around four specialised sessions dealing with the aspects listed above. The Workshop concluded with a plenary discussion session aimed at summarising the most important points arising from the specialised sessions, reaching conclusions with respect to the way in which future agricultural development is likely to impact upon soil quality in the CEEs, and making recommendations for future work aimed at helping relevant policy and decision makers in the CEEs. Although the Workshop was held just over two years ago it has not been possible to publish the proceedings before now, principally because of a serious illness to one of the Editors. Nevertheless, despite this time gap it is thought that much of information reported here is still both topical and timely and will be of general interest to a wide audience both in the West and in the CEE countries.

Finally, the editors wish to thank the NATO Science Committee for supporting the Workshop so generously and for the help and guidance received both before and after the event.

M. J. Wilson,
Macaulay Land Use Research Institute,
Aberdeen, UK.
November, 1999

B. Maliszewska-Kordybach,
IUNG,
Pulawy, Poland

Soil Quality, Sustainable Development and Environmental Security.
Introduction to General Concepts

M. J. Wilson
Macaulay Land Use Research Institute,
Aberdeen, UK

1.General Background

Recent definitions of "soil quality" emphasise that the term encompasses the capacity of the soil to promote crop growth in a safe and healthy manner, but at the same time in a way that is not detrimental to the resource base or to the wider environment. In other words, soil quality is now thought of in a way that stresses functionality in a broad context, involving soil not only as a medium for crop production but also as an important reservoir for water storage, as a buffer for the filtration, transformation and neutralisation of pollutants and as a habitat for plants and animals so protecting genetic biodiversity. It is evident, therefore, that the preservation of soil quality links directly to general issues of environmental security.

The concept of "sustainablity" or "sustainable development" is now well-established and the Brundtland definition is widely accepted, namely "development that meets the needs of the present without compromising the ability of future generations to meet their own needs". Thinking in terms of sustainable agricultural development in particular, it is evident, in any given country, that the satisfaction of the primary need for food and fibre production requires to be balanced by several other considerations. These would include the protection and, if possible, the enhancement of overall environmental quality and of the natural resource base, most immediately the soil itself. In addition, sustainable agricultural development should attempt to use non-renewable resources efficiently, as well as natural biological cycles and controls where these are effective. Finally, it is vital that sustainable agriculture should be economically viable and that it should enhance the quality of life both for the farmer and for society as a whole. These concepts were embodied in a bill presented to the US Congress in 1990.

From the above discussion it is evident that the preservation of soil quality is a pivotal link in the development of sustainable agriculture. The execution of whatever management strategy is chosen to promote sustainablity in agriculture will inevitably have to take protection of soil quality into account. But in order to do this effectively it is necessary to understand the principal factors that control soil quality, as well as how these factors inter-relate one with another. In general, soil quality and associated agricultural development will relate strongly to prevailing climatic conditions and to the nature of the land surface on which the soil occurs. There will also be a relationship with how Man has used these soils in the past, particularly in terms of management practices. These factors will all combine to impart to the soil a variety of attributes of a physical, chemical and biological nature which collectively may be thought of as defining, or as a means of measuring, the quality of the soil, irrespective of the soil function being considered.

Although it would be unrealistic to suppose that any single soil attribute could be used as a universal index of quality, nevertheless, in the context of a defined threat to soil quality in relation to a given function, there will be some measureable attributes that are more useful than others, and these attributes are now briefly reviewed.

2. Soil Quality; Attributes, Threats and Strategies for Maintenance

Although the various attributes that bear directly on soil quality will be described under the conventional headings of physical, chemical and biological, it is important to realise that there are strong linkages and interactions between each of these spheres that should never be discounted.

Among the more important physical attributes that that may be considered as useful for assessing soil quality may be included:

(a) Soil Depth
(b) Water Holding Capacity
(c)Aggregate Stability
(d) Bulk Density
(e) Infiltration Rate
(f) Hydraulic Conductivity

These attributes relate mainly to threats to soil quality from erosion due to water and wind and to compaction. Degradation due to crusting and water-logging should also be borne in mind. Clearly, these attributes may link to each other, for example bulk density, infiltration rate and hydraulic conductivity, as well as to other attributes in different groups. Thus, aggregate stability and water holding capacity often relate to organic matter content. So far as useful soil chemical attributes are concerned, the following are widely accepted as being important to measure:

(a) pH
(b) Base Saturation
(c) CEC
(d) Nutrient Availability (Major and Minor)
(e) Electrical Conductivity
(f) Potentially Toxic Elements or Compounds

These attributes may be associated with several readily identified threats to soil quality, some of which are inter-related in the same way as the above properties.

Thus, measurements of soil pH, base saturation, CEC and nutrient availability must all be measured in attempting to assess soil degradation by increased acidification and nutrient depletion. There is clearly a direct relationship between the measurement of potentially toxic elements and compounds and contamination of the soil by radionuclides, heavy metals and organic pollutants. Measurement of electrical conductivity is related to assessment of problems of soil salinisation and possibly to over-fertilisation, although it might be anticipated that the latter is not of any great significance in Central and Eastern European countries (CEECs).

Finally, a number of soil biological attributes have been suggested as being of direct use as soil quality indicators although they are not all of equal weight so far as their utility is concerned. These attributes are as follows:

(a).Soil organic matter content
(b). Soil respiration
(c) Microbial biomass carbon
(d). Nitrification
(e) Enzyme assays
(f). Soil micro-organisms (population/community)
(g). Soil invertebrates

Probably, measurement of soil organic matter content is most important as it relates strongly to loss of soil quality as assessed by measurement of physical and chemical properties, in addition to threats of a more direct biological nature, such as reduced soil microbial activity and reduced biological diversity. For this reason soil organic matter content is often regarded as a key indicator of soil quality. The other biological attributes listed above are probably not quite so useful as general indicators of soil quality at present, although in the light of ongoing research they may be in the future. Thus, soil respiration measurements are subject to a high degree of spatial and temporal variability, particularly when measured in the field, as are those attributes relating to carbon and nitrogen cycling. Again, quantification of microbial populations and communities may be difficult and there may be problems with methodology in the assay of soil enzymes.

Most soil and crop management strategies aimed at combating the threats to soil quality outlined above are well-known, although practices will vary according to area, soil type, climate and socio-economic constraints. Clearly, the application of chemical fertilisers will directly address problems of nutrient depletion of the soil and, similarly, the use of lime will keep the problem of soil acidification in check. Again, the use of organic wastes will help to build up soil organic matter and together with the use of conservation tillage systems of various kinds will reduce the likelihood of soil erosion through the improvement of soil physical properties like aggregation, porosity, bulk density and permeability. The growth of cover crops will also help to build up soil organic matter in addition to reducing leaching of nitrogen and potassium from the soil. However, although such strategies and practices are quite obviously helpful in maintaining soil quality, ultimately socio - economic pressures and constraints will undoubtedly be decisive in their employment by farmers in CEECs. Throughout this area, there has been a sharp fall in agricultural output since 1989, reflecting the depressed state of agriculture and the generally lower levels of inputs since that time. This will clearly have had an impact upon natural capital stock, including various aspects of soil quality.

3. Specific issues to be addressed

The broad issues addressed at this Workshop concern definitions and concepts of soil quality, sustainable agriculture and environmental security, characterisation of the present quality of agricultural soils in CEE countries, identification of the management practices that would be most useful in maintaining and enhancing soil quality in CEECs and prediction of the likely impact of future policies and socio-economic changes in these countries on soil quality.

Definitions and concepts are extensively dealt with in the first two papers of this volume but perhaps it would be useful to briefly preface the papers dealing with the present situation concerning soil quality in CEE. Concern about deteriorating soil quality is widespread and is considered by many in the environmental science community to be global in scale. For the CEE countries, it is still not clear, at least in the West, as to what is the true situation with regard to changes in soil quality. Certainly, there is a perception that the soils of CEE are seriously contaminated by heavy metals and other pollutants but it remains to be shown that this is actually true. For example, many papers have been written showing that certain Polish soils contain very high amounts of heavy metals but it is unclear whether this is simply a local problem that is confined to certain metal smelting areas rather than a national one. Indeed, it was not clear at the commencement of this Workshop as to what are the major threats to the soil quality of the CEECs and this was amongst the specific questions identified that were addressed. It was thought to be very important that any conclusions should be based upon an objective approach involving analysis of real samples rather than on extrapolation involving few samples and mathematical models. The latter approach, which has been widely used in the work on critical loads of acidity to soils, may have led to an unnecessarily bleak view of the damage that has already been done, and that will occur in the future in the light of current plans for the abatement of acid emissions, to terrestrial ecosystems in Europe. The major questions concerning soil quality in the CEECs may be summarised as follows:

What is the current status of the soils of the CEECs with respect to quality?

What are the major threats to soil quality in the different countries of CEE?

What is the most effective means of monitoring soil quality in the CEECs?

The way in which soils and crops are managed is evidently of vital importance to the maintenance of soil quality in the CEECs and to the ability to minimise threats in this area. In particular, information is needed on the effectiveness of the management practices that are currently used in CEECs, as well as those that are currently used or being developed in western countries and which could be transferred to and utilised in the CEECs. Consideration of this latter aspect, however, inevitably involves socio-economic and policy issues and raises as well the question as to how the changes in the economic systems that have occurred in the CEECs during the last 10 years have impacted upon soil management practices and on soil quality. The specific questions to be addressed in this area , therefore, include:

What are the most effective soil/crop management practices for sustaining or enhancing soil quality in the CEECs?

Will socio-economic constraints in the CEECs allow such management practices to be used?

How have socio-economic/policy changes in rural areas of the CEECs since 1989 impacted upon soil management practices and consequently on soil quality in general?

This brings to the fore consideration of the likely impact of future agricultural policies and practices in the CEECs, especially those related to sustainable agriculture and the accession to wider markets. It seems inevitable that many dilemmas will emerge within different countries and regions, stemming from the fact that priorities for the development of the agricultural sector need somehow to be balanced with the wider perspective brought about by globalisation trends in agriculture and the increasing emphasis accorded to overall environmental protection. Future socio-economic and policy changes in the CEECs could clearly bring about an adverse impact on national soil quality and advice needs to be available to policy makers as to how such an impact can be minimised. In this connection there is an obvious need to identify priorities for research into this general area. Specific questions to be addressed by the Workshop therefore include:

What socio-economic/policy changes are likely to occur in rural areas in the CEECs in the immediate future and what will be the likely impact upon soil quality and management practices?

What information should be given to policy makers that could help to protect soil quality in the CEECs given that drastic socio-economic changes will occur in the rural areas of the CEECs?

What are the priorities for research into the inter-related aspects of soil quality, sustainable development and environmental security in the CEECs?

PART I

OVERALL CONCEPTS OF SOIL QUALITY, SUSTAINABLE DEVELOPMENT AND ENVIRONMENTAL SECURITY

Synopsis and Overview

Although the three concepts of soil quality, sustainable agricultural development and environmental security are widely understood in a general sense, further consideration reveals the need for more precise definition and deeper understanding. *Davis and Miller* seek to clarify these issues in the opening paper of this volume. While their analysis proceeds mainly from American experience, there seems little doubt of its applicability to other countries in general and to the countries of central and eastern Europe in particular. Starting from the notion of national security, *Davis and Miller* point out that up until recently this has been perceived almost entirely in a military/political context, with threats to security being seen as ultimately external in origin and essentially conflictual in nature. However, there are sound reasons for adopting a much broader perspective and for incorporating into the notion of national security, internally derived threats such as resource depletion and environmental degradation. Closer examination of the national security concept, even from the traditional military/political point of view, reveal that it incorporates three essential elements, namely predictability, flexibility and control. Whether at a national or an individual level, security inevitably involves a feeling that the world is predictable, that there is the flexibility to make choices and that it is possible to maintain reasonable control. But these same elements also underlie what we would regard as healthy ecosystems. Environmental degradation often proceeds, initially at least, in unpredictable and imperceptible ways, leading to decreasing flexibility of response and ultimately to a loss of control. *Davis and Miller* point out that healthy ecosystems are characterised by the predictablity of the various controlling factors, of which the soil is one, by the ability of systems to respond flexibly if one of the ecosystem's components is damaged or destroyed, and by the possession of a high degree of internal integrity and control, making them less vulnerable to external forces and damage. Healthy agroecosystems may have similar characteristics, for example, the soil system should be predictable and flexible enough to support a variety of cropping systems, although human activities, in one form or another, are evidently essential in maintaining the internal integrity of these systems, emphasising their susceptibility to damage. On the other hand, human activities not necessarily connected with agriculture have the potential to inflict deleterious effects upon all types of ecosystems and particularly on the integrity of the soil ecosystem In other words, the quality of the soil is adversely affected. The extent to which this has happened in central and eastern Europe remains one of the key questions to be addressed at this meeting. It is concluded that soil degradation tends to reduce national security by rendering farming operations more unpredictable, increasing the need for costly control measures, such as fertilisation and irrigation, and reducing flexibility in the choice of crops.

M.J. Wilson and B. Maliszewska-Kordybach (eds.),
Soil Quality, Sustainable Agriculture and Environmental Security in Central and Eastern Europe, 1–2.

Várallyay shows that the soil resource and the maintenance of its quality are important elements of sustainable agricultural development of a national economy and emphasises that the concept of soil quality can only be understood in the context of multifunctionality. The main characteristics and functions of soil in the biosphere are that it is a renewable resource on condition that it is subjected to prudent maintenance and conservation; that it reacts, transforms and integrates the combined inputs of other natural resources from the atmosphere, hydrosphere, lithosphere and biosphere, thus forming a "life medium" essential to the development of ecosystems; that it is the most important medium for biomass production and is the primary nutrient source of the biosphere; that it is a major store of heat and water as well as plant nutrients and other elements; that it acts as a natural filter and detoxification system so preventing or inhibiting the migration of various pollutants to deeper horizons and ground and surface waters; that it moderates and buffers to a high degree the various stresses caused by environmental factors and human activities; and that it acts as a gene reservoir in the biosphere and thus constitutes an important element of biodiversity. The soil resource is, therefore, of particular significance both in terms of sustainable agricultural development and environmental protection. *Várallyay* reiterates that the modern concept of soil quality must reflect its diversity of functions, as well as how these functions may vary in space and time. In the utilisation of the soil resource, failure to appreciate the true nature of soil quality may result in misguided management and over-exploitation, leading to soil degradation and eventual serious environmental deterioration. The ability of soil to perform its functions is related to the impact of various soil properties, which are themselves the result of soil processes. *Várallyay* advocates that maintenance of soil quality should be based upon control and evaluation of soil processes, followed by risk assessments and analyses of sensitivities and impacts. Using such a comprehensive approach, soil quality can be assessed from both the agricultural and the environmental points of view, thus constituting an important element of an economically viable, environmentally sound and socially acceptable agricultural development.

SOIL, SUSTAINABILITY AND SECURITY

The Importance of Ecosystem Integrity

C. B. DAVIS
F. P. MILLER
The Ohio State University
School of Natural Resources
2021 Coffey Road, 210 Kottman Hall
Columbus, OH 43210 USA

I. Introduction

These are disquieting times, as the nations of the world attempt to adjust to the new political, economic and cultural realities of a post-Soviet Union world. This is especially true in Russia, other former Soviet Republics, and Central and Eastern European Countries (CEECs) where change since 1989 has been truly revolutionary. The people of this region have seen the demise of Communism as the definer of social, economic, and political order and its abrupt replacement by democracy movements, capitalism and market economies. Gone are the centralized, top--down bureaucracies that restricted participation in decision making and constrained dissent. Gone are repressive controls on speech, movement and assembly. Gone is the "Cold War" with all of its uncertainties and insecurities. Former enemies are now friends. Former Soviet Republics are now independent, free, and equal, and the countries of the former Warsaw Pact are setting their own courses in international affairs.

At a different scale, citizens in these emerging market democracies are enjoying their newfound individual rights and freedoms, the right to real choice among political parties and candidates for office, the right to own property and to make a profit, the right to accumulate capital and possibly get rich, and the freedom to choose what is best for one's self and one's family. Choice also extends to the consumer arena where the variety and quality of consumer goods has expanded considerably. Industrialists can choose what they will manufacture and farmers can choose what they will produce. People can choose what they will purchase and who they will select to lead them. Does this not seem like the best of worlds?

M.J. Wilson and B. Maliszewska-Kordybach (eds.),
Soil Quality, Sustainable Agriculture and Environmental Security in Central and Eastern Europe, 3–16.

For the newborn capitalists of the CEECs and the former Soviet Republics with their newfound freedoms and access to resources and financing, this may very well be the best of worlds. But for many, if not most, citizens of these new democracies, the times are not so good. Gone are the old regimes that restricted freedoms but provided security, the security of jobs, pensions, cradle to grave health care and subsidies for agriculture and industry. Gone is the certainty of being able to afford to feed and house yourself and your family even if only modestly. For these citizens, the new world is in many ways less predictable with fewer real choices than the old one had been. Freedom is a primary human goal, but freedom without the resources required to exercise security it is useless, and freedom without security is meaningless.

It is within this context of time and place that we came to Poland to examine one small but important ingredient of this process of social change -- the quality of soil resources in the region - and to consider how those resources ultimately affect security in the region.

2. Security -- Predictability, Flexibility and Control

The concept of security can be examined at several scales. At the lowest level is personal security, and psychologists and marketers make good livings dealing with our individual insecurities. At a higher level, we devise mechanisms to provide for social security, and at the level of the state we have notions of national security, notions that may be vaguely defined but are non-the-less effective in directing national and international policies and actions. Alliance formation among nation states may be pursued in the interest of collective security. And, more recently, we have begun to grapple with the concepts of global or comprehensive security and the relatively new idea of environmental security.

The English word security is derived from the Latin securus meaning "free from care." When we are secure, we are free from care. We can examine the foundations of security by looking at some notions of "national" security. As we will see, the foundations that we identify for national security are also the foundations for all other manifestations of security; security is the same regardless of the social scale examined. The definitions of national security offered here are derived from the American experience. Definitions from other cultures or countries will differ in wording and emphasis, but we contend that the foundations of the concept are universal across scales and across cultures.

Berkowitz and Bock (1968) define national security as "the ability of a nation to protect its internal values from external threat." A related definition holds that national security is "the ability to protect national interests and boundaries while preserving existing political ideologies and governments. Former U.S. Secretary of Defense Harold Brown (3) suggested that a nation is secure if it can determine its own future without being influenced by what happens outside its own borders. In 1989, participants in a conference on environmental security suggested that all security considerations center on the notions of predictability and control (16). To these, Davis (7) added the notion of flexibility. To be secure, a nation must have predictability, flexibility and control. The same can be said for security at other levels.

For a CEEC farmer to be secure, he must believe that his world and life are at least minimally predictable. He must have predictable access to land, a predictable climate, predictable soils, reasonably predictable markets, etc. Without such predictability, he would have trouble deciding where he will plant; when he will plant and harvest; how much seed, fertilizer, pesticide, etc. he will have to buy and apply and how much product he might expect to sell and at what price. Predictability provides him with a modicum of security.

Flexibility means having choices and the ability to act upon those choices. Our CEEC farmer requires flexibility so that he can optimize for predicted conditions, select the most promising crops, pursue the most promising markets and prices and so on. He must also have flexibility in the face of adversity. He must know that if things don't go as predicted, he has options. With such flexibility, his security is further enhanced. He will survive even if things go wrong.

Finally there is control, not control in the sense of power over others, but control in the sense of being in command of one's situation. It is a control built of skill, experience and wisdom, a control based on having choices and the ability to decide and act. It is the state of believing that one's abilities and actions count, that one is not at the mercy of outside forces. It is the belief that we can control our own lives and fortunes.

When one lives in a predictable world, has options, and control, one is secure, whether the one in question is our CEEC farmer or his country. For many in the CEEC today, farmers and others, security is fragile at best. Guarantees of employment, health care, social security, subsidies for agriculture, etc. have disappeared or have been significantly reduced since 1989. The predictability of the old system is gone, and with only limited options for employment, many turn to the land for subsistence or to even less desirable pursuits. With declining predictability and decreasing options, control is also lost. Anything that threatens predictability, flexibility and control threatens security.

3. Environmental Degradation and Security

Early notions of security focused on political and economic security. During the last decade, however, we have become increasingly aware that degradation of ecological systems and resource bases pose a growing threat to security at all levels (7). Pollution of

---

[1]The masculine gender is used throughout this chapter for stylistic reasons. It is recognized that farmers include both men and women.

our rivers and drinking water supplies and conflict over who has access to the waters in our major rivers reduce security throughout the world. Pollution of the atmosphere and concerns about ozone holes and global warming have been the focus of international meetings and of negotiations, often heated, among nations. In an important 1994 paper, ecologist Peter Vitousek (21) brought our attention to another form of environmental degradation -- Landscape Alteration -- noting that approximately half of the ice-free terrestrial surface of the earth has been transformed by human activity. These changes have resulted in impacts such as increases in albedo, increases in atmospheric methane and N_2O, habitat loss and fragmentation, changes in biodiversity, watershed modifications, soil degradation through erosion, salinization, desertification and contamination, and the increased use of fragile lands for production. Indeed, he suggests that land degradation is the most critical environmental problem facing us today.

Food and fiber are fundamental human needs, making agriculture a fundamental necessity in any economy whether the goal is subsistence, national self-sufficiency or competitiveness in regional and global markets. The soil, as Várallyay has pointed out in this volume, is 'the most important medium for biomass production.' Soil degradation leads to reduced production. Reduced production erodes security at all levels, from the poor farmer trying to eke out a subsistence living from the land to the nation trying to compete in international markets such as the European Union.

Further, insecurity can lead to degradation of the soil. Dalton (this volume) suggests that current economic conditions in the CEEC may lead to increased soil degradation at the hands of the farmers attempting to survive without access to off-farm employment or the financial or material resources to care for the land. He notes that in such cases, production may occur at the cost of 'natural capital stocks' such as soil fertility. Thus, land degradation and security interact in complex ways that can lead to positive feedback that accelerates environmental decline.

4. Soil Quality

The central objective of this workshop was to examine soil quality in the CEEC and its impact on sustainable development and security in the region. Discussion of such a topic must recognize that concepts of quality are inherently normative in that they deal with notions of good and bad, desirable and undesirable, acceptable and unacceptable. Quality can only be measured against some standard, and that standard is established by humans, based on their needs and wants, on their utility. As Kay noted, "The physical and biological sciences can describe and even predict changes in the biosphere, but they alone cannot determine which change is better." (13). Science may be able to tell us how to achieve and maintain soil quality or how and why quality declines, but it cannot tell us what quality is.

There are various definitions of soil quality, but all have three ideas in common. First, quality is a measure of a soil's capacity to produce crops, assimilate pollutants, retard flooding or provide some other product or service to humans. Thus, notions of soil quality are both normative and utilitarian. Second, it is expected, as we have already noted, that quality soils will provide such products and services 'on an ongoing basis,' e.g. that quality soils will remain quality soils unless they are disturbed or abused. Third, soil quality is derived from certain soil attributes or characteristics such as texture, tilth, aggregation, organic matter content, nutrient status, water holding capacity and depth. In this view, structure or soil components define a soil and determines its quality.

It might be argued that the last two ideas are based on a fundamental misunderstanding of how soil and other ecological systems function, arising, perhaps, from two misconceptions. The first is what might be called the assumption of constancy. We have a tendency to believe that what we would call a quality ecosystem - - be it a mature forest, a high quality wetland, or a high quality soil system -- is a permanent feature of the landscape, that it is there because it is adapted to local environmental conditions, and that it will remain there unless it is disturbed by some severe natural event or by human actions. This is what ecologists call the equilibrium view of ecosystems. More will be said about this in the section on Ecosystem Dynamics.

The second misconception is based on the assumption that things (e.g. morphological-chemical-biological attributes) are more important than processes. We can describe soils at any place and time in terms of their structure, their parts and the states of those parts. But this is a static view. An alternative view is that soil is more than a set of attributes, more than the sum of its components. It is a living, functioning system, an ecological system. Systems are made up of components that interact with one another. It is these interactions, these processes that determine what an ecosystem will be and how it will respond to environmental changes, exploitation and stress. It is these processes that determine its renewability and sustainability. It is these processes that determine its quality -- how well it serves our needs now and in the future. Therefore, if our goal is to maintain or enhance soil quality, we must understand how ecosystems function.

5. Ecosystem Dynamics

In a healthy ecosystem such as a soil ecosystem of high quality, internal processes will be more important than external events in determining ecosystem behavior. External events or forces are certainly important, but they are mediated in such ecosystems by internal processes (10). We might say that such "healthy" ecosystems have a high degree of internal control. When perturbed, they are able to cope, weathering the perturbation if it is of limited duration or intensity or adjusting to new external conditions if the perturbation is sufficiently severe or prolonged (18). In general, as ecosystems develop, internal control

---

[2]The term healthy is used here in the general, intuitive sense of Ehrenfeld (1992).

increases. Resources are garnered and energy is dissipated as the ecosystem builds structure and order; nutrient mobilization and retention increase; species diversity increases, at least for a while, as unused resources are exploited and, finally, resources become increasingly allocated for system maintenance as the ecosystem approaches a stable state.

One view of this process, a view that has dominated ecological, conservation and management thinking for decades, is that there is only one optimal stable state for a given ecosystem, and that once an ecosystem attains that stable state, it will tend to remain their in steady-state or dynamic equilibrium. Because of this view of ecosystem dynamics, management and conservation strategies have focused largely on maintaining ecosystems in their supposed ideal or normal equilibrium states or on facilitating their recovery to equilibrium whenever they diverge. Quality, in this context, is defined in terms of an ecosystem's ability to remain at equilibrium in the face of disturbance or to return quickly to that equilibrium when driven from it for any reason. The focus is on constancy and change is considered to be an aberration.

To many ecologists today, this equilibrium view of ecosystem dynamics is oversimplified and misleading (see for instance:2, 12 and 6). Alternatively, they suggest that change is a normal component of ecosystem development (2) and should not be viewed as aberrant or undesirable. They offer a non-equilibrium view of ecosystem dynamics that considers ecosystem development to be open ended. That is, ecosystems can exist in any number of stable states (10), and internal control enables functional ecosystems to "self-organize" in response to changes in external conditions. Here the focus is not on constancy but on variability. Ecosystems characterized by high levels of variability have flexibility; they can respond in a variety of ways to changes in their environment. Some components or attributes may be lost and some new ones gained in this process, but the ecosystem will self-organize to form a new stable state that will persist for some period of time until events or conditions once again break it down.

Holling (10) offers a model of ecosystem dynamics that follows this non-equilibrium view. In his model, ecosystems cycle through a series of stages that he calls "exploitation, conservation, creative destruction, and renewal." The exploitation stage is characterized by opportunistic exploitation of space and resources by pioneer species that allocate much of their energy to reproduction and dissemination (r strategists). These r strategists excel in getting to a site or a resource first and becoming established, but they are generally not very competitive. They will dominate a site until replaced by longer-lived, more competitive K strategists that put more of their energy into maintenance than reproduction. K strategists are there for the long term. Their success is dependent upon their ability to persist on a site. This conservation stage may persist as a stable community for a considerable period of time, giving rise to the false impression of constancy discussed above.

Ultimately, however, disturbance, senescence or other factors will disrupt that stability, mortality will increase and the ecosystem will decline. Holling calls this third stage "creative destruction," because the forces causing the destruction also release resources that had been accumulated in the biomass during the first two stages, and these resources will fuel the "creative renewal" of the ecosystem. The fourth stage involves the dual functions of mobilization of the energy and minerals bound in the dead biomass and retention of those resources until they can be used again. This stage takes place largely in the soil or on its surface. The space newly opened by creative distruction and the newly available resources provide conditions that will support the development of a new community or ecosystem. Pioneer species move in to exploit those resources and the process of self-organization begins anew.

This cycle can occur at any scale of space and time. It can occur on the scale of an entire forest or it can occur in a forest opening created by the death of one tree. Generally, small or intermediate scale disturbance that occurs frequently relative to the life expectancies of the longest-lived members of the community have a salutary effect on the community by rejuvenating it on an ongoing basis. Management schemes that try to eliminate disturbance factors in an effort to "protect" a desirable ecosystem are often counter-productive in that they interfere with this internal rejuvenation process that is disturbance dependent. As a result, "protected" ecosystems can become increasingly vulnerable to catastrophic destruction.

Thus, ecosystems are dynamic. Their defining feature is not stability or some idealized notion of climax or steady state. Their defining quality is their renewability, their capacity for continual self-organization in the face of ever-present and sometimes essential disturbance. The success of this renewal capability rests with ecosystem functions that provide it with internal control and the ability to build and maintain variability and flexibility. The ecosystem changes constantly, but it maintains its integrity.

6. Ecosystem Integrity

In recent years, some ecologists and resource managers have proposed ecosystem integrity as a goal for conservation and a target for sustainable development. The concept of ecosystem integrity embodies non-equilibrium explanations of ecosystem dynamics in that it focuses primarily on ecosystem processes, on internal control and flexibility, on self-organization and on the possibility of multiple stability states. "An ecosystem exhibits integrity if, when subjected to disturbance, it sustains a self-organizing, self-correcting capability to recover toward an end-state that is 'normal' and 'good' for that system ." (19) In other words, an ecosystem has integrity if it responds to Holling's creative destruction phase or to any stress by self-organizing to a mature stage that is optimal under prevailing environmental conditions ("normal" and "good") and also retains the ability to respond to future change.

This idea of ecosystem integrity is not entirely a scientific concept. Like the concept of "quality," notions of "integrity" include social, economic, political and even psychological considerations. Kay (12), for instance, defines ecosystem integrity as "the state in which we want ecosystems to be." Ecological, biological, chemical and physical factors operating in and on an ecosystem of high integrity will lead to repeated self-organization as described above, each leading to one of any number of possible stability states. As with notions of quality, scientists can tell us how this process works and maybe even predict outcomes in given situations, but scientists alone cannot determine which outcome is better. "Integrity can only be defined distinctly for specific ecosystems, in the context of the humans which are an integral part of the ecosystem ."(12)

Regier (19) points out that "end states other than the pristine or naturally whole may be taken to be 'normal' and 'good.'" Agricultural ecosystems are certainly not pristine and naturally whole. But, they are ecosystems and as such, if left alone (withdrawn from human management), they exhibit all of the features of non-equilibrium systems. This can be demonstrated simply by observing what happens when a piece of farm land is abandoned and allowed to return to nature. It passes through what has been called old field succession or what we here are calling the exploitation and conservation stages of ecosystem development. Actually, the concept that the old field is returning to nature is a misconception. In fact, it has always been nature, nature maintained at an early stage of ecosystem development by human actions. Earlier we noted that soil quality is defined at least partly as 'the ability to produce crops on an ongoing basis.' Similarly, we can say that an agricultural ecosystem that can produce crops on an ongoing basis is both 'normal' and 'good.' To be able to produce crops on an ongoing basis, an agricultural ecosystem must have ongoing soil fertility, water-holding capacity, etc., and these features are products of processes that take place on and in the soil. Agricultural practices such as crop rotation, fallowing and minimum tillage are implemented to allow natural processes to restore and renew the soil component of the ecosystem. Beyond that, farmers may use various amendments such as fertilizers, irrigation, drainage, liming, etc. to maintain or renew soil quality. In doing so, farmers become participants in the Hollingian ecosystem cycle. Regier (19) calls such involvement "cultural integrity." He defines cultural integrity as "human capability, individually and through institutions, to complement the integrity of a modified natural ecosystem." Thus, to Regier and others, ecosystem integrity has two interacting and complementary components, natural integrity and cultural integrity. If we are serious about sustainable development, we must learn how to balance natural and cultural integrity, and this must be done on a site-by-site basis.

7. Ecosystems, Stress and Sustainability

We have said here that change and even disturbance are natural components of ecosystems and that disturbance can actually be essential to ecosystem self-renewal. But, we must also be concerned about stress that is not natural, but human in origin and how such stress might

---

[3]Human dominated or managed ecosystems or landscapes are sometimes called " cultural landscapes."

affect integrity and sustainability. Regier (19) points out that what he calls cultural stresses differ in kind and degree from natural stresses. He notes that it is "often assumed, tacitly or explicitly, that the disintegrative effect of such exotic/cultural stress will be similar to that of some natural stress." But, he goes on to cite findings from his research (17) that "this is not usually the case, at least if the cultural stresses achieve any measure of intensity. Instead the consequent disintegration has generally been abnormal or pathological." Thus, before we can address questions of sustainability, we must look briefly at how ecosystems respond to stress.

In their much cited paper, Rapport et al. (18) looked at five types of stress: stress associated with harvesting of renewable resources; stress related to pollutant discharges into ecosystems; stress associated with physical restructuring and land-use changes; stress associated with the introduction of exotic species into ecosystems; and stress associated with extreme natural events to which ecosystems cannot accommodate, and they examined their impacts on ecosystems. They found that such stress is often manifested in much the same ways in different ecosystems:

1) *Changes in nutrient cycling:* Downward leaching and lateral transport of nutrients is widespread in disturbed terrestrial and aquatic ecosystems.
2) *Changes in primary productivity:* Stresses other than eutrophication generally lead to decreases in primary productivity. Further, the ratio of productivity to biomass almost always declines.
3) *Changes in species diversity:* One of the most widespread signs of ecosystem distress is a reduction in species diversity. Exceptions occur when the stress increases habitat diversity or when the stress is not too severe.
4) *Retrogression*: Stress alters the character of a community in the direction of less mature systems. Communities are simplified, niches are opened, the nutrient inventory is at least partially lost, the community becomes less stable, etc.
5) *Changes in size distribution of species*: There is a general reduction in the average size of dominant biota, at least temporarily.
6) *Other signs of distress*: Other less well documented effects of stress include alterations in disease incidence and changes in amplitude of fluctuations in component populations.

Rapport et al. (18) note further that by the time these symptoms of distress are recognized, the ecosystem is often in an advanced stage of response, and they offer a model of progressive response to aid early detection of problems. They call this model an "Ecosystem General-Adaptation Syndrome."

> *Alarm Reactions* -- "The earliest indications of ecosystem response to stress may be found in abnormal fluctuations in sensitive populations or communities, abnormalities in reproduction, changes in the distribution of sensitive species, and, in the case of contaminant stress, biochemical or histological abnormalities within

populations and organisms. Such early-warning symptoms manifest themselves primarily at the reproductive and species-specific levels." For example, pollutant contamination of soils might manifest itself early by "the accumulation of litter which suggests impaired decomposer activity," or by "increased nutrient leaching."

Coping Mechanisms -- "It seems clear that ecosystems have cybernetic mechanisms which operate to counteract the impact of stressors, at least to some extent. The ecosystem's ability to cope with specific stresses appears, in part, to be a function of its evolutionary history. Ecosystems that have evolved in relatively unstable environments are more likely to resist moderate stresses from human activity, especially those which mimic natural stresses. The precise mechanisms whereby ecosystems resist or cope with stress vary. One mechanism involves the replacement of more stress-sensitive species with functionally similar, but more resistant species." The components change but processes are maintained. "Another mechanism involves external feedback loops." Here ecosystems are able to tap external sources of energy and materials to maintain productivity when stress impairs their ability to utilize internal sources effectively. In such cases there is a loss of internal control, but the system continues to function because of the successful garnering of external supplies. Of course, this coping mechanism is effective only when adequate external supplies are available. Another coping mechanism is the "conversion of toxic compounds to less deleterious forms. Heavy metals are chelated in soil organic matter, and gaseous SO_2 is converted to sulfate."

Ecosystem Breakdown -- At some point, if the stress factors are not moderated or removed, the coping capacity of the ecosystem will become overwhelmed. Homeostatic or, more precisely, homeorhetic mechanisms will cease to be effective and ecosystem breakdown will occur. Once this threshold is reached, the self-organizing capabilities of the ecosystem may be irreversibly damaged such that the ecosystem will not recover even if the stress agent is removed. The ecosystem will degrade spontaneously and restoration will require extraordinary measures that are usually costly and time consuming.

As ecosystems pass through such a degradation process, internal control and flexibility are progressively depleted. In the early stages, minor and relatively inexpensive cultural measures, including the removal of serious stresses, will often suffice to restore an ecosystem's self-organizing capabilities. As deterioration progresses, however, more cultural integrity will be required to counter the loss of natural integrity if we are to maintain the ecosystem as "what we want it to be." Compromising on cultural integrity will compromise ecosystem integrity, and we will not get the productivity or other services that we want and need. Further, such distressed ecosystems will be more vulnerable to stresses to which they would have been able to respond when healthy. Such ecosystems are less productive and less predictable than those with high integrity. Thresholds occur when the stressed ecosystem loses internal control and flexibility. Such ecosystems are at the mercy of external forces and are highly unpredictable and unreliable.

Sustainability of ecosystems, including soil ecosystems, requires that we learn to recognize ecosystem distress at an early stage and take remedial action by removing the stress factors and applying cultural integrity measures sufficient to counter the impact of lost natural integrity. Only then will ecosystems maintain their critical capabilities of self-organization. Only then might they be sustainable.

8. Sustainable Development

Sustainable development is another vaguely defined concept. Probably the most common notion of sustainable development would be development that does not diminish the capacity for future development. To ecologists and environmentalists, this usually means development without damaging the ecosystems that make development possible. Economists and developers, on the other hand, may interpret this broad definition to mean economic development that does not threaten future economic development or development that does not destabilize or otherwise threaten local, regional, or global economies. Politicians might see sustainable development in terms of maintaining political stability or, in a more immediate sense for them, sustaining current political systems and regimes. For these reasons, one's interpretation of sustainable development is tied to what it is that is to be sustained.

We suggest that these ecological, economic and political definitions are all valid ways of looking at sustainability. But, sustaining ecological systems is fundamental. Unless we find ways of sustaining ecosystems, or what economists call natural capital, we will not be able to sustain economic or political systems. We further suggest that we should begin addressing this problem of sustainable development by addressing ecosystem integrity as a dynamic interplay of natural and cultural integrity.

As Kay (13) noted, ecosystem integrity is what we want an ecosystem to be. Advocates of ecosystem preservation call for protecting natural integrity and maintaining ecosystems in pristine states. At the opposite extreme, many development economists believe that natural and human (cultural or social) capital are 100 % interchangeable, that it is alright to eliminate natural capital if it is replaced with human capital. We would suggest that both views have merit. Where we decide that we want an ecosystem to be pristine, as in the case of nature reserves, sacred groves, biosphere reserves, etc., it makes sense to focus totally on natural integrity to bring about sustainability through Hollingian cycles of ecosystem self-organization. If, however, we have decided that an ecosystem, for any reason, should or must depend mostly or even entirely upon cultural integrity, it is legitimate to convert natural capital largely to human capital. Cities are cases in point. As David Orr (14) has pointed out, cities, especially megalopolises, will never be sustainable in a natural sense. They will always be dependent upon imports of food and materials and much of their energy needs. Further, they will never be able to assimilate their own wastes. As Orr points out, "Urban concentrations must be justified on their contributions to intellectual, economic, and cultural life, not on their natural integrity. Therefore, such ecosystems will have to be subsidized by society if they are to be sustainable, e.g. they will be highly dependent upon cultural integrity for their sustainability.

The major challenges of sustainable development lie between these two extremes. On most sites around the world, sustainability will be achieved only through appropriate blending of natural and cultural integrity. On each site we must determine what we want the ecosystem to be and what mix of natural and cultural integrity will sustain that ecosystem.

We suggest that the major goal of sustainable development should be to maintain the self-organizing capabilities of local and regional ecosystems, subsidizing natural integrity with cultural integrity as required. We caution, however, that as natural integrity is lost, maintaining ecosystem integrity by cultural inputs becomes increasingly expensive and decreasingly efficient and effective.

9. Soil Quality and Security

Finally, we come full circle to the question of security and how that relates to soil quality, soil integrity and the dynamics of soil ecosystems. As we noted at the beginning of this chapter, our sense of security at all scales of human existence depends upon three factors: predictability, flexibility and control. Security in matters of the land, of agriculture, of forestry, of food, fiber, and shelter rests on those three factors. Our CEEC farmer wants and needs security. He needs predictability from his land. He also needs to know that his land provides him with choices or flexibility, so he can optimize his yields and the quality of his life or so he can accommodate to the unexpected. And he needs to know and understand how his land maintains internal control and how this enhances his own control. Degrading land, land out of ecological control, means farmers with decreased control. As we can see by looking at degraded lands around the world, lands without internal control are lands sooner or later abandoned.

The CEECs want and need security. They need predictability from their agricultural sectors, from their farmers and their land. They need flexibility in land and agricultural capabilities so they can optimize their positions in European and global markets. And they need to be able to count on consistency over time from their agricultural production sectors and the land upon which they depend, e.g. they need a soil/agriculture system that has considerable internal control.

In matters of the land, security derives from ecosystem integrity, the self-organizing, self-renewing capabilities of healthy ecosystems, including and probably most importantly soil ecosystems. As we have seen, ecosystems with high ecological integrity have significant internal control. They also have considerable variability which makes them flexible in the face of changing conditions, disturbance or even stress. And, because of this internal control and flexibility, these ecosystems tend to be relatively predictable. Thus, ecosystem integrity provides us with predictability, flexibility and control without which our security is diminished.

10. References

1. Berkowitz, M. and Bock, P.G. (1968) National security, in D.L. Sills (ed.), International Encyclopedia of the Social Sciences, Vol. 11, Macmillan/Free Press, New York, pp.40-45.

2. Botkin, D.B. (1990) Discordant Harmonies: A New Ecology for the Twenty-first Century, Oxford University Press, New York.

3. Brown, H. (1983) Thinking About National Security: Defence and Foreign Policy in a Dangerous World, Westview Press, New York.

4. Clark, W.D. and Munn, R.E. (eds.) (1986) Sustainable Development of the Biosphere., Cambridge University Press, Cambridge.

5. Costanza, R, Norton, B.G. and Haskell, B.D. (Eds.) (1992) Ecosystem Health: New Goals for Environmental Management., Island Press, Washington, DC.

6. Crossley, J.W. (1996) Managing ecosystems for integrity: Theoretical considerations for resource and environmental managers, *Society & Natural Resources*, **9**, 465-481.

7. Davis, C.B. (1993) Needed behavioural change: Steps towards environmental security, in N. Polunin and J. Burnett (eds.), Surviving With The Biosphere, Edinburgh University Press, Edinburgh, pp. 397-416.

8. Ehrenfeld, D. (1992) Ecosystem health and ecological theories, in R. Costanza et al. (eds.) Ecosystem Health: New Goals for Environmental Management., Island Press, Washington, DC, pp. 135-143.

9. Holling, C.S. (1973) Resilience and stability of ecological systems, in R.F. Johnston et al. (eds.) *Annual Review of Ecology and Systematics*, Annual Reviews, Inc., Palo Alto, CA, pp. 1-23.

10. Holling, C.S. (1986) The resilience of terrestrial ecosystems: local surprise and global change, in W.C. Clark and R.E. Munn (eds.), Sustainable Development of the Biosphere, Cambridge University Press, Cambridge, UK, pp. 292-320.

11. International Peace Research Institute, Oslo/United Nations Environment Programme (cited as PRIO/UNEP (1989). Environmental Security: A Report Contributing to the Concept of Comprehensive International Security, Oslo, Norway.

12. Kay, J.J. (1991) A nonequilibrium thermodynamic framework for discussing ecosystem integrity, *Env. Mgmt.* **15** (4), 483-495.

13. Kay, J.J. (1993) On the nature of ecological integrity: some closing comments, in S. Woodley, et al. (eds.) Ecological Integrity and the Management of Ecosystems., St. Lucie Press, St. Lucie, FL, pp. 201-212.

14. Orr, D.W. (1992) Ecological Literacy: Education and the Transition to a Postmodern World, State University of New York Press, New York.

15. Polunin, N. and Burnett, J. (eds.) (1993) Surviving With The Biosphere, *Proceedings of the Fourth International Conference on Environmental Future*, Budapest, April, 1990, Edinburgh University Press, Edinburgh.

16. PRIO/UNEP (1989) - see International Peace Research Institute, Oslo/United Nations Environment Programme.

17. Rapport, D.J. and Regier, H.A. (1992) Disturbance and stress effects on ecological systems, in B.C. Patton (ed.) Complex Ecology: the Part-Whole Relation in Ecosystems, Prentis-Hall, New York.

18. Rapport, D.J., Regier, H.A. and Hutchinson, T.C. (1985) Ecosystem behavior under stress, *The American Naturalist* **125** (5), 617-640.

19. Regier, H.A. (1993). The notion of natural and cultural integrity, in S. Woodley et al. (eds.), Ecological Integrity and the Management of Ecosystems, St. Lucie Press, St. Lucie, FL, pp.3-19.

20. Sills, D.L. (ed.), International Encyclopedia of the Social Sciences, Vol. 11, Macmillan/Free Press, New York.

21. Vitousek, P.M. (1994). Beyond global warming: Ecology and global change, *Ecol.* *75*(7), 1861-1876.

22. Vos, C.C. and Opdam, P. (eds.) (1993) Landscape Ecology of a Stressed Environment, Chapman & Hall, New York.

23. Woodley, S., Kay, J. And Francis, G. (Eds.) (1993) Ecological Integrity and the Management of Ecosystems, St. Lucie Press, St. Lucie, FL.

24. World Commission on Environment and Development (1987) Our Common Future, Oxford University Press, New York.

SOIL QUALITY IN RELATION TO THE CONCEPTS OF MULTIFUNCTIONALITY AND SUSTAINABLE DEVELOPMENT

G. VÁRALLYAY
Corr. Member of Hung. Acad. Sci., research professor
Research Institute for Soil Science and Agricultural Chemistry of the Hungarian Academy of Sciences
H-1022 Budapest, Herman O. 15. Hungary

1. Introduction

In spite of the great differences in the definition of the "quality of life" there is a world-wide agreement on three criteria: healthy food (sufficient quantity, high quality); clean water; and pleasant environment. All three are closely related to soil quality, rational land use and proper soil management.

Soils represent a considerable part of the natural resources of central and eastern Europe. Consequently, rational land use and proper soil management, ensuring normal soil functions, are important elements of *sustainable agricultural development*, having particular significance both in the national economy and in environmental protection.

The basic aim of *agriculture* is the production of high-quality products with low costs (inputs) and without undesirable or harmful environmental side-effects in the biosphere of a given area at the present time or in the future. The relative importance of these criteria (quantity, quality, efficiency, profitability, environmental impacts) has changed considerably during the history of agriculture, depending on socio-economic conditions and according to political decisions.

Actions towards sustainable agricultural development are joint tasks of the state, policy- and decision-makers at various levels, land owners, land users, and, to a certain extent, each member of society. Only their joint efforts can be successful towards an efficient, economically viable, socially acceptable and environmentally sound *sustainable* development.

2. Sustainable Development, Responsibilities of Agricultural Environment Management

The term „*sustainable development*" was not mentioned at the UN Conference on Human Environment (Stockholm, 1972). Twenty years later it became the most fashionable and most frequently used term at the World Summit on the Environment (Rio de Janeiro, 1992). The two parts of the expression

M.J. Wilson and B. Maliszewska-Kordybach (eds.),
Soil Quality, Sustainable Agriculture and Environmental Security in Central and Eastern Europe, 17–33.

("sustainable" and "development") can be a potential point of conflict in itself. It was interpreted and translated in different ways in various countries, depending on their environmental and socio-economic conditions, historical background and even political situation, sometimes laying more emphasis either on "development" or on "sustainability". Consequently, the realisation of AGENDA-21 will require global compromises and joint efforts instead of regional or national economic and/or social confrontation.

FAO adopted the following definition of sustainable agricultural development in 1988. "It is the management and conservation of the natural resource base, and the orientation of technological and institutional change in such a manner as to ensure the attainment and continued satisfaction of human needs for present and future generations. Such sustainable development (in agriculture, forestry and fisheries sectors) conserves land, water, plant and animal genetic resources, is environmentally non-degrading, technically appropriate, economically viable, and socially acceptable." [5]. The concept was adopted in Hungary within the „AGRO-21" Programme best suited to the country's natural endowments and socio-economic circumstances [1].

Sustainable development includes efficient agricultural production, using environment-friendly, energy- and material saving technologies, and paying special attention to quality and socially acceptable rural development. Both these latter factors represent a challenge to modern soil science, rational land use and soil management. The problems and tasks are schematically illustrated in Figure 1.

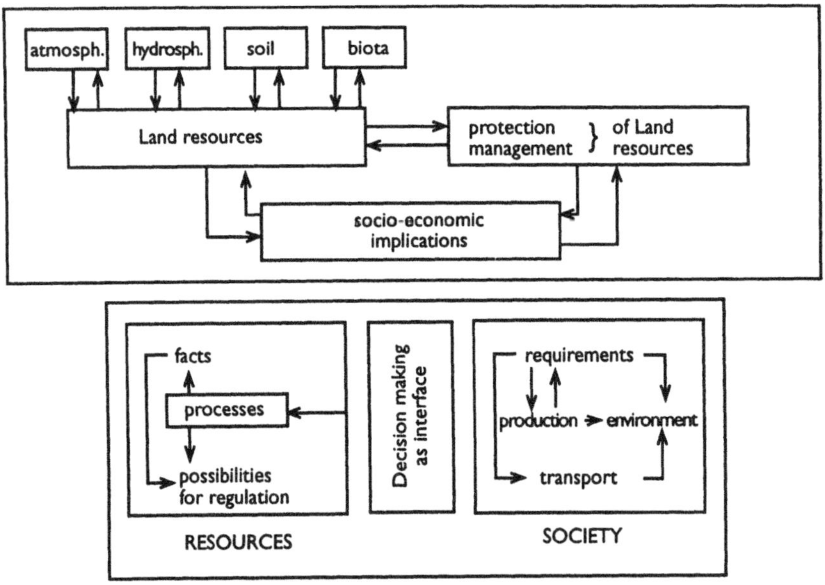

FIGURE 1. Relationships between resources and society

A given set of land resources, including geological formations, relief, atmosphere, surface and subsurface water resources, soil, biota, vegetation etc., is used, managed and hopefully protected by society according to its requirements and depending on the prevailing socio-economic conditions. These conditions may be modified by the historical background and formulated by decision-makers at various levels.

In the decision-making "machinery" certain questions have to be answered in an appropriate way, such as:

- What are the potential land use alternatives under given conditions, taking into consideration what is expected from the land? This would include biomass production for food, fodder, industry, energy; meaningful work for the local rural population; exploitation of mineral resources; locations for building construction and infrastructure; drinking water supply; places for recreation and sports; conservation of pleasing landscape and biodiversity, etc.
- What are the potential and actual efficiencies of the various alternatives, based on a comprehensive and realistic cost/benefit analysis?
- What are the predictable ecological and economic risks (risk analysis), the environmental consequences and potential side effects (impact analysis)?

The efficient realisation of these new tasks requires new conceptual approaches and adequate expertise. Consequently, research, education, innovation and extension services will play a prominent role in future sustainable development in all countries and particularly those of central and eastern Europe.

3. Soil Functions and Soil Quality

The main functions of soil in the biosphere are as follows [20, 22]:

(a) Soils are a most significant, *conditionally renewable* natural resource. During rational biomass production they do not change irreversibly and their quality does not decrease unavoidably and fundamentally. But their "renewal" based on soil resilience does not occur automatically. Soil conservation, which includes the maintenance and increase of soil fertility, requires ongoing activities, such as sustainable land use, application of agricultural technology, remediation and reclamation.

(b) Soil is both a *reactor* and *a transformer,* integrating the combined influences of other natural resources, such as solar radiation, the atmosphere, surface and subsurface waters, deeper geological strata and biological resources. Biogeochemical cycles in soils develop a "life-medium" for microbiological activities, as well as habitats for natural vegetation and cultivated crops.

(c) Soil is the most important medium for biomass production and primary food.

(d) Soil is a major *natural store* of heat, water, plant nutrients and other elements, including wastes and potentially harmful elements or chemical compounds. The stored water and plant nutrients ensure water and nutrient uptake by plants for shorter or longer periods, without continuous additional water or nutrient supply.

(e) Soil is an efficient "natural filter" and detoxification system that may prevent the deeper horizons and subsurface waters from being contaminated as a result of the deposition or addition of various pollutants to the soil.
(f) Soils represent a high capacity buffer medium of the biosphere that, within limits, may moderate the various stresses caused by environmental factors and/or human activities.
(g) Soil is a significant gene reservoir of the biosphere and an important element of bio-diversity.

These functions are all equally important but society has used them in different ways throughout history, depending on the given natural conditions and socio-economic circumstances. In many cases the character of a certain function was not taken into consideration during the utilisation of the soil resources. Such cases resulted in over-exploitation, decreasing the efficiency of one or more of the soil functions, and eventually to serious environmental deterioration [7].

The modern concept of "soil quality" has to reflect the multifunctionality of soil and consequently requires a new, complex, multidisciplinary, "fit for purpose" approach [3,4,6,12,16,19]. Soil functions express the soil's ability to serve or limit Man's various needs, directly or indirectly, and even in cases where this is not immediately obvious, such as declining water and air quality, animal or human health hazards. According to this concept, soil quality expresses the ability of soil to fulfil its various functions. Consequently, there is not an overall, general term for the definition, quantification and evaluation of "soil quality" at the present time, only "purpose-dependent" or "function-dependent" soil quality interpretations, with their own and specific rating or evaluation systems. These specific "soil quality" evaluations have to be dynamic, expressing the changing needs of a society.

Different "soil quality" ratings are necessary for:

- agricultural production areas;
  drinking water supply;
- industrial sites and urban areas;
- various land use practices;
- various crops (suitability classes);
- fertiliser application;
- tillage;
- irrigation and drainage,

and are assessed according to different criteria Before the wide-spread use of fertilisers, soils with good natural nutrient supply were classified as good quality soils. Again, before mechanised tillage, soil workability was an important quality criterion, and before irrigation and drainage the natural soil moisture regime was the main factor of soil productivity. The criteria expressing "soil quality" for high input farming differ from those of low-input agricultural or bio-farming systems..

For these reasons, a unanimously accepted definition for "soil quality" does not exist or criteria for its quantification, evaluation and, especially, standardisation [11].

4. Soil Processes and Their Control

The multifunctionality of soil is determined by the combined influences of soil properties, which are the results of soil processes (mass and energy regimes, abiotic and biotic transport and transformation and their interactions) under the combined influences of soil forming factors. Any soil-related human activity influences the soil through these processes. Consequently, the control of soil processes is a great challenge and the main task of soil science and soil management in sustainable development [2, 7, 22]. The control should include the

- registration of facts and consequences;
- evaluation of potential reasons (definition and quantification of soil processes, analysis of influencing factors and their mechanisms);
- a comprehensive risk assessment, sensitivity and impact analysis;
- development of efficient technologies for the "best" control alternatives.

Concerning soil process control four fundamental questions have to be answered:

Q1. Is it necessary to control soil processes?

With the exception of some unusual cases (e.g. strictly protected areas) a certain degree of control is always necessary. Usually even the soils of protected ecosystems require an appropriate process control for the prevention of undesirable environmental side- effects (e.g. soil acidification due to atmospheric deposition). Abandoned (set-aside) land cannot find its way back to its "original" condition without human intervention or the regeneration process will take an unacceptably long time. For example, set-aside arable land will never revert to forest, species-rich grassland or wetland with high bio-diversity as it was before deforestation, ploughing or water regulation. Restoration needs well-defined human measures and permanent "control activities".

Q2. What are the main purposes of the control of soil processes?

The general answer cannot be more specific than to ensure or guarantee "normal" or "desirable" soil functions." The definition of "normal" or "desirable" is purpose-specific and depends on the criteria of "priority functions", which are determined by given conditions and requirements.

The most important objectives of soil process control are as follows:

- maintenance (stabilisation) of the present situation. This is not meaningful if the present state is not desirable.
- prevention of unfavourable changes.
- re-establishment of a previous "original" or more natural situation. Here, the definition and the time criteria of "original" are rather questionable);
- modification of the present conditions in a positive direction (e.g. improvement of one or more soil properties, soil amelioration);

- recultivation of disturbed lands, remediation of polluted sites, or rehabilitation of degraded ecosystems. Sometimes the re-establishment of a previous situation, or a certain soil improvement where deterioration is not fully reversible.

Q3. To what extent is soil process control necessary?

From the practically "O-control" in strictly protected areas, to the "full control" in, for example, soils in greenhouses, there is a wide spectrum of control levels. The "optimum", "desirable", "required" or "tolerable" rate of soil process control is determined by the main objective, costs, efficiency and impacts, whether positive and negative economic and ecological consequences.

Q4. How can the various soil processes be controlled?

The conceptual model of the rational strategy for the control of soil processes is schematically illustrated by Figures 2 and 3 [20, 22].

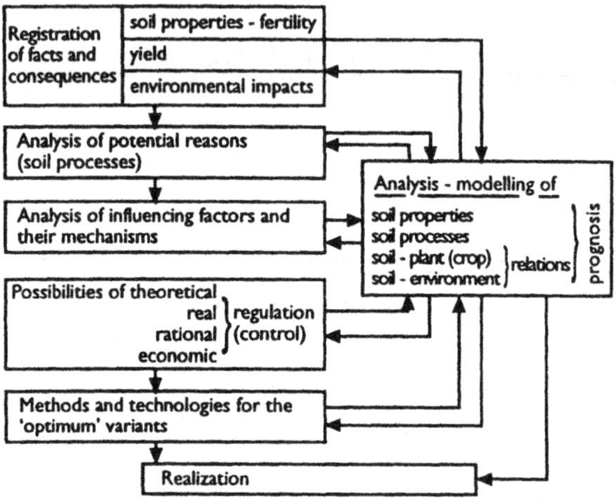

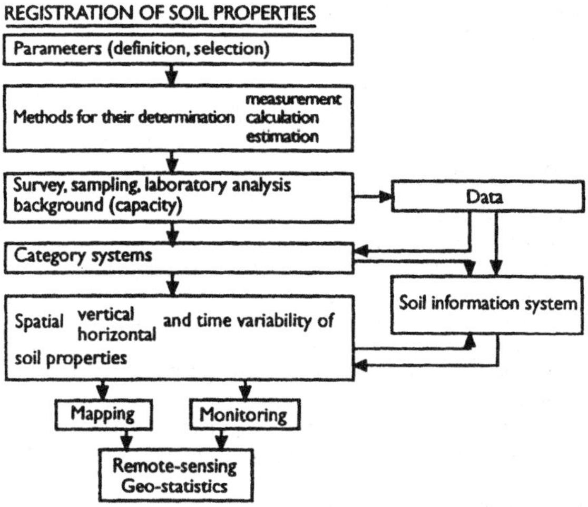

FIGURE 2. Control of soil processes

The scientifically based planning and implementation of sustainable land use and rational soil management to ensure desirable soil functions require adequate soil information: exact, reliable, "detectable" (preferably measurable) and accurate, quantitative territorial data on well-defined soil and land properties. This would include the characterisation of their spatial (vertical, horizontal) and temporal variabilities as well as pedo-transfer functions relating to soil processes and biogeochemical cycles, including the factors and mechanisms that influence and determine them.

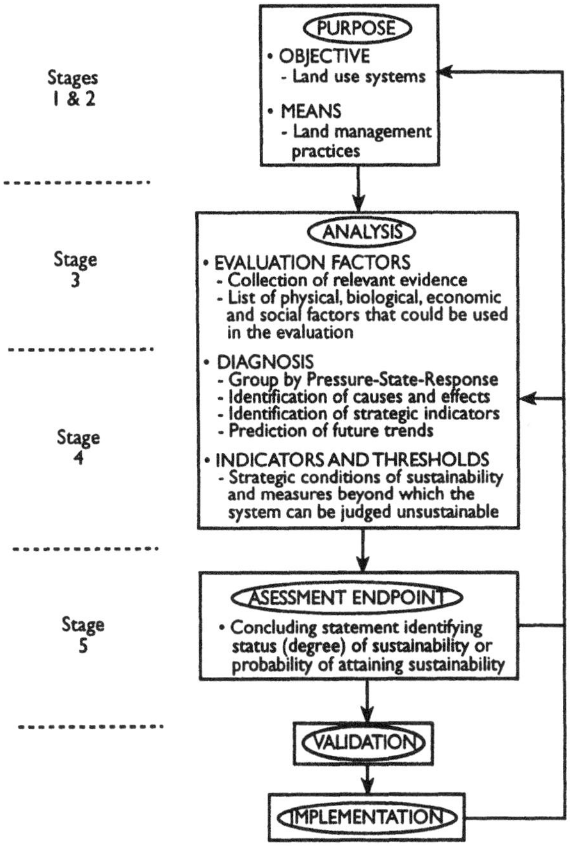

FIGURE 3. Flowchart showing the stages of soil process control

5. Main Elements of Rational Land Use and Soil Management for Sustainable Agricultural Development

(1) Territorial co-ordination of agro-ecological conditions (land-site characteristics) and requirements of cultivated crops, taking into consideration both production and environmental aspects in short-, mid- and long-term scales:

- rational land use and optimisation of cropping pattern according to natural conditions [10];
- selection of appropriate crops (in crop rotations) and cultivars according to the given natural circumstances (selection and breeding of varieties or even genotypes tolerant to certain ecological constraints, such as frost, soil acidity, salinity/alkalinity, drought, moisture surplus, extreme soil texture, soil compaction, etc.);
- improvement of soil ecological conditions according to the crop requirements (amelioration of land, reclamation/improvement of soil, soil conservation, irrigation and drainage, appropriate agricultural technology).

(2) Rationalisation of the structure of agricultural fields, perhaps under a new ownership structure:

- optimisation of field size according to given physiographic conditions (development of more homogeneous fields, uniform agrotechnical measures, taking into consideration a rational level of biodiversity);
- development of appropriate infrastructure (rational territorial arrangement of fields, roads, canals, buildings, forest shelter belts, etc., taking into consideration landscape preservation requirements).

(3) Reduction to a minimum of production wastes, including plant residues, animal excrement, harvest wastes and wastes from food processing, with efficient recycling with no harmful environmental side-effects.

(4) Improvement of existing agro-ecological conditions (amelioration, reclamation) and prevention, or at least moderation, of undesirable soil degradation processes.

The main limiting factors of soil fertility in Hungary are as follows:

- extremely coarse textures (low amount of inorganic and organic colloids);
- acidity;
- salinity/alkalinity particularly in the deeper layers;
- extremely heavy textures (high amounts of clay and swelling clay minerals);
- waterlogging (or regular over-moistening);
- water and wind erosion.

These soil degradation processes are summarised in Table 1 [14].

The necessity and rationale for the reclamation of soils with limited fertility depends on economic/cost-benefit and ecological considerations. The radical

No.	Limiting factor of soil fertility	Area, 1000 ha	%
1.	Extremely coarse texture	746	8.0
2.	Soil acidity - combined with erosion - combined with shallow depth	1200 (348) (67)	12.8 (3.7) (0.7)
3.	Salinity/alkalinity	757	8.1
4.	Salinity/alkalinity in the deeper layers	245	2.6
5.	Extremely heavy texture	630	6.8
6.	Peat formation (waterlogging)	161	1.7
7.	Soil erosion - combined with acidity	1455 (348)	15.6 (3.7)
8.	Shallow depth - combined with acidity	217 (67)	2.3 (0.7)

No.	Soil degradation processes
1.	Soil erosion: - by water - by wind
2.	Soil acidification
3.	Salinisation/alkalisation
4.	Physical soil degradation - structure destruction - compaction - surface sealing
5.	Extreme moisture regime -overmoistening,waterlogging - drought sensitivity
6.	Biological degradation - decrease of organic matter - deterioration of soil biota
7.	Unfavourable changes in the Nutrient regime - leaching - biotic and abiotic immobilisation
8.	Decrease of the buffering capacity, soil pollution, "toxicity"

TABLE 1. Limiting factors of soil fertility and soil degradation processes in Hungary

amelioration of salt-affected soils, sandy soils or peatlands requires such expensive and complex measures that it is uneconomic. At the same time saline lakes and soils, wetlands and sand regions are, in many cases, protected ecosystems and habitats of special plants and animals and, consequently, have particular environmental value. These areas should be kept in "original" condition and their reclamation is not advisable, although this was sometimes advocated in the last decades. On the contrary, the improvement of soils with moderate limitations (e.g. liming of acid soils, loosening of compacted soils, etc.) can be an efficient and economic tool for increasing biomass production for food, fodder, raw material for industry, energy or for other purposes.

Soil degradation is usually a complex process in which several component features of can be recognised, leading to the loss of land or its productive capacity, to the limitation of normal soil functions, and/or to the decrease in soil fertility due to unfavourable changes in soil processes and, consequently, soil properties. Soil degradation may be the result of natural factors and/or human activities.

In Hungary the considerable extension of the various undesirable soil degradation processes represents serious crop production and environmental problems both in the directly affected territories and in their surroundings [17].

Recognizing the world-wide importance of the prevention, control and „management" of soil degradation processes, UNEP initiated a world programme for the Global Assessment of Human-induced Soil Degradation (GLASOD), resulting as a first step in the complilation of a World Map at a general scale of 1:5,000,000.

Soil degradation is *not* an unavoidable consequence of intensive, but rational, agricultural production and social development. Most soils are resilient to some extent, and most of the degradation processes and their unfavourable consequences can be efficiently prevented, or at least moderated [7]. But it needs ongoing actions and appropriate soil- and water conservation practices need to be widely adopted [17, 18]. The key word is *prevention*.

For successful *prevention* satisfactory *prognosis* is necessary. It can be rationally based on a comprehensive

- sensitivity analysis: evaluating the susceptibility of soils to various soil degradation processes, such as water and wind erosion [13], acidification, salinisation/ alkalisation [15], structure destruction or compaction [24], susceptibility to nitrate leaching; vulnerability to various chemical pollutants [23];
- impact analysis: evaluation of the "positive" and "negative" impacts of various human activities.

Between 1960 and 1970 considerable efforts were made in Hungary to prevent soil degradation processes based on the results of comprehensive soil survey, including soil testing and soil analysis, and involving large scale soil mapping programmes. There was a high rate of state subsidy on the main soil reclamation practices and soil degradation control measures [8]. In the 70's these well co-ordinated activities decreased dramatically. This tendency continued during the (re)privatisation of the 90's and has not been able to balance the further extension of soil degradation processes due to the inappropriate and practically uncontrolled land diversification, ignorance of "off-site" effects, the short-term profit orientation, improper land use, involving improper cropping patterns, soil management and agricultural technology Undoubtedly, other countries in central and eastern Europe have faced, and are facing similar soil degradation problems. In Hungary, the "sustainable development" concept is reflected only in the latest legislation documents, giving hope for the future.

(5) Improvement of the efficiency of agricultural water management and soil moisture control [18]. The main objective here is to increase the water storage within the soil in plant available form without any unfavourable environmental consequences, thus aiming

- to reduce evaporation, surface runoff and filtration losses of water, both atmospheric precipitation and irrigation water;
- to increase the available moisture range of the soil so as to help infiltration into the soil, increase the water storage capacity and reduce the immobile moisture content;
- to improve the vertical and horizontal drainage condition of the soil in a given area, so preventing over-saturation and waterlogging.

The highly variable moisture regimes of soils may sometimes necessitate a special "double-faced" soil moisture control so as to ensure efficient drainage or irrigation, as and when required. Both actions are costly and faced with serious limitations. Efficient drainage is inhibited by poor permeability of soil, a limited capacity of drainage canals and drain water reservoirs and salinity problems. Irrigation is limited by factors relating to relief; limited and still decreasing water resources available for crop production and increasing water demand from other sectors. Consequently, all efforts have to be taken to improve agricultural water use efficiency by proper soil management. Most of the measures for soil moisture control incorporate elements of environment protection.

In recent years a comprehensive soil survey/analysis/categorisation/mapping/monitoring system was developed in Hungary for the exact characterisation of hydrophysical properties, modelling and forecasting of water and solute regimes of soils. The system may serve as a scientific basis for soil moisture control and it is efficiently used for practical soil water management both for crop production and environmental protection [18].

(6) Precision plant nutrient management is a relatively recent development and involves

- rational use of fertilisers adequate to the crop requirements, according to the dynamics of their nutrient uptake, soil conditions and agroclimate of a given site;
- efficient recycling of crop residues and production wastes, including organic manures;
- utilisation of other wastes with plant available nutrient but without potentially harmful chemical compounds.

Before World War II the plant nutrient status of Hungarian soils was rather poor, due to a negative nutrient balance [25]. From 1955 there was a rapid increase in fertiliser consumption. This trend resulted in substantial yield increases and a significantly improved nutrient status of many agricultural soils. But in spite of these developments there were serious problems and inadequacies in fertiliser application technology involving improper N-P-K ratios, lack of Ca, Mg and micro-nutrients, a limited variety of fertilisers, problems with fertiliser storage, time of application, means of distribution, etc. The main problem, however, was a tendency for fertiliser application to be polarised between:

(a) better soil → rich farm → higher rate of fertiliser application (in spite of the lower requirements ← better nutrient status of soils) → overdosage;

(b) poor soils → poor farms → lower rate of fertiliser application (in spite of the higher requirements ← lower nutrient supply of soils) → underdosage.

Over-generalization and the uncritical acceptance of the "maximum-concept" led to false conclusions, diminished the effectiveness and efficiency of mineral fertilisation, and resulted in undesirable environmental side-effects. These included:

- soil acidification, due to inadequate type of fertiliser, lack of simultaneous lime application with consequences for mobilisation of toxic element, fixation of some of the nutritive elements, etc.;
- increased loads to surface waters by P compounds and contamination of subsurface drinking water resources by nitrates (see later);
- accumulation of harmful toxic elements in the various stages of the "food chain".

Most of these side-effects, however, are not inevitable and could be prevented, or at least reduced, by precision nutrient management, based on the nutrient requirements and nutrient uptake dynamics of cultivated crops, and by taking into account the nutrient status and other properties of soils, as well as agroclimate and hydrology characteristics of a given site. All of these factors were taken into consideration in the development of a new plant nutrition advisory system in Hungary which has been efficiently used on several hundreds of hectares year by year [21, 25].

In Hungary, as well as in many other CEECs, after the political and economic changes of 1990 the high-rate state subsidy for fertilisers was withdrawn and this resulted in a dramatic decrease in fertiliser application: 300 kg/ha in 1985 → < 10 kg/ha in 1992. This was partly due to confused expression of government policy, and partly because some new farmers thought that the same (or similar) yields could be obtained without applying P and K fertilisers. This is true, however, only for soils with good or luxury P and K supply, topped up by cheap, state-subsidised P and K fertilisers in the late 80's, and only for a certain "exploitation" period, after which the consequence will be a dramatic yield decrease. An understanding of these facts was reflected by a certain increase in fertiliser application in 1994 and 1995 (→ 30-40 kg/ha).

(7) Control of soil pollution.

Most of the elements occurring on Earth can be found in the soil. Their quantity, quality, solubility, mobility; availability for micro-organisms, plants, animals and human beings show an extremely wide spectrum. Most of these elements are essential for the living organisms, but the same elements can be harmful, or even "toxic" for the same organisms, when occurring over a certain threshold value.

The occurrence and accumulation of these elements (e.g. heavy metals and other potentially toxic elements such as Al, As, F, organic pollutants; water soluble salts; nitrates, P and S compounds; etc.) can be due to natural sources, such as:

- air (NO_x, SO_x, etc.) via wet and dry atmospheric deposition;
- water (B, Na, N, etc.) via irrigation water or groundwater;

- soil and geological deposits (P, K, Ca, Mg, Na, Fe, Al, Mn, As, Co, Cu, Ni, Se, Zn, etc.) via local weathering and soil formation processes;

or it can be the consequence of various industrial, urban or agricultural activities. In addition to the increasing quantity of these elements, the sudden, and sometimes surprising, mobilisation of previously immobile pollutants following changes in the soil environment (e.g. soil acidification, salinisation/ alkalisation, destruction of soil structure. Transformation of clay minerals, decrease of organic matter content and buffer capacity, limitations in the „filter function" of the soil, etc.) is even more harmful. This non-linear, time-delayed effect of potentially harmful chemical compounds typifies the "*chemical time bomb*" (CTB) problem.

For the evaluation of the status, regime and potential environmental impacts of these elements (compounds) in soil the following determinations are recommended as shown in Figure 4. [9, 23]:

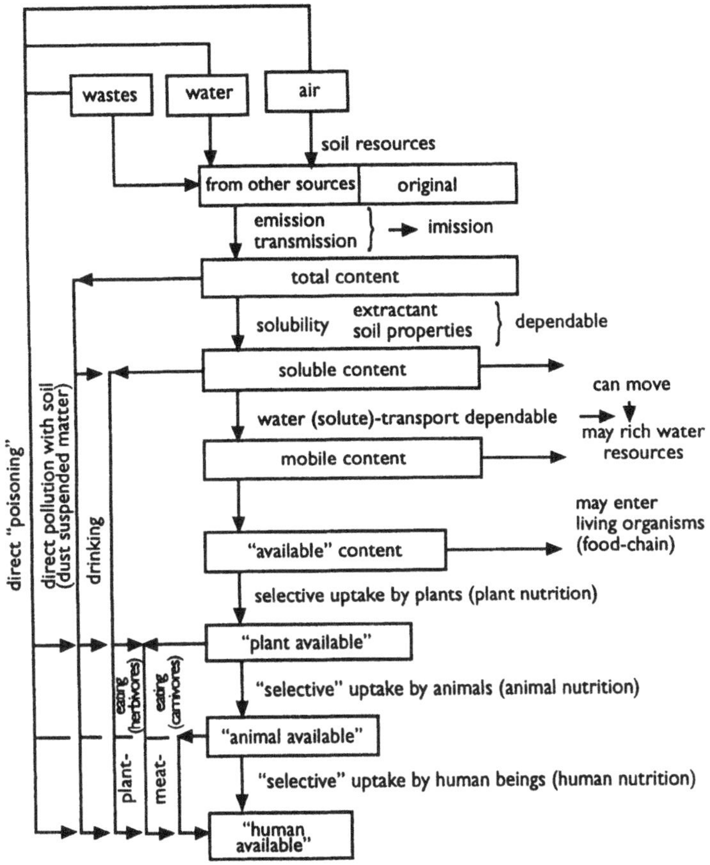

FIGURE 4. Sources and pathways of soil pollution

(a) Determination of Total Content.
(b) Determination of Soluble Content. This quantity is highly solute specific. The "soluble content" depends on the chemical characteristics of the given compound and on the prevailing soil properties, such as the soil's reaction and carbonate status, texture, clay content, clay mineral association, organic matter content and quality, absorption capacity, base saturation, exchangeable cation composition, moisture content, redox potential, microbial activity, etc. The "in situ" soluble content depends to a great extent on plant-root activities. Non-soluble compounds may be

- immobile: cannot be transported by the liquid flow, so that there are limited possibilities of contaminating surface and subsurface water resources;
- not available for plants and, consequently not toxic for plants, animals and human-beings.

(c) Determination of Mobile Content. Only the mobile fraction of the various elements or compounds can be transported and translocated to other places by leaching, accumulation or migration processes, and may reach water resources or plant root surfaces.
(d) Determination of Plant Available Content. This is a plant dependent characteristic because of their selective ion uptake. Only the "available fraction" of a given element can be phytotoxic, resulting in harmful (or even lethal) changes in plant metabolism. Non plant-available elements cannot enter the "food chain", except when they are deposited and accumulated on the surface of plants (e.g. Pb on leaves, Cr and Hg in roots). Some plants can take up and accumulate some non-phytotoxic elements, as exemplified by the accumulation of Cd and Al in the leaves of spinach and lettuce, leading to toxicity problems for plant-eating animals or humans.
(e) Determination of Animal Available Content. Potentially harmful elements taken up by plants or deposited on plant surfaces may enter plant-eating organisms and may result in serious health problems or even death. In some cases, the element may be non-toxic, or at least not lethal, for plant eaters, but after a quick or slow translocation it may accumulate in particular organs (e.g. in fish liver etc.,) and may cause health problems when carnivores eat that organ.
(f) Determination of Human Available Content. Elements toxic to humans may be ingested by man enter either directly through air, water or dust, from plant surfaces, or through the food-chain and may result in various health problems or even death.

The main potential possibilities of *soil pollution control* are schematically illustrated in Figure 6. Its main elements are [23]:

- emission reduction aimed at preventing or reducing the quantity of pollutants deposited or transported to the soil surface or into the soil;
- prevention of the mobilisation of potentially harmful chemical compounds or elements which are already present in the soil but in a temporarily immobile form;
- decrease of the susceptibility/vulnerability of soil against various pollutants so that a higher critical load of pollutants may be tolerated, consequently reducing the "exceedance risk" with concomitant unfavourable ecological consequences.

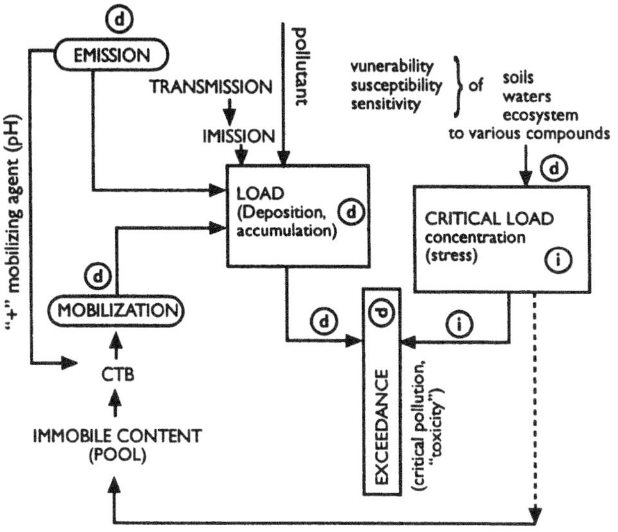

FIGURE 5. Strategy for pollution control (i: increase; d: decrease)

For a comprehensive assessment of the status and regime of pollutants in soil and for an evaluation of their ecological impacts and environmental hazards it is necessary to identify and quantify

- the character of the contaminants (their total, soluble, mobile, plant-, animal-, and human-available and toxic quantities); and
- their sources and pathways (atmospheric wet and dry deposition, vertical movement within the soil; horizontal transport, such as surface runoff, seepage in the unsaturated zone, groundwater flow; abiotic and biotic transformation, etc.)

Soil pollution is sometimes perceived in the West as being a serious problem in all CEECs. However, in recent years considerable efforts have been made in Hungary to assess the true situation along the above lines and similar efforts are taking place in neighbouring countries. At the same time, thematic maps have been prepared on the susceptibility/vulnerability of soils against various pollutants, their critical loads calculated and used as guidelines for non-harmful waste, waste-water and sewage sludge disposal.

6. Conclusions and Future Tasks

The modern concept of "soil quality" needs a complex, multidisciplinary, "fit for purpose" approach based on the multifunctionality of soils. The definition, quantification and interpretation of soil quality has to express the ability of the soil to perform various functions, and has to reflect its territorial and historical variability, as determined by natural conditions and socio-economic circumstances. The neglect of the "multifactorial" and "multifunctional" character of soil quality may result in over-exploitation and misguided management of soil resources, and may lead to serious environmental deterioration.

On the basis of these assessments, soil quality can be interpreted from both production and environmental points of view. Agricultural production is economy driven. In contrast, the maintenance of the multifunctionality of soils and the protection of the natural environment are not completely economy-dependent elements of sustainable development, but rather may be regarded asimperative tasks for human society aimed at improvement of the *"quality of life"*.

Both production and environmental aspects have to be taken into consideration for efficient, economically viable, socially acceptable and environmentally sound sustainable development, which should include the following main elements:

1) Territorial co-ordination of agro-ecological conditions and land-site characteristics, as well as the agro-ecological requirements of cultivated crops, taking into consideration both production and the environmental aspects in the short-, mid- and long-term scales.
2) Rationalisation of the structure of agricultural fields, particularly under new ownership structures.
3) Reduction to a minimum of production wastes with promotion of efficient re-cycling with no harmful environmental side-effects.
4) Prevention of unfavourable soil degradation processes.
5) Improvement of the efficiency of agricultural water management and soil moisture control. Increase the water storage capacity of the soil in a plant available form with no unfavourable environmental consequences.
6) Precision plant nutrient management.
7) Prevention and control of soil pollution.

These actions are joint tasks for the state, decision makers at various levels, land owners, land users and, to a certain extent, each member of society. Only by such joint efforts can sustainable agricultural development be efficiently harmonised with successful environment protection, thus ensuring a pleasant environment and a promising future.

7. References

1. AGRO-21, 1995. Future view of the agriculture. "AGRO-21" Brochures. 10. 5-26. Budapest.
2. Arnold, R. W., Szabolcs, I., and Targulian, V. (eds.) (1990) *Global Soil Change*. IIASA. Laxenburg.-Budapest.
3. Bouma, J., and Bregt, A. K. (eds.) (1989) *Land Qualities in Space and Time*. Pudoc, Wageningen.
4. Doran, J. W., and Jones, A. J. (eds.) (1996) *Methods for Assessing Soil Quality*. Soil Sci. Soc. Am. Special Publ. No. 49.
5. FAO (1997) *Land Quality Indicators and Their Use in Sustainable Agriculture and Rural Development*. FAO Land and Water Bull. No. 5. Rome.
6. Fresco, L. O., Stroosnijder, L., Bouma, J., and Keulen, J. van (eds) (1994) *The Future of the Land (Mobilising and Integrating Knowledge for Land Use Options)*. John Wiley and Sons, Chichester.
7. Greenland, D. J. and Szabolcs, I. (eds.) (1993) *Soil Resilience and Sustainable Land Use*. CAB Inter-national, Wallingford.

8. Hinricksen, D. and Enyedy, Gy. (eds.) (1990) *State of the Hungarian Environment*. Hung. Acad. Sci. - Min. for Environment and Water Management - Central Statistical Office, Budapest.
9. Kádár, I. (1995) *Contamination of the Soil-Plant-Animal-Human Foodchain with Chemical Elements in* Hungary (In Hungarian) KTM-MTA TAKI, Budapest.
10. Láng, I., Csete, L. and Harnos, Zs. (1983) *The Agro-ecological Potential of Hungarian Agriculture in 2000*. (In Hungarian) Mezõgazd. Kiadó. Budapest.
11. Nortcliff, S. (1977) Standardisation of soil quality - what progress? *ISSS Bulletin*. No. 92. 1997/2. 24-26.
12. Soil Survey Workshop: Perspectives and Strategies for the 21st Century. *ITC Journal*. 1993/1. 1-101.
13. Stefanovits, P. and Várallyay, Gy. (1992) State and management of soil erosion in Hungary. In: *Proc. Soil Erosion Prevention and Remediation Workshop*, Budapest, 1992. Vol. I. 79-95.
14. Szabolcs, I., and Várallyay Gy. (1978) Limiting factors of soil fertility in Hungary. (In Hungarian) *Agrokémia és Talajtan*. 27. 181-202.
15. Szabolcs, I., Darab, K. and Várallyay, Gy. (1969) Methods for the prognosis of salinization and alkalization due to irrigation in the Hungarian Plain. *Agrokémia és Talajtan*. 18. Suppl. 351-376.
16. Thomasson, A. J., and Jones, R. J. A. (1989) Land evaluation at regional scales. In: Bouma, J., and Bregt, A. K. (eds.) *Land Qualities in Space and Time*. 231-240. Pudoc, Wageningen.
17. Várallyay, Gy. (1989a) Soil degradation processes and their control in Hungary. *Land Degradation and Rehabilitation*. 1. 171-188.
18. Várallyay, Gy. (1989b) Soil water problems in Hungary. *Agrokémia és Talajtan*. 38. 577-595.
19. Várallyay, Gy. (1989c) Land evaluation in Hungary - Scientific problems, practical applications. In: Bouma, I., and Bregt, A. K. (eds.) *Land Qualities in Space and Time*. 241-252. Pudoc, Wageningen.
20. Várallyay, Gy. (1993) Soil data-bases for sustainable land use: Hungarian case study. In: Greenland, D. I. and Szabolcs, I. (eds.) *Soil Resilience and Sustainable Land Use*. 469-495. CAB International. Wallingford.
21. Várallyay, Gy. (1994) Precision nutrient management - impact on the environment and needs for the future. *Commun. Soil Sci. and Plant Anal.* 25. (7-8) 909-930.
22. Várallyay, Gy. (1995) Sustainable development, a challenge for rational land use and soil management. Proc. Conf. "From Soil Survey to Sustainable Farming"; Vedecké práce, Vysk. ustav podnej urodnosti, Bratislava, 19/II. 227-241.
23. Várallyay, Gy. (1996) Soil pollution management in Hungary. In: *Remediation of Soil and Groundwater. Opportunities in Eastern Europe*. (Eds.: McBeam, E. A., Balek, J., Clegg, B.) NATO ASI Series 2. Environment. Vol. 17. Kluwer Academic Publishers, Dordrecht - Boston - London. Part II. II/5. 87-112.
24. Várallyay, Gy. and Leszták, M. (1990) Susceptibility of soils to physical degradation in Hungary. *Soil Technology*. 3. 289-298.
25. Várallyay, Gy., Buzás, I., Kádár, I. and Németh, T. (1992) New plant nutrition advisory system in Hungary. *Commun. Soil Sci. Plant Anal.* 23. (17-20) 2053-2073.
26. Várallyay, Gy., Szûcs, L., Zilahy, P., Rajkai, K. and Murányi, A. (1985) Soil factors determining the agroecological potential of Hungary. *Agrokémia és Talajtan*. 34. Suppl. 90-94.

PART II

SOIL QUALITY STUDIES IN CENTRAL AND EASTERN EUROPEAN COUNTRIES

Synopsis and Overview

Although the desirability of the wide-ranging approach to the assessment of soil quality proposed in Part I of this book may be agreed, nevertheless from a practical point of view many studies, both in the past and currently, focus on a particular threat or upon the impact and behaviour of a single group of pollutants. For example, there have been many reports and papers on the heavy metal contamination of soils in the countries of central and eastern Europe and the perception has grown, especially in the West, that there is a serious and widespread problem. Whether perception always corresponds to reality is another matter. The only way to resolve the issue is to take a comprehensive and thorough approach to the matter on a national scale. This was the background of the paper by *Terelak and Motowicka-Terelak* for the agricultural soils of Poland with regard to their heavy metal and sulphur status. The study involved the sampling and analysis of 40,000 soils distributed over the whole of the national territory. It was found that 79% of the soils had heavy metal contents at background levels, whilst a further 18% contained slightly elevated levels of heavy metals. The only limitation on these soils is that they should be excluded from cultivation of plants which would be directly consumed by children. Only 0.3% of the soils analysed are classed as heavily or very heavily polluted with heavy metals and in this case they should be excluded from agricultural production. A general conclusion was that the heavy metal status of agricultural soils in Poland was lower than that of most other countries in Europe and in the world. With regard to sulphur, only 3.9% of Polish agricultural soils could be regarded as heavily sulphate polluted, although this does not necessarily mean that soil acidification is not a problem.

Similarly, in a study of organic contaminants of agricultural soils in Poland, focusing particularly on PAHs which are known for their strong mutagenic and carcinogenic properties, *Maliszewska-Kordybach* found that although "hot spots" existed near urban areas and emission sources, in general only background levels were detected in the great majority of samples. In fact, from the data reported it appears that the average level of PAHs in the soils of certain central European countries, namely Poland, Slovakia and the Czech Republic, is similar to that found in Germany, Switzerland and the UK.

In a study of soil quality of the agricultural soils of Hungary, *Murányi* presents results from comprehensive national soil monitoring systems that have been in place for some time. The results for heavy metal pollution are rather similar to those found for Poland, with background levels being recorded for the most part, but with large deviations in certain hot spots indicating serious heavy metal contamination at a local level. In general though, it was concluded that the soil cover of Hungary is not

M.J. Wilson and B. Maliszewska-Kordybach (eds.),
Soil Quality, Sustainable Agriculture and Environmental Security in Central and Eastern Europe, 35–36.

polluted with heavy metals. Average values recorded were Cd 0.5ppm, Co 9 ppm, Cr 18 ppm, Cu 19 ppm, Pb 18 ppm, Zn 49 ppm and Hg 0.05 ppm. However, soil monitoring has brought other problems to light. For example, there is a continuing trend towards soil acidification and soil erosion is a significant problem on 24% of the total land area of Hungary. *Murányi* describes and compares approaches to soil pollution problems based upon Environmental Risk Assessment and Environmental Monitoring and concludes that an integrated concept is the best solution. He further concludes that in the measurement of soil quality over the long term, it is essential to take into account both the chemical and microbiological characteristics of the soil.

Turning to a study of soil degradation and pollution in the Czech Republic, *Podlešáková et al* describe the results of a comprehensive national approach to ascertain the seriousness of the threats resulting from soil erosion, soil acidification, loss of organic matter and pollution with heavy metals and organic compounds. It is evident that erosion is a serious problem with water erosion causing deterioration of 17% of the country's land base and wind erosion affecting 10.4% of all arable land. Surprisingly, it was found that although soil acidification is important in areas of dying forests in the CR, the data for other forest soil do not show convincingly that recent soil acidification has occurred. The results for pollution of soils with heavy metals and organic compounds are similar to those described above, with serious contamination being confined to certain areas which are identified and described in detail.

Finally, *Andronikov* reviews the current status of the soils in Russia on the basis of work done at the V. V. Dokuchaev Soil Institute. Soil erosion is again identified as a serious problem and with the drastic reduction in organic inputs, mineral fertilisers and lime since the early 1990s, there has been a corresponding decline in overall soil fertility. In certain areas, which may be extensive, salinization, water-logging, overgrazing and desertification problems have continued to increase. As far as soil pollution is concerned, whilst the problem is most serious in the vicinity of the main industrial centres, it nevertheless seems that extensive areas of arable lands, particularly those on chernozems near the border with Ukraine may be badly affected. A specific problem with radionuclide pollution of soils occurred as a result of the Chernobyl accident and it is estimated that 4.9 mln. ha have been contaminated.

In conclusion, it worth emphasizing that policy makers and scientists in many countries in Central and Eastern Europe have long been aware of the potential seriousness of environmental problems related to soil quality and have been active in promoting and pursuing studies aimed at understanding, identifying, characterising and ameliorating such problems on their national territories. This is reflected in the extensive data bases that exist in many of these countries as well as in the establishment and support of national monitoring schemes. There are, of course, continuing problems with lack of resources and the depressed state of agriculture generally, a theme that will be discussed particularly in Part IV of this book.

THE HEAVY METALS AND SULPHUR STATUS OF AGRICULTURAL SOILS IN POLAND

HENRYK TERELAK, TERESA MOTOWICKA-TERELAK

Institute of Soil Science and Plant Cultivation
24-100 Puławy, ul. Czartoryskich 8, Poland

1. Introduction

The protection and proper use of soil, which is one of the most important elements of the biosphere, is a major focus of human activities. The soil is characterised by defined physical, chemical and biological properties developed under the influence of natural and anthropogenic factors of the soil forming process. These properties are in a state of dynamic equilibrium. Human activities, particularly industrial ones, can destroy the natural biogeochemical balance existing in the soil and in these circumstances it is often very difficult to bring a degraded soil back to its natural stage and in extreme cases the degradation can be irreversible [6].

The intensive development of industrial and agricultural activities in some areas of Poland has caused serious degradation of soils [6, 10, 11]. Many publications and reports published in different media have created much interest in data concerning the quality of the environment and some reports have lead to a perception that the majority of agricultural soils in Poland are considerable polluted with various chemical elements and substances [2]. In addition, on the world map of soil degradation published by UNEP/ISRIC [12] in 1990, the area of Poland was presented as strongly degraded. This opinion, even though it seems doubtful that there is enough scientific evidence to draw such an extreme conclusion, was based on inadequately documented or fragmentary data from some specific regions and generalised for the whole country. Propagation of such an unreliable information has caused serious social and economic problems, manifested by sale limitations of some Polish agricultural products on the foreign and domestic markets. This situation lead the Ministry of Agriculture and Food Economy to establish in 1991 a country wide soil and crop monitoring program to evaluate the true situation with regard to the state of Polish soils.

This article shows the total contents of Cd, Cu, Ni, Pb, Zn and the concentrations of S-SO_4 in the 0-20 cm soil layer of agricultural land in Poland and the degree of soil pollution with these elements.

M.J. Wilson and B. Maliszewska-Kordybach (eds.),
Soil Quality, Sustainable Agriculture and Environmental Security in Central and Eastern Europe, 37–47.

2. Methods and Materials

Since 1991 the Regional Agro-Chemical Laboratories (OSChR) and the Department of Soil Science and Land Protection of the Institute of Soil Science and Plant Cultivation (IUNG) in Pu³awy have been carrying out a research programme with the aim of evaluating the ecological state of the agriculturally used soils in Poland. About 45 000 soil samples collected throughout the country from a depth of 0-20 cm were analysed by the end of 1997. The content of heavy metals (Cd, Cu, Ni, Pb, Zn) is determined in the soil solution after soil decomposition in aqua regia and in the case of sulphur ($S-SO_4$) nephelometrically after extraction of sulphates from the soil with calcium phosphate solution - $Ca\ (H_2PO_4)_2 \cdot H_2O$ (Ensminger's method).

Granulometric composition, humus content, pH and texture, the criteria that influence the content of heavy metals and sulphur in soils and that may control their transfer to the food chain [4], have been also determined in the soil samples. Evaluation of the degree of soil pollution with heavy metals and sulphur is based on standards [4] which take into consideration their content in the soil as well as the above mentioned soil properties. The degree of soil pollution with heavy metals [4] was classified according to a system consisting of 6 classes: no pollution - background level of metals (0°), elevated content of metals (I°), slight (II°), medium (III°), heavy (IV°), very heavy (V°) pollution [4]. To assess the soil pollution with sulphur [4] the scale consisted of 4 classes depending on sulphate content ($S-SO_4$ mg/100 g soil): 0° - natural (background) content (≤ 1.5), 1° - slightly increased content (1.6-2.5), 2° - weak pollution (2.6-5.0), 3° - heavy pollution (> 5.0).

On the base of the results obtained, synthesis of the content of heavy metals and sulphur in soils [10, 11] as well as maps of soil pollution by these components [8, 9] were prepared. The research programme carried out will supply data on soil properties that are indispensable for the evaluation of the ecological condition of the agriculturally used soils in Poland and for the information system on that subject that is being created in the Institute of Soil Science and Plant Cultivation in Pu³awy.

The report has been based on the analytical results of 40,500 bulk soil samples taken from the 0-20 cm soil layer of agricultural land in Poland during the period 1991-1996. A bulk soil sample consists of 10-15 single soil samples collected from the area of 10 x 10 m (100 m^2) and represents an area of about 460 ha.

3. Results and Discussion

3.1. CADMIUM

Cadmium does not fulfil any physiological functions in an organism. Even though it is found in rocks and soils in a considerably scattered form, it is very mobile and easily enters the food chain.

Background concentrations of cadmium reported in the world soils literature range from 0.2-1.05 mg/kg of soil with a mean of about 0.5 mg/kg of soil [3]. The content of cadmium in soil depends on the geological origin of soil parent material, soil texture, intensity of weathering processes, organic mater content in the soil, application of sewage sludges, wastes, high doses of phosphorus fertilizers etc. A serious source of soil pollution with cadmium is dust emission from non-ferrous metallurgical plants, as well as dust coming from mine waste dumps.

The results of our studies, based on analysis of over 40,000 soil samples showed that the mean cadmium content in agricultural soils in Poland is 0.22 mg/kg, with a range from 0.01 to 49.73 (Table 1). The expected range, however, obtained by excluding 5% of extreme values is 0.10-0.46 mg/kg of soil. The expected range, as suggested by Dudka [1] gives the approximate actual content of soil metals in a given area and expresses how dangerous it is from the contamination point of view for the soil environment. The data presented above showed that the mean cadmium concentration and its variability in agricultural soils in Poland is not substantially different than that in soils of other countries [1, 3].

TABLE 1. Content of Cd, Cu, Ni, Pb and Zn in the 0-20 cm layer of agricultural soils and degree of their pollution with these elements

Metal	Geometric mean mg/kg	Range mg/kg		Percentage of agricultural soils within degrees of pollution					
		found	expected	0°	I°	II°	III°	IV°	V°
Cadmium	0.22	0.01-49.73	0.10-0.46	88.8	9.6	1.1	0.3	0.1	0.1
Copper	6.5	0.2-725.0	3.1-13.7	96.5	3.0	0.3	0.1	0.1	0.0
Nickel	6.3	0.1-328.3	2.6-15.0	95.1	4.4	0.4	0.1	0.0	0.0
Lead	13.7	0.1-5000.0	7.4-25.2	96.9	2.5	0.4	0.2	0.0	0.0
Zinc	32.5	0.5-5754.0	16.2-65.1	87.9	10.6	1.3	0.2	0.0	0.0

The highest concentrations of cadmium were found in agriculturally used soils in southern Poland, especially in the highly industrialised Silesia region. Accumulation of cadmium in these soils is undoubtedly partially connected with dust emission from the metallurgical industry in this region. However, it should be stressed that a high content of cadmium in the soils of this area can be also related to outcrops of non-ferrous metal ores or soil parent rocks rich in heavy metals [3].

The data presented in Table 1 show that 88.8% of the agricultural land in Poland is characterised by natural (0°) and in 9.6% by slightly increased (I°) contents of cadmium. The total percentage of soils on which all plants species can be grown (0°+I°) with some limitations (I°) concerning raw materials for baby and children food production is as high as 98.4%. Slightly polluted soils (II°) represent 1.1% of the agricultural land in Poland. Danger related to an increased cadmium pollution level (III° + IV° + V°) concerns only 0.5% of the agriculturally used soils, but locally, in regions such as Silesia, these soils can represents as much as 17% of the agricultural land. In general, the cadmium content of Polish soils does not pose a particular environmental risk as far as food quality is concerned, except in the heavily industrialised regions.

3.2. COPPER

Copper in soils occurs usually in the form of poorly mobile compounds and may be associated with carbonates, sulphates, organic matter and clay materials [3]. Acidic reactions increase copper phytotoxicity in the soil but neutral and alkaline reactions decrease it.

The copper content of the unpolluted soils of the world ranges from 1 to 140 mg/kg and depend strongly on the content of the clay fraction in the soil.

The results of our investigations (Table 1) showed that the mean copper content in agricultural soils in Poland is 6.5 mg/kg with a wide range of found (from 0.2 to 725.0 mg/kg) and expected (3.1-13.7 mg/kg) values. The average concentration of copper in soils in Poland (6.5 mg/kg) is similar to the background level reported for other countries [3]. The highest copper content in agricultural soils in Poland was found in the southern and south-west part of the country (the region of copper production and metallurgical engineering, as well as in the regions in which soils with natural accumulations of this element occur). Soil contamination with copper poses a risk only in areas with copper mining and smelting industries Even in the Silesia area, the average concentration of copper in soils (11.2 mg/kg) is much lower than the upper value of the expected range obtained for the whole country (Table 1).

In the area of Poland, 96.5% of the agricultural soils show natural (0°), and 3.0% increased (I°) copper contents (Table 1). The latter soils are mainly located in the copper mining areas within Legnica Province. In practice, soils weakly polluted with copper (II°) do not occur outside this area, and make up only 0.3% of the total agricultural land of the country. The medium (III°) and strongly (IV°) copper contaminated soils are located at certain points in some small areas and constitute 0.2% of the agricultural land. On a national basis, the percentage of soil contaminated with copper on a greater scale (III°-V°) is very small. Therefore, it can be concluded that copper is not an element that limits output of high quality of agricultural products, except in small areas affected by the copper industry and in soils formed from parent rock material containing high amounts of this metal.

3.3. NICKEL

Nickel is geochemically connected with iron and cobalt and this connection controls its distribution, appearance and behaviour in the soil environment. High concentrations of iron hydroxides and clay minerals in soils limit nickel mobility in the environment.

The average nickel content in soils of the world is strongly differentiated. It is commonly accepted that a value of 100 mg Ni/kg of soil is the maximum permissible level [3]. The average nickel content in agricultural soils in Poland is 6.3 mg/kg with the found range from 0.1 to 321.3 mg/kg (Table 1). It is significant that the upper level of the expected range for nickel (15.0 mg/kg) throughout the country does not exceed the threshold value. Similarly, as in the case of cadmium and copper, high concentrations of nickel were found in soils of the southern part of Poland. High contents of nickel in soils of this area are not limited to some places polluted with this element but are found over larger areas and are related to the type of soil parent rocks (flysch, heavy clays).

The data presented in Table 1 show that 95.1% of the agricultural land in Poland is characterised by a natural content of nickel (0°) and 4.4% by a slightly increased (I°) content of this element. Many soils with elevated nickel levels are very likely associated with the use of nickel in smelting and galvanising processes [10]. Soils slightly polluted (IV°) with nickel appear sporadically in Poland, but the area they occupy does not exceed 0.4% of the total agricultural land of the country (Table 1). The data presented in Table 1 demonstrate that soils belonging to the higher degrees of contamination with nickel (III° + IV° + V°) are of very limited occurrence. On the basis of the data obtained, it can be concluded that nickel concentration in agricultural soils in Poland does not significantly affect agricultural production and crop quality, even in polluted regions such as Silesia.

3.4. LEAD

Lead is characterised by poor mobility in soils due to the low solubility of the minerals in which it occurs. The lead content in soils depends on the mineralogical and granulometric composition and origin of the soil bedrock. In soils, an acidic reaction, poor sorption properties and low content of humus favour excessive lead uptake.

The lead content in soils of the world ranges from 25 to 40 mg/kg [3], and in the agricultural soils in Poland it is 13.7 mg/kg of soil with a range of 0.1-5000.0 mg/kg of soils (Table 1). The extreme values reported were found in a number of places in Silesia and are associated with lead and zinc mining [10, 11]. The expected range of lead content in the soils studied (7.4-25.2 mg/kg of soil) is much more narrow than the found one (0.1-5000.0 mg/kg soil). This confirms an insignificant contribution of soils with very high concentrations of this element. Larger areas of soils with a high lead content occur in the southern part of Poland, especially in district of Katowice. The lead content in soils of this area is strongly related to metallurgical engineering, mining industry and heavy car traffic.

The data presented in Table 1 show that 96.9% of the agricultural soils in the country exhibit a natural content (0°) of lead, and 2.5% an increased level (I°) - Table 1. Only a small proportion of the agricultural land in Poland is characterised as slightly (II°) - 0.4% or medium (III°) - 0.2% polluted with lead. These soils only occur locally and are, in practice, of no danger to the quality of plant materials produced. Higher degrees of soil pollution with lead are of little importance, except in some small areas in Silesia.

3.5. ZINC

Zinc is indispensable for the metabolic processes of living organisms. Zinc availability for plants is high, and an acidic reaction favours its mobility in the soil environment.

The mean zinc content in soils of the world ranges from 27 to 235 mg/kg [3]. The natural zinc content in soils in Poland should not exceed 40 mg/kg [3].

The average concentration of zinc in agricultural soils in Poland is 32.5 mg/kg (Table 1) with a found range of 0.5-5754.0 and an expected range of 16.2-65.1 mg/kg of soil. The above data show that both low and high zinc concentrations are not so common in Polish soils. Nevertheless, the upper limit of the expected range (65.1 mg /kg of soil) considerably exceeds the zinc concentration considered as natural background - 40 mg/kg of soil [3]. High zinc concentrations occur mainly in soils of southern Poland and is to a high degree connected with zinc emission into the environment by the smelting industry of non-ferrous metals located in Silesia. The origin of parent rock material also plays an important role, particularly in areas which are not directly exposed to industrial pollution.

The data presented in Table 1 show that 87.9% of farm land in Poland do not show zinc pollution (0°). Slightly elevated concentrations of zinc (I°) are found in 10.6% of soils. Only small areas (1.3%) of agricultural soils show weak (II°) and 0.2% medium (III°) zinc pollution levels. Soils polluted with zinc appear in small areas over the whole area of Poland but particularly in the Katowice district and in the adjacent regions.

3.6. HEAVY METAL POLLUTION OF AGRICULTURAL SOILS IN POLAND

Soil pollution with all metals studied as presented by data in Table 2 and graphically in Figure 1 show that 78.9% of the agricultural land in Poland is characterised by a natural content (0°) of all heavy metals studied. These soils can be used for all kinds of field and orchard farming, including vegetables. About 18% of the agriculturally utilised soils in the country show a slightly elevated level (I°) of heavy metals. These soils should be excluded from the cultivation of plants for direct consumption by children. Despite some limitations these soils are very valuable for agricultural production. Slightly polluted soils (II°), where there is potential danger of chemical plant contamination occupy 2.3% of agricultural land in Poland. No horticultural plants for direct consumption should be grown there. The only crops allowed in this case are cereals, root crops and forage crops. Since the soils showing medium (III°) pollution with heavy metals in Poland occupy only about 0.5% of the total area of the agricultural land they are of little consequence for the overall quality of agricultural production. Cereals, forage crops and root crops can be grown on these soils if they are periodically checked for heavy metal contents of edible plant parts and fodder material. The contribution of heavily (IV°) and very heavily (V°) heavy metals polluted soils to the total area of agricultural land does not exceed 0.3%. These soils should be excluded from the agricultural production but they can be used for growing industrial plants, production of sowing material, afforestation etc.

TABLE 2. Pollution of agricultural soils with all heavy metals (0-20 cm layer)

Area	Percentage contribution of agricultural soils within degrees of pollution					
	0°	I°	II°	II^I°	IV°	V°
Poland	78.9	18.0	2.3	0,5	0.2	0.1

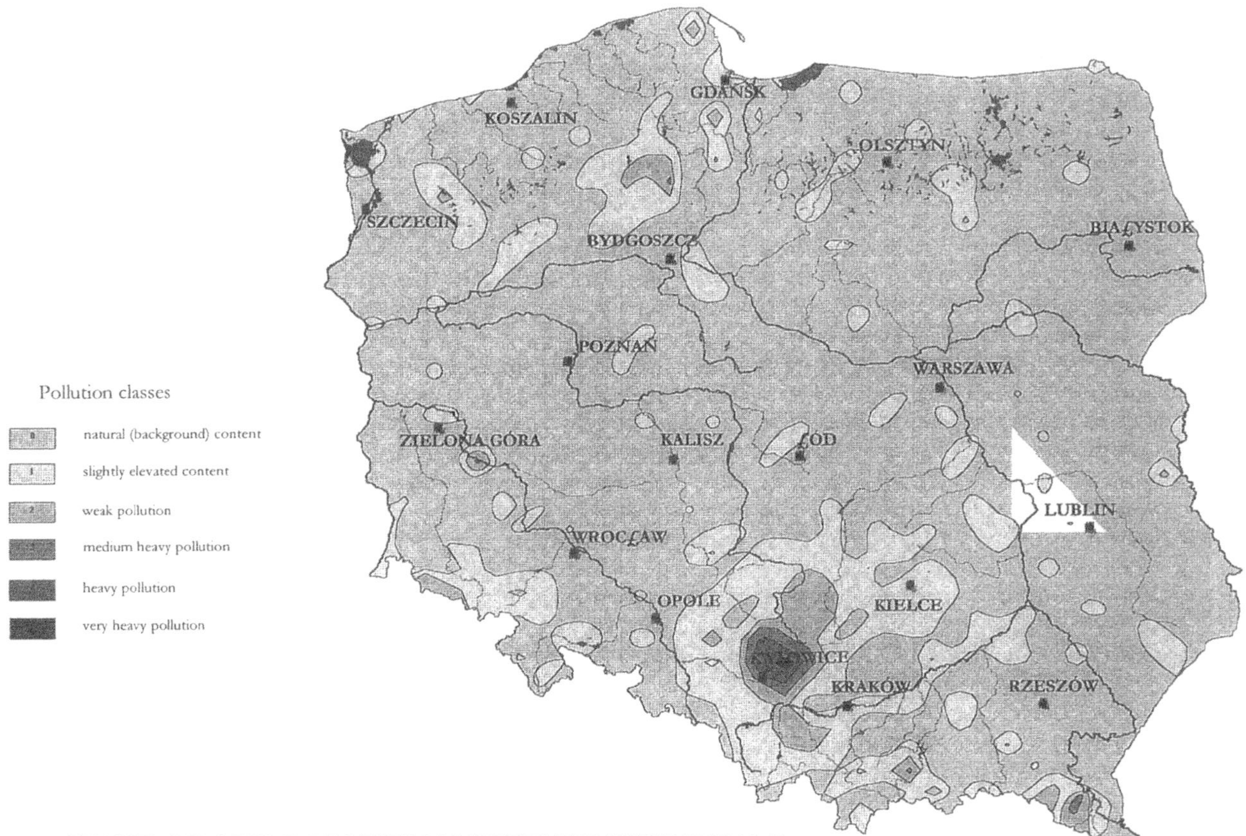

Fig. 1. STATE OF SOIL POLLUTION WITH HEAVY METALS
(see also coloured plate at end of book)

3.7. SULPHUR

Sulphur, in addition to N, P, K and Mg is an essential macroelement for plant growth. The concentration of sulphur in soils depends on the organic matter content, level of organic and mineral fertilisers applied, and on the input of this element associated with air pollution. Both excess and deficiency of sulphur in soils can have an adverse effect on plant growth. Excessive concentrations of sulphur in plants are considered as one of the important factors affecting the nutrition value of food and foodstuff, including impact on the quality of the processed food [5].

Chemical degradation of soils through acidification and significant increase of sulphate concentrations ($S\text{-}SO_4$) are the primary negative effects of soil pollution with sulphur caused by anthropogenic factors. Among a number of sulphur forms, sulphate sulphur is the most important fraction of this element in the context of crop production and environmental protection [5, 7]. Sulphates are a direct source for uptake by plants. The concentration of sulphates can be also considered as an index of soil pollution with sulphur.

There is a wide range of sulphate sulphur concentrations ($S\text{-}SO_4$) in the A horizons (0-20 cm) of mineral soils in Poland. In general, sandy and low organic matter soils contain smaller amounts of sulphur while heavy soils, black earth and alluvial soils in particular [7], contain much more.

The data presented in Table 3 show that the average concentration of sulphate ($S\text{-}SO_4$) in agricultural mineral soils in Poland is 1.82 mg/100 g with a range from 0.01 to 50.00 mg/100 g of soil. It was found that 57.0% of soils are characterised by a natural content of sulphate (0°) - ≤ 1.5 mg/100 g of soil, but 3.9% is heavily polluted (3°) with this element - > 5.0 mg/100 g of soil. A slightly increased content of sulphate (1°) - 1.6-2.5 mg $S\text{-}SO_4$/100 g of soil, and weak pollution (2°) with this element - 2.6-5.0 mg/100 g of soil were found in 25.4 and 13.7% of agricultural land, respectively. The largest areas of soils polluted with sulphur are located in the south and south-west regions of Poland (Figure 2).

TABLE 3. Content of sulphur ($S\text{-}SO_4$) in the 0-20 cm layer of agricultural soils and degree of their pollution with this element

Area	Range mg/100 g	Mean value mg/100 g	Percentage of agricultural soils within degrees of pollution			
			0°	1°	2°	3°
Poland	0.01-50.00	1.82	57.0	25.4	13.7	3.9

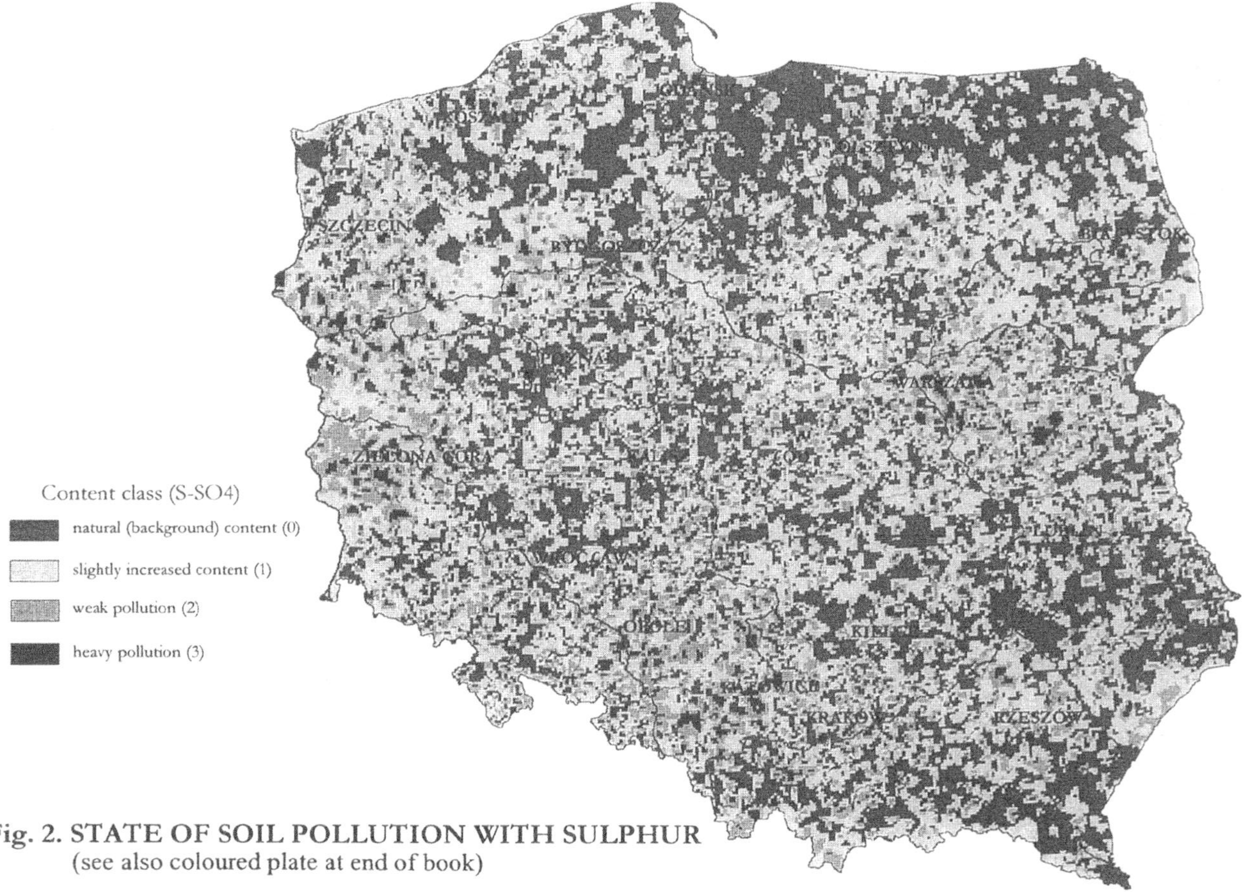

Fig. 2. STATE OF SOIL POLLUTION WITH SULPHUR
(see also coloured plate at end of book)

4. Conclusions

1.The mean content of cadmium, copper, nickel, lead and zinc in the 0-20 cm layer of agricultural soils in Poland is respectively: 0.22, 6.5, 6.3, 13.7 and 32.5 mg/kg of soil. These values are lower than in most other countries of Europe and the world.
2. More than 78% of the agricultural soils in Poland show a natural (0°) content of heavy metals. Soils with an increased (I°) content of heavy metals constitute 18.0%. Soils that guarantee production of raw plant materials with quality parameters occupy 97% of the agricultural land on total. About 3.0% of all soils are polluted with heavy metals to various degrees (II°-V°). These soils do not create any major problem for Polish agriculture and do not significantly influence the quality of agricultural products in Poland.
3. Mineral soils of agricultural land in Poland contain on average 1.82 mg $S\text{-}SO_4$/100 g.
4. About 57% of the agricultural soils in Poland show a natural (0°) content of sulphate ($S\text{-}SO_4$), whereas heavy sulphate polluted soils (3°) contribute only 3.9%. Soils with slightly increased (1°) sulphate content and weakly polluted with this element occupy, respectively 25.4 and 13.7% of the agricultural land.
5. The perception that Polish agricultural soils are heavily polluted by heavy metals and sulphate ($S\text{-}SO_4$) is not confirmed by this study. In fact, this perception is directly contradicted by the results presented here wich are based on the most thorough sampling exercise undertaken in Poland (40.500 samples). Some areas of agricultural soils with various degree of pollution by heavy metals and sulphate do occurs mainly in the industrialised areas of southern and south-western parts of Poland.

5. References

1. Dudka, S. (1992) *Total content of main and trace elements in the surface layer of Polish soils* (in Polish). IUNG, Pu³awy, Seria R (293).
2. Gliñski, J. and Turski, R. (1991) Chemical pollution of Polish soils, in N.H. Batjes and E.M. Bridges (eds.) *Soil Vulnerability to Pollution in Europe,* ISRIC, Wageningen, pp. 99-104.
3. Kabata-Pendias A. and Pendias H. (1993) *Biogeochemistry of trace elements* (in Polish), PWN Warszawa.
4. Kabata-Pendias, A., Motowicka-Terelak, T., Piotrowska, M., Terelak, H. and Witek, T. (1993) *Assessment of soil and plant pollution with heavy metals and sulphur - Guidelines for agriculture,* IUNG Pu³awy, Seria P (53a).
5. Motowicka-Terelak, T. and Dudka, S. (1991) *Chemical degradation of soils polluted with sulphur and its effects on cultivated plants* (in Polish), IUNG, Pu³awy, Seria R (284).
6. Siuta, J. (1995) *Soils* (in Polish), Instytut Ochrony ŒErodowiska, Warszawa.
7. Terelak, H., Motowicka-Terelak, T., Pasternacki, J. and Wilkos, S. (1988) Sulphur contents in mineral soils of Poland (in Polish), *Pam. Puł. Supl. 91.* 1-59.
8. Terelak, H., Motowicka-Terelak, T.and Budzyñska, K. (1994) *Content of sulphur in soils. Map in scale 1:500 000.* Okr. Przeds. Geod.-Kartogr., Bia³ystok.
9. Terelak, H., Piotrowska, M. and Budzyñska, K. (1994) *State of soil pollution with heavy metals. Map in scale 1:500 000.* Okr. Przeds. Geod.-Kartogr., Bia³ystok.

10. Terelak, H., Piotrowska, M., Motowicka-Terelak, T., Stuczyñski, T. and Budzyñska, K. (1995) The content of heavy metals and sulphur in soils of agricultural land and the degree of their pollution with these elements (in Polish), *Zesz. Probl. Post. Nauk Rol. 418, vol. 1.* 45-60.

11. Terelak, H., Piotrowska, M., Motowicka-Terelak, T., Stuczyñski, T., Budzyñska, K., Pietruch, Cz. and Sroczyñski, W. (1997) Chemical properties of soils and heavy metals and sulphur contents in soils and plants (in Polish). Report prepared for the Ministry of Agriculture and Food Economy, IUNG, Pu³awy.

12. *World map on status of human induced soil degradation* (1990), UNEP/ISRIC, Wageningen.

ORGANIC CONTAMINANTS IN AGRICULTURAL SOILS IN CENTRAL AND EAST EUROPEAN COUNTRIES AS COMPARED TO WEST EUROPEAN COUNTRIES; EXAMPLE OF PAHs.

B.MALISZEWSKA-KORDYBACH
Institute of Soil Science and Plant Cultivation
24-100 Pulawy, Poland

1. Introduction

Polycyclic aromatic hydrocarbons (PAHs) - know for their strong mutagenic, carcinogenic and toxic properties - belong to the group of persistent organic compounds (POPs) i.e. organic contaminants which are relatively resistant to biodegradation, can remain in the environment for long periods and have the potential to cause adverse environmental effects. There are thousands of PAH compounds in the environment but in practise PAHs analysis is restricted to the determination of 6 to 16 compounds. Individual PAHs differ substantially in their physical and chemical properties. Generally, the high molecular weight compounds (> 3 aromatic rings) are less water soluble, less volatile and more lipophilic than lower molecular weight ones [1,19,32]. The best know model compound from this group is the highly carcinogenic benzo/a/piren (BaP). PAHs are listed by the United States Environmental Pollution Agency and the European Commission as priority pollutants.

PAHs are multimedia contaminants which occur in all compartments of the environment; atmosphere, inland and sea waters, sediment, soils and vegetation [19]. The widespread occurrence of PAHs is largely due to their formation and release in all processes of incomplete combustion of organic materials - Table 1. At the present time natural sources of PAHs do not contribute significantly to overall PAH emission which derives mainly from anthropogenic sources. PAHs emitted to air can be transported over long distances before they are deposited with atmospheric precipitation on soils, vegetation or sea and inland waters [8,31]. The last century of industrial development and higher utilisation of coal and petroleum products caused significant increase of PAHs concentrations in the natural environment; from about four fold in the plough layer of soils from a semi-rural area in Rothamsted, UK [14], ten fold in lake sediments in the US [9] and about 50 times in ice from Greenland [15].

The majority of PAHs released to the atmosphere concentrate finally in the terrestrial environment [26,32,40], which is the main sink for most of the hydrophobic organic contaminants - Table 2.

M.J. Wilson and B. Maliszewska-Kordybach (eds.),
***Soil Quality, Sustainable Agriculture and Environmental Security in Central and Eastern Europe*, 49–60.**

TABLE 1. Sources of PAHs in the environment

NATURAL	ANTHROPOGENIC
	Pyrogenic origin
• Volcanic activity • Natural fires	• Combustion of materials for energy supply. Stationary Sources; Mobile Sources; - industry - vehicles - domestic heating - airplanes - power and heat generation - ships • Combustion of materials for waste minimalization. • Other miscellaneous. - unregulated fires - cigarette smoking
	Biogenic origin
• Crude oil formation • Plant pigments and resins	

It is estimated [26,40], that more than 90 % of the total burden of PAHs resides in the subsurface soils though soil profile data show that these pollutants may extend into much deeper layers [12,37,41].

The main cause of contamination of the terrestrial environment with PAHs is atmospheric deposition [2,6,11-14] although only a fraction of it is deposited directly on the soil whereas the other part is initially intercepted by plants and incorporated to soil with vegetation at the end of the growing season [30,31]. The other sources of PAHs to agricultural land are sewage sludges and composts applied as fertilisers, road runoffs, accidental fuel spills and leakages [2,11,13,14,32,39,40].

TABLE 2. Partition of POPs in the environment

Part of the environment	PAHs[1]	PCB[2]
	Percent of total	
Soil	**93.9**	**89.2**
Sediments /freshwaters/	4.5	9.9
Wastes and sludges	1.4	-
Air	0.1	0.5
Vegetation	0.1	-
Water	< 0.01	0.3
Biota	< 0.01	0.1

[1])Wild and Jones, 1995 [40]; [2])Paterson and Mackay, 1988 [26]

PAHs incorporated into the terrestrial environment are not static and recalcitrant but may be subject to several processes. The main mechanisms include biodegradation, chemical transformations, volatilisation, leaching, photolysis, sorption to the soil solid phase and transfer to plants and grazing animals [2,4,11,13,19,39,40]. The overall loss or dissipation of organic contaminants from soil is dependent on the properties of the chemicals, environmental factors including soil characteristics, temperature or humidity and anthropogenic factors such as cultivation or drainage [2,11,13,17,19,32,40].

TABLE 3. Human exposure to PAHs

Source of PAHs	Exposure to PAHs[1] (non smokers)	PAH burden[2]
	% of total	
Food	**93.0**	< 0.01
Air	4.9 - 18.0	0.1
Water	0.2	< 0.01
Soil (accidental ingestion)	1.9	**94.0**

[1]) Menzie *et al.*, 1992 [25]; [2]) Wild and Jones, 1995 [40]

PAHs present in soil may exhibit toxic activity towards different biological elements of the soil environment such as plants, microorganisms, invertebrakes, *etc.* [10,16,19,32]. Pollution of agricultural ecosystems with PAHs may lead to an increase of PAH content in food products. It is estimated [25,26] that most human risk associated with PAHs is related to food consumption - Table 3. This information relates to the fact that soils account for about 90 % of total PAH environmental burden [40] and indicates that the problem of pollution of agricultural soils - which are the basis for food production - with these contaminants is crucial in assessing the degree of human risk. Nevertheless, information on the content of PAHs in soils from agricultural regions both in Europe and in other parts of the world, is very limited.

TABLE 4. Estimated emission of PAHs in Poland (CEC) and in the UK (WEC).

Source	Tonnes/year		Percent of total	
	UK[1]	Poland[2]	UK[1]	Poland[2]
Industrial processes	19	67	2.6	11.8
Stationary sources	610	464	85.4	82.0
Incineration/fires	6	-	0.8	-
Vehicles	80	35	11.2	6.2
Total	712	566	100	100
kg/year/km^2	2.9	1.8	-	-

[1]) Wild and Jones, 1995 [40], [2]) Environment 1998 [6]

The aim of the paper is to evaluate the content of PAHs in agricultural soils in CE countries as related to WE countries. Two countries, for which most comprehensive data sets are available; Poland (Central Europe) and the UK (West Europe) were chosen as examples. Total PAHs emission in Poland is about 20 % lower than in the UK, but in both countries most of the PAHs derive from stationary sources (domestic heating, power stations) - Table 4. In Poland, industry is responsible for higher PAH emissions and transport for lower emissions than in the UK. However, the observed discrepancies between the countries may be due not only to real differences in PAH emissions but also to different estimation methods, the amount of the potential sources included and the number of PAH compounds taken into account.

2. Materials and Methods

Evaluation of the level of PAHs in agricultural soils in Poland was based on the results of 2-years studies (PASMP2) within the 3-year Polish Arable Soils Monitoring Programme (PASMP) sponsored by the Polish Ministry of the Environment and started in 1995 in the Institute of Soil Science and Plant Cultivation in Pulawy, Poland. The area under study covers different regions from typical rural territories in Eastern and North-Western Poland to the highly industrialised Upper Silesia district (Katowice voievodeship) – Fig.1.

FIGURE 1. Poland - the area under study (218 000 km^2)

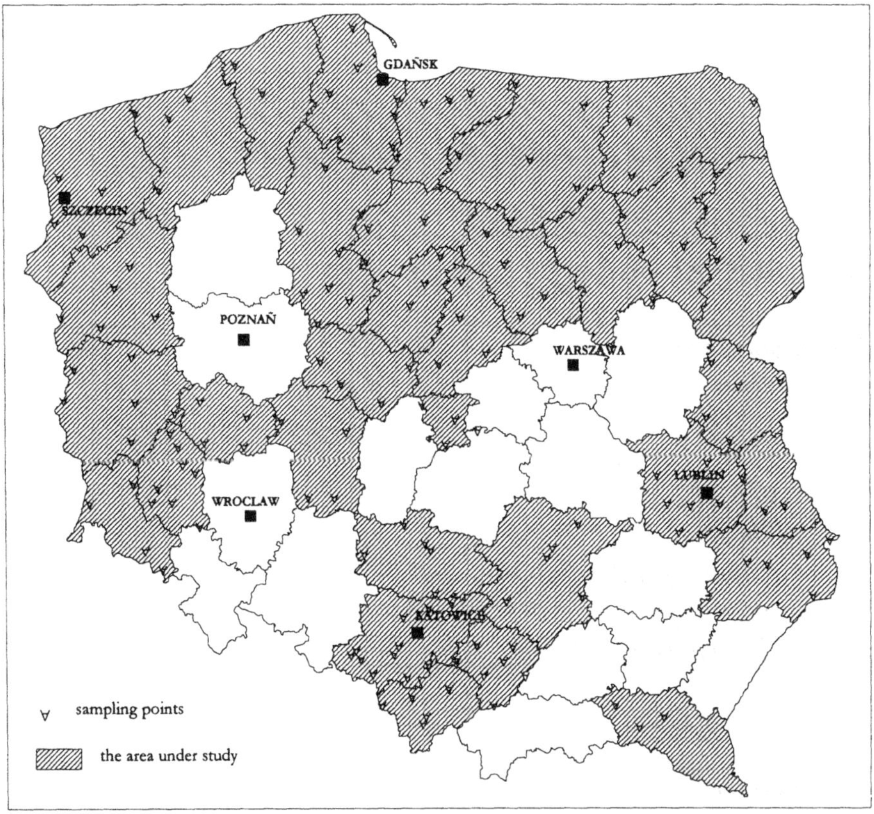

The soil samples (n=151) were taken (1995-1996) from the 0-20 cm soil layer from the area covering about 65% of the territory of the country - Figure 1. The samples were air-dried, ground and screened <2 mm. The sieved fractions were stored in the dark at ambient temperatures. Subsamples of the soils (5-10 g) were Soxhlet-exctracted with methylene chloride followed by cleaning-up on Silica micro-columns. 13 PAH compounds (US EPA List excluding naphthalene, acenaphtylene and acenaphtene) were determined by HPLC analysis (UV detection at 254 nm) of the extracts according to the procedure given elsewhere [20].

The evaluation of the content of PAHs in the UK soils was based on the results published by Cousin *et al.* [4] from Lancaster University, UK. In this study 45 soil samples taken from 10 regions widely distributed over the UK (remote, rural or semirural locations) were analysed for the content of 12 PAH compounds by means of HPLC method with fluorescent detection. About 85 % of the samples were taken from the pastures, natural grassland and other grazing areas. The rest included scrubland, woods and land in common use. In this case arable soils were not included in the study.

3. PAH Content in Agricultural Soils in Poland.

The soils under investigation - Table - 5 were mostly light in texture (median value for fraction <0.02 mm content - 20%), slightly acidic (median value for pH - 5.4), with low organic matter content (median content - 1.73%) and corresponded very well to general characteristics of Polish soils [27]. The variability of the results was rather low; in 90% of the samples organic matter (OM) content was below 2.66 %, pH value was lower than 6.5 and fraction <0.002 mm content did not exceed 14 %.

TABLE 5. Characteristic of the soils (PASMP2; n=151)

Soil property	Statistical parameters					
	Average	Median	Stand. dev.	90 percentile	Range	
					Measured	Interquartile
OM content (%)	1.93	1.73	0.80	2.66	0.8 - 5.75	1.42 - 2.16
pH_{KCl}	5.4	5.4	0.9	6.5	3.6 - 7.2	4.7 - 6.1
fraction < 0.02 mm content (%)	25	20	16	44	2 - 88	13 – 36
fraction < 0.002 mm content (%)	7	6	7	14	1 - 52	3 – 8

Statistical evaluation of the content of ΣPAH as well as the most known PAH compound, benzo(a)pyrene, in Polish arable soils - according to the results received in this study - is given in Table 6. The range of the measured ΣPAH concentrations was 89 - 11 391 μg/kg but in 90 % of the soil samples ΣPAH content did not exceeded 1150 μg/kg and the median was 327 μg/kg - Table 6. Due to the large area covered by the studies (about 200 000 km^2) the variability of the results was rather high: standard deviation for ΣPAH - 1 027 μg/kg. The median content of the most known PAH compound - benzo(a)pyrene - was 31 μg/kg.

TABLE 6. Statistical evaluation of the content of ΣPAHs in arable soils in Poland (PASMP2; n=151).

Parameter	Σ13PAH	Σ13PAH/OM 2 % OM	Σ13PAH/OM 10 %OM	BaP
	μg/kg			
Average	579	505	2 525	62
Median	327	315	1 575	31
Std .deviation	1 028	837	4 185	132
90 percentile	1 150	891	4 455	75
Range: measured	89 - 11 391	67 - 9 076	335 - 45 380	3 - 1 420
interquartile	194 - 683	187 - 588	935 - 2 940	16 - 68

Simultaneously, to enable the assessment of the level of soil pollution with PAHs according to the existing regulations [5,36] the content of PAHs was adjusted (ΣPAH/OM) to standardised soils with 2 % and 10 % OM content.

TABLE 7. The results of the studies on the content of PAHs in agricultural soils in Poland

Region	n	Σ PAH (μg/kg) Average	Median	St.dev	90 perc.	Range	Reference
Poland /MPPAS1/ ~100 000 km^2 (arable lands)	80	520	346	549	1 061	110 - 4 122	Maliszewska-Kordybach and Terelak, 1997 [24]
Poland /MPPAS2/ ~200 000 km^2 (arable lands)	151	579	327	1 027	1 150	89 - 11 391	this study
Pulawy district ~ 300 km^2 /Lublin rural region/ (arable + grasslands)	136	248	196	358	387	70 - 2 787	Maliszewska-Kordybach and Smreczak, 1998 [22]
Lodz district ~ 1 000 km^2 /semirural region/ (arable +grasslands)	26	189	172	110	385	60 - 657	Stuczynski *et al.*, 1992, [35]
Tarnowskie Góry district ~ 40 km^2 /Upper Silesia industrial region / (arable + grasslands)	68	1 852	817	2 911	4 943	62 - 12 760	Maliszewska-Kordybach, 1993 [17]

In Table 7 the data received during this study are compared with the results of earlier investigations [17,22,23,24,35] on the content of PAHs in Polish agricultural soils. The median concentration of PAHs (as determined in PASMP2) in Polish agricultural soils is about twice as high as in the typical rural district of the country (180 μg/kg in Lublin province [21-23]) and about half that of in the most industrialised Upper Silesia region of Poland (817 μg/kg) [17] .

4. Comparison of the Content of PAHs in Polish Soils with the Level of these Contaminants in Soils in West European Countries (WEC).

Information on the content of PAHs in West European soils - as well as for the Central or East European ones - is very limited. Most of the published data refer to highly contaminated soils in urban/industrial areas, military sites, on roadsides, *etc.* (point source pollution) [3,5,7,28,29,33,40,42]. Agricultural soils in rural and semi-rural areas, subjected mainly to diffuse contamination, are characterised usually by much lower PAHs content [7,13,17,21-24,34,37]. More information on the content and the fate of PAHs in the terrestrial environment was given recently by Jones *et al.* [11], Wild and Jones [40] and Maliszewska-Kordybach [19]. The last work [19] includes accessible data on the content of PAHs in soils in some CE and WE countries as affected by regional type, land use and soil characteristic.

The most comprehensive general (country-wide) data set on the content of PAHs in WEC soils is available for the UK from the work of Cousin *et al.* [4]. Statistical evaluation of this data is given in Table 8.

TABLE 8. Statistical evaluation of the content of ΣPAHs in soils in the UK (n=45). (Cousin *et al.* [4])

Parameter	Σ12PAH	Σ12PAH/OM 2 % OM	Σ12PAH/OM 10 % OM
	μg/kg		
Average	1 100	136	680
Median	460	75	375
Std .deviation	1 500	186	930
90 percentile	2 760	347	1 735
Range: measured	20 - 7 400	2 - 809	10 - 4 045
interquartile	200 - 1 100	28 - 157	140 - 785

The median content of PAHs in Polish soils (327 μg/kg - Table 8) is slightly lower than concentration of these contaminants in the UK soils (460 μg/kg - Table 6). It corresponds to the concentration of PAHs in soils in Wales in the UK (302 μg/kg) [13] and in Bavarian soils in Germany (321 μg/kg) [34]. The variability (standard deviation value 1 500 μg/kg) of the data covering all UK territory is about the same as in PASMP2 studies in Poland (1 028 μg/kg).

For further evaluation of the data on PAH contents in Polish soils and in the UK soils, six levels of soil contamination with these compounds were distinguished - Table 9.

TABLE 9. The groups of soil contamination with PAHs in Poland (MPPAS2) and in the UK[1]

PAHs content (μg/kg)	Poland	UK	PAHs content (μg/kg)	Poland	UK
	Percent of total			Percent of total	
		ΣPAH			
< 200	31	27	< 200	31	27
200 - 1 000	56	37	< 1 000	87	64
1 000 - 2 000	10	18	< 2 000	97	82
2 000 - 8 000	1	16	< 8 000	99	98
> 8 000	1	2	> 8 000	1	2
		ΣPAH/2%OM			
< 200	31	82	< 200	31	82
200 - 1 000	60	18	< 1 000	91	100
1 000 - 2 000	7	0	< 2 000	98	100
2 000 - 8 000	1	0	< 8 000	99	100
> 8 000	1	0	> 8 000	100	0

[1]) Cousin *et al.* [4]

The lowest limit value (200 μg/kg) corresponds to the background value proposed for PAHs in Polish agricultural soils [17,21] and to the background value for ΣPAHs according to Dutch regulations [36] (value 1 000 μg/kg for the soil with 10 % of OM recalculated for the soil with 2 % OM). The value 1 000 μg/kg is equal to Canadian limits for PAHs in agricultural soils [2,11] and the highest value of 8 000 μg/kg corresponds to Dutch "intervention" value (40 000 μg/kg, soil with 10 % OM) [36] recalculated for the soil with 2 % of OM.

As can be seen from Table 9, in both countries (Poland and the UK) the least contaminated soils group (ΣPAH <200 μg/kg) includes about 30 % of total samples and the most polluted one (ΣPAH >8 000 μg/kg) - accounts for a very small percentage (1 - 2 %) of the total. A more distinct difference was observed in the group 200 - 1 000 μg/kg which includes 56 % of Polish soils and only 37 % of the soil samples from the UK [4] as well as in the group 1 000 - 2 000 μg/kg; 10 % and 18 % correspondingly

Discussing the results received in the two countries (PASMP2 and the work of Cousin *et al.* [4]) it has to be remembered that soil samples from both countries differ significantly. The Polish studies were on very light, low organic matter (median 1,7 %)

soils from arable lands while the UK soils [4], collected mainly from grasslands were rather heavy (mostly gley or loamy) and characterised by high (median 12,4 %) OM content. Land use as well as soil properties may influence accumulation and losses of PAHs in the soil environment [19]. Earlier works [7, 22, 23, 34]indicate that the content of PAHs in the soils from grasslands is about 20% higher than in arable soils from the same area, which corresponds to the relation between the level of these pollutants in English and Polish soils under discussion. Although there is not much information on the relationship between soil characteristics and the level of PAHs, it is known that soil organic matter content is the most important factor controlling PAHs sorption processes [2,4,11,32,40]. Significant correlation between PAHs and OM content was found to be true only for the soils from the relatively small areas having similar histories of atmospheric deposition [13,17,22,23]. Nevertheless the level of soil pollution with PAHs is evaluated on the basis of PAHs concentration normalised with respect to organic matter content of the soil in Dutch [36] and Swiss [5] regulations (10 % of OM) or in Polish propositions (2 % of OM) [18,21]).

Relating ΣPAH value to soil OM content (ΣPAH/2%OM and ΣPAH/10%OM) - Tables 6, 8 and 9 - changes totally the relationship between the levels of soil pollution with PAHs in both countries (as compared to the evaluation on the base of ΣPAH value). Now the soils in the UK appeared to be much „cleaner". *E.g.* if the soil with 2 % OM is applied as the standard then 82 % of soil samples from the UK and only 31 % of soils in Poland are in the lowest class < 200 μg/kg (Table 9) with the median content of PAHs in UK soils being 75 μg/kg (Table 8) while in Poland - 315 μg/kg (Table 6).

The above results indicate that the evaluation of the level of soil contamination with PAHs may be open to various interpretation. The existing soil quality criteria for organic contaminants vary widely in making allowances for functionality of the soil, the levels (background, clean-up, investigation, full agricultural use, plant protection, *etc.*) and the soil type (OM content) [2,5,11,19,36,38]. Criteria applied for the evaluation of agricultural soils contamination refer to very different PAHs limits such as background value of 1 000 μg/kg (for the standard soil with 10 % OM) in Dutch regulations [36], 1000 μg/kg for agricultural use of the soil in Canadian regulations [2,11], background value of 500 μg/kg and "plant protection level" of 8 000 μg/kg in German regulations [38], "threshold level"(when no further reduction is necessary) of 50 000 μg/kg and "action level" (when land ought not to be used for that purpose unless it has first been remedied) of 500 000 μg/kg in the UK regulations [2,11] and background value of 200 μg/kg and the „uncontaminated soil" limit of 1 000 μg/kg (for the standard soil with 2% of OM) in Polish propositions [18,21]. More information on soil quality criteria for PAHs is given in the works of Beck *et al.* [2], Jones *et al.* [11], Maliszewska-Kordybach [19] and Wild and Jones [40].

The above results show that the content of PAHs - model organic contaminants from POPs group - in soils from rural or semi-rural locations in two countries, characterised by similar PAHs emission and representing Central Europe (Poland) and West Europe (the UK), is on the same level (median value of 327 - 460 μg/kg). This suggests diffuse contamination, mainly by atmospheric deposition, of the soils under study. Much higher concentrations of these contaminants can be expected near local PAHs pollution sources.

The concentrations of PAHs in the soils in Poland and the UK are low and most of the studied area can be classified as unpolluted with these compounds. However, an evaluation of the level of contamination of the soils with PAHs depends on the regulatory standards being applied. A specific problem with soil quality regulations for organic pollutants is the influence of environmental conditions (*e.g.* soil characteristic) on the significance of the concentration of the contaminant. Current soil quality criteria in some countries vary very widely in many respects emphasizing the need for more uniform international cooperation in this field .

5. References

1. ATSDR- US Agency for Toxic Substances and Disease Registry (1990) Public health statement: polycyclic aromatic hydrocarbons.
2. Beck, A.J., Wilson, S.C., Alcock, R., and Jones, K. (1995) Kinetic constrains on the loss of organic chemicals from contaminated soils: implications for soil-quality limits, *Crit.Review Environ. Sci. Technol.* **25**, 1-43.
3. Bradley, L.J.N., Magee, B.H., and Allen, S.L. (1994) Background levels of polycyclic aromatic hydrocarbons (PAH) and selected metals in New England urban soils, *J. Soil Contam.* **3** (4), 349-361.
4. Cousins, I.T., Kreibich, H., Hudson, L.E., Lead, W.A., and Jones, K. (1997) PAHs in soils: contemporary UK data and evidence for potential contamination problems caused by exposure of samples to laboratory air, *Sci. Total Environ.***203**, 141-156.
5. Enggist, A., Liechti, H., and Borer, F. (1994) Untersuchungen der Schadstoffbelastung von Boden und Vegetation entlang von Kantonsstrassen sowie von Strassenwischgut, Vokswirtschaftsdepartement des Kantons Solothurn, Amt für Umweltschutz. Berichte Nr.6. Solothum, Januar 1994.
6. Environment 1998 (1998), Polish Central Statistical Office, Warszawa.
7. Fränzle, O., Krinitz, J., Schmotz, W., Delschen, T., and Leisner-Saaber, J. (1995) Harmonisierung der Untersuchungsverfahren und Bewertungsmastäbe für den Bodenschutz mit der Russischen Föderation. Report No UBA-FB 95-048. Texte 60/95. Terytze K. (ed.), Umweltbundesamt, Berlin.
8. Gustafsson, O., Gschwed, P.M., and Buesseler K.O. (1997) Using ^{234}Th diequilibria to estimate the vertical removal rates of polycyclic aromatic hydrocarbons from the surface ocean. *Marine Chem.* **57**, 11-23.
9. Hites, R.A., Laflamme, R.E., and Farrington, J.W. (1977) Sedimentary polycyclic aromatic hydrocarbons: the historical record, *Science* **198**, 829-831.
10. Jensen, J. and Folker-Hansen, P. (1995) Soil quality criteria for selected organic compounds. Danish Environmental Protection Agency. Working Report No.47.
11. Jones, K.C., Alcock, R.E., Johnson, D.L., Northcott, G.L., Semple, K.T., and Woolgar, P.J. (1996) Organic chemicals in contaminated land: analysis, significance and research priorities, *Land Contam.& Reclam.* **4**, 189-197.
12. Jones, K.C., Stratford, J.A., Tidridge, P., and Waterhouse, K.S. (1989) Polynuclear aromatic hydrocabons in an agricultural soil: long-term changes in profile distribution, *Environ. Poll.* **56** 337-351.
13. Jones, K.C., Stratford, J.A., Waterhouse, K.S., and Vogt, N.B. (1989) Organic contaminants in Welsh soils: polynuclear aromatic hydrocarbons, *Environ. Sci. Technol.* **23**, 540-550.

14. Jones, K.C., Stratford, J.A., Waterhouse, K.S., Furlong, E., Giger, W., Hites, R., Schaffner, C., and Johnston, A.E. (1989) Increases in the polynuclear aromatic hydrocarbons content of an agricultural soil over the last century, *Environ. Sci. Technol.* **23**, 95-101.
15. Kawamura, K., Suzuki, I., Fujii, Y., and Wanatabe, O. (1994) Ice core record of polycyclic aromatic hydrocarbons over past 400 years, *Naturwissenschaften* **81**, 502-505.
16. Kummerova, M., Slovak, L., and Holoubek, I. (1995) Phytotoxicity studies of benzo(a)pyrene with *Lactuca Sativa. Toxicol. Environ. Chem.* **51**, 197-203.
17. Maliszewska-Kordybach, B. (1993) Trwalosc wielopierscieniowych weglowodorow aromatycznych w glebie. 82 pp. IUNG, H(4), Pulawy.
18. Maliszewska-Kordybach, B. (1996) Polycyclic aromatic hydrocarbons in agricultural soils in Poland; preliminary proposals for the criteria for the evaluation of the level of soil pollution, *Environ. Geochem.*, **11**, 121-127.
19. Maliszewska-Kordybach, B. (1999) Persistent organic contaminants in the environment; PAHs as a case study. In: J.C. Block, V.V. Goncharuk and P.Baveye (eds.), *Bioavalibility of organic xenobiotics in the environment.* NATO ASI Series, Kluwer Academic Publishers, Dordrecht/Boston/London - in press.
20. Maliszewska-Kordybach, B. and Oleszek W. (1994) The use of high performance liquid chromatography for the determination of polycyclic aromatic hydrocarbons (PAHs) in soil samples. *Acta Chromat.* **3**, 84-93.
21. Maliszewska-Kordybach, B. and Smreczak, B. (1997) Zawartosc wielopierscieniowych weglowodorow aromatycznych w glebach uzytkowanych rolniczo na terenie woj. lubelskiego. *Roczniki Gleboznawcze,* **48,** 95-110.
22. Maliszewska-Kordybach, B. and Smreczak, B. (1998) Polycyclic aromatic hydrocarbons (PAH) in agricultural soils in the Eastern Poland, *Toxicol. Environ. Chem.* **66,** 51-56.
23. Maliszewska-Kordybach, B. and Terelak, H. (1996) Evaluation of the content of polycyclic aromatic hydrocarbons in soils from rural region of Poland (Lublin province), in: *Proceedings Third International Symposium and Exhibition on Environmental Contamination in Central and Eastern Europe,* Warsaw, Poland, pp. 81-83.
24. Maliszewska-Kordybach, B. and Terelak, H. (1997) Wielopierscieniowe weglowodory aromatyczne w glebach polnocnej i wschodniej Polski, in "*Zwiazki organiczne w srodowisku i metody ich oznaczania*" Biblioteka Monitoringu Srodowiska, Warszawa, pp. 169-170.
25. Menzie, C.A., Potocki, B.B., and Santodonato, J. (1992) Exposure to carcinogenic PAHs in the environment, *Environ.Sci.Technol.* **26,** 1278-1284.
26. Paterson, S. and Mackay, D. (1989) A model illustrating the environmental fate exposure and human uptake of the persistent organic chemicals, *Ecol. Modelling,* **47**, 85-114.
27. Pondel H., Sadurski, W., and Wilkos S. (1985) Zawartosc substancji organicznych w glebach Polski. *Pamietnik. Pulawski,* **85,** 5-27.
28. Sanka, M., Strnad, M.., Vondra, J., and Paterson, E. (1995) Sources of soil and plant contamination in an urban environment and possible assessment methods, *Intern. J. Environ. Anal. Chem.* **59,** 327-343.
29. Siepak, J. and Andrzejewska M. (1997) Zanieczyszczenie wielopierscieniowymi weglowodorami probek gleb poddanych antropopresji z terenu miasta Poznania. In: *Mat. IV Konferencji Naukowej „Geochemiczne, hydrochemiczne i biochemiczne zmiany srodowiska przyrodniczego na obszarach objetych antropopresja."*, Krakow, Poland.
30. Simonich, S.L. and Hites, R. (1994) Vegetation-atmosphere partitioning of polycyclic aromatic hydrocarbons, *Environ. Sci. Technol.* **28,** 939-943.
31. Simonich, S.L. and Hites, R.A. (1994) Importance of vegetation in removing polycyclic aromatic hydrocarbons from the atmosphere. *Nature* **370,** 49-51.
32. Sims, R.C. and Overcash, M.R. (1983) Fate of polynuclear aromatic compounds (PNAs) in soil-plant systems, *Residue Reviews* **88,** 1-68.
33. Slebioda, M.., Kot, A., Wodecki, Z., and Kolodziejczyk, A. (1996) Impact of motor traffic on PAHs contamination of soils, in "*Third Intern. Symposium TOCOEN'96 and Satellite Workshops*", Luha ovice, Czech Republik, pp. 292-294.
34. *Soil background and reference values in Germany* (1995) Bund-Länder Arbeitsgruppe Bodenschutz Report, Ed. Bayerisches Staatsministerium für Landesentwicklung und Umweltfragen.

35. Stuczynski, T., Motowicka-Terelak, T., Piotrowska M., and Maliszewska-Kordybach B. (1992) *Stan chemizmu gleb woj. lodzkiego z uwzglednieniem wystepowania metali ciezkich, siarki i wielopierœcieniowych wêglowodorów aromatycznych*, Raport IUNG, Pulawy, Poland.
36. Swartjes, F.A. and van den Berg, R. (1993) Remediation of contaminated soil and groundwater: Proposals for criteria and priority settings, *Mat. of Workshop on Contaminated Soils.*, Stockholm.
37. Tebaay, R.H., Welp, G., and Brümmer, G.W. (1993) Gehalte an Polycyclischen Aromatischen Kohlenwasserstoffen (PAK) und deren Verteilungmuster in unterschiedlich belasteten Böden, *Z.Pflanzenernähr. Bodenk.* **156**, 1-10.
38. Trenck, K.T., Ruf, J., and Flittner, M. (1994) Guide values for contaminated sites, *Environ Sci. and Pollut. Res.* **1**, 253-261.
39. Wild, S.R. and Jones, K.C. (1992) Organic chemicals entering agricultural soils in sewage sludges: screening for their potential to transfer to crop plants and livestock, *Sci. Total Environ.* **119**, 85-119.
40. Wild, S.R. and Jones, K.C. (1995) Polynuclear aromatic hydrocarbons in the United Kingdom environment: a preliminary source inventory and budget, *Environ. Poll.* **88**, 91-108.
41. Wilson, S.C., Duarte-Davidson, R., and Jones, K.C. (1996) Screening the environmental fate of organic contaminants in sewage sludge applied to agricultural soils: 1. The potential for downward movement to groundwaters, *Sci. Total Environ.* **185**, 45-57.
42. Zerbe, J., Sobczynski, T., and Siepak, J. (1995) Zanieczyszczenie gleby w ogrodach dzialkowych metalami ciezkimi i wielopierscieniowymi weglowodorami aromatycznymi, *Przyroda i Czlowiek*, **z. 5**, 133-141.

QUALITY AND CONTAMINATION OF AGRICULTURAL SOILS IN HUNGARY AS INDICATED BY ENVIRONMENTAL MONITORING AND RISK ASSESSMENT

A. MURÁNYI
Research Institute for Soil Science and Agricultural Chemistry of the Hungarian Academy of Sciences
H-1022 Budapest, Herman Ottó 15. Hungary

1. Environmental Monitoring

The main reason for environmental monitoring is to detect changes in the state and functioning of ecosystems at a stage such that timely counteractive measures can be initiated, developed and evaluated. It is stressed that in environmental control, monitoring should be applied as an instrument and not as an objective itself. Sampling is only the first step in the monitoring process. It should be followed by the interpretation and evaluation of the monitoring results, and concluded with a timely reporting of the achieved results. The period between sampling and reporting is often considerable, thereby devaluing the monitoring results for their intended use. Traditional routine monitoring activity can be defined as long-term, standardised measurement, observation, evaluation and reporting of the environment in order to define status and trends. In the case of soil, rational soil management so as to ensure normal soil functions, requires adequate information. This was the reason why the Soil Fertility Monitoring System was initiated in Hungary in 1978.

1.1. SOIL FERTILITY MONITORING SYSTEM

In the Soil Fertility Monitoring System the most changeable soil characteristics (nutrient content, pH, etc.) were measured in the topsoil of about 100 000 agricultural fields covering near to 5 million hectares, in 3-year cycles. The program started in 1978 and stopped before completing the third cycle. The results were used efficiently to monitor changes in the main soil parameters as well as to make recommendations of nutrient supply for the farmers. 7142 fields representing almost 0.5 million ha have been tested three times. The interval between the 1st and 3rd cycles was 6 years on the average.

Soil Acidification. This monitoring program proved to be very useful in the evaluation of the rate of soil acidification by comparing the results of the 1st and 3rd cycles (Table

M.J. Wilson and B. Maliszewska-Kordybach (eds.),
***Soil Quality, Sustainable Agriculture and Environmental Security in Central and Eastern Europe*, 61–77.**

1). The data clearly indicated that the proportion of non-calcareous soils (3647 fields) with $pH_{KCl} < 5.5$ had increased by 6 %, from 60.7 % to 66.7 % at the expense of the soil areas with higher pH. At the same time the percentage of the fields without carbonate increased from 51 % to 54 %, the territory of calcareous soils decreased [18].

The results of this monitoring program brought the process of soil acidification into sharper focus, particularly its reasons and consequences. Fast soil acidification is caused mainly by improper fertiliser application, by decreasing amelioration (state subsidies on liming were decreased) and by acid deposition.

TABLE 1. Distribution of tested non-calcareous soils in the various pH_{KCl} categories

pH_{KCl}	Testing cycle 1st cycle 1977 - 1980	 3rd cycle 1983 – 1985
	% of tested area	
< 4.0	3.5	5.8
4.1 - 4.5	15.3	15.9
4.6 - 5.0	20.3	22.2
5.1 - 5.5	21.6	22.8
< 5.5	*60.7*	*66.7*
5.6 - 6.0	22.4	20.6
6.1 - 6.5	16.9	12.7
5.6 <	*39.3*	*33.3*

The susceptibility of soils to acidification was classified on the basis of parent material, pH and carbonate status, texture, organic matter content and soil depth. Soil acidification is one of the most important degradation processes in Hungary, since 46 % of the total area is susceptible to acidification: 14 % is highly susceptible, 5 % is susceptible, 23 % is moderately susceptible and 4 % is slightly susceptible. Significant pH decreases could be observed in the south-western, eastern and north-eastern parts of Hungary, where the soils are susceptible to acidification [19].

As a consequence of soil acidification the fertility of soils decreased. Soil acidification resulted in significant effects on fertiliser efficiency and crop yield. Table 2 clearly indicates that winter wheat, corn and sunflower produced the highest yields between pH_{KCl} 6.5 - 7.5, where the NPK fertiliser input was the most efficient as well. Decreasing pH corresponded with lower yields and higher NPK fertiliser needs. Degradation of soil quality by acidification was closely related to the decline of the soil fertility function.

1.2. SOIL INFORMATION AND MONITORING SYSTEM

A Soil Information and Monitoring System was developed in Hungary since it was realised that the scientifically-based planning and implementation of sustainable land use and rational soil management, ensuring normal soil functions, require more information on the soils.

1236 representative observation points were selected: 865 points on agricultural land, 183 points in forests and 189 points in environmentally threatened "hot spot" regions (representing 12 different types of environmental hazards or particularly sensitive areas). The representative sampling sites were selected by regional experts.

The basic observation points are characterised in great detail. Not only the basic soil properties (pH, $CaCO_3$ content/acidity, humus content, CEC, exchangeable cations) and nutrient status (N forms, available plant nutrients) but also potentially toxic elements (total content, mobile content), soil physical characteristics (particle size distribution, pF curve, hydraulic conductivity) and biological activity (cellulose test, dehydrogenase activity, CO_2 production) are determined. The chemical, physical, biological status and pollutant- and nutrient content of the soils are characterised at the same time. Some soil parameters are measured every year, others every 3 years or every 6 years, depending on their stability as a function of time. The first sampling was in 1992.

TABLE 2. Yield and NPK fertiliser dose (kg NPK for 1 t yield/ha) of the dominant crops in Hungary depending on soil pH

soil pH_{KCl}	winter wheat yield t/ha	winter wheat NPK kg/t	corn yield t/ha	corn NPK kg/t	sunflower yield t/ha	sunflower NPK kg/t
< 4.50	3.76	82.7	5.58	62.7	1.66	164.5
4.51 - 5.50	4.14	77.3	6.23	59.9	1.85	149.7
5.51 - 6.50	4.64	69.4	6.71	57.2	1.92	133.8
6.51 - 7.50	*5.33*	*65.2*	*7.07*	*56.0*	*2.14*	*130.8*
7.51 <	4.45	76.8	6.87	55.0	1.92	145.8

Heavy Metals in Soils. This complex soil database can be analysed and interpreted in many ways. For instance the average heavy metal content of Hungarian soils are shown in Table 3 [10]. The data were sorted according to the soil texture and soil type.

The average heavy metal contents exhibit a close relationship with the soil texture. According to the genetic soil types, the average concentrations are the highest in the case of Zn (51 mg/kg), while Cr, Cu and Pb are in the same range (18 - 21 mg/kg).

The average heavy metal contents in Table 3 clearly demonstrate that the soil cover of Hungary is not polluted. However, the average concentration of 1196 samples hides the high concentrations of polluted soils [11]. For instance the Cd content is higher than 1.0 mg/kg in 39 soil samples (the maximum is 13.4 mg/kg), the Pb concentration is higher than 42 mg/kg in 100 soils (the maximum is 2372 mg/kg) and the Zn content is higher than 102 mg/kg in 67 samples (the maximum is 618 mg/kg). That means that only 3 %, 8 %, 6 % of the samples had twice the average Cd, Pb, Zn content. A minority of the soil samples measured for heavy metal content can be considered as representing polluted sites.

Soil erosion. Erosion is also an important soil degradation process in Hungary, greatly influencing the quality of soils. The erosion status of soils is shown in Table 4.

TABLE 3. Average heavy metal content of Hungarian soils according to the soil types and texture in mg/kg

	As	Cd	Co	Cr	Cu	Hg	Mo	Pb	Zn
Soil types									
Skeletal soils	4	0.3	5	8	13	0.02	0.7	11	27
Lithosols	8	0.7	8	19	17	0.06	1.3	24	49
Brown forest soils	6	0.4	9	19	17	0.04	0.7	17	43
Chernozem soils	6	0.5	10	21	21	0.03	1.5	16	51
Salt-affected soils	5	0.4	11	20	20	0.04	1.9	26	54
Meadow soils	7	0.5	10	23	24	0.05	1.7	23	62
Bog soils	6	0.5	5	15	19	0.05	1.3	15	32
Alluvial and sedimentary soils	5	0.5	9	21	25	0.05	1.1	23	69
average (n=1196)	*6*	*0.5*	*9*	*18*	*20*	*0.05*	*1.1*	*21*	*51*
Soil texture									
Sandy soils	3	0.3	3	6	10	0.02	0.6	9	19
Loamy soils	7	0.5	8	17	18	0.03	0.9	16	44
Clayey soils	6	0.6	12	26	26	0.05	1.6	26	71

The GIS map of soil erosion was compiled in the RISSAC GIS Lab on the basis of the original map, edited by Stefanovits and Duck (1:500 000) and used to determine the areas of the different categories [16, 20]. It can be seen that erosion is significant on 24.2 % of the total area of Hungary.

TABLE 4. Soil erosion in Hungary

	Area in 1000 ha	% of the total area of Hungary
Strong erosion	562	6.0
Medium erosion	864	9.3
Slight erosion	831	8.9
Non eroded areas	4407	47.4
Sedimentation areas	1689	18.2
Forests	951	10.2
Total area	9304	100.0

1.3. PROBLEMS OF ENVIRONMENTAL MONITORING

There are certain identified problems with the existing state of environmental monitoring which will only increase in the future. In general terms, these problems concern: (1)

the information gained from monitoring; (2) the high cost of monitoring, and (3) the harmonisation of monitoring throughout the EU Member States.
The existing approach is limited in that complete information about soil/water quality and toxicity is not being collected. This is in a large part a strategic problem, in that the design and operation of monitoring systems are characterised by uncertainty. In many cases, goals are unclear.
Traditional environmental monitoring is frequently characterised by

- being data rich, but information poor;
- occurring without information objectives;
- more is better; money in - data out;
- using high tech must be an improvement;
- know how is better than "know why".

In addition the following technical problems are encountered:

- many compounds are overlooked. - The number of potentially polluting substances is enormous. Of the 100 000 chemicals identified, only 30 - 40 are regularly monitored. Thus, terrestrial and aquatic contamination and toxicity may be underestimated.
- no accounting for interactions between pollutants. Most polluting loads consist of a complex mixture of toxicants. With a chemical-specific approach, chemical interactions and combined toxicity effects cannot be accounted for.
- unknown environmental fate and toxicity. Information on specific chemicals with respect to their long-term toxicity and environmental fate is relatively scarce. Thus, the 'health' of a terrestrial/aquatic ecosystem as a whole cannot be evaluated.
- difficulty in analysis. Individual substances in a mixture may be difficult to detect, identify and quantify [21].

2. Environmental Risk Assessment

Paying more and more attention to the effects of pollution sources resulted in the elaboration of the principles of Environmental Risk Assessment (ERA). The best formal definition of risk is a probability distribution of possible (future) frequencies of harmful consequences. The main reason for risk assessment is the protection of our environment. The risk assessment procedures may be applied for classification of new and existing substances, as well as for the characterisation of contaminated sites.
An Environmental Risk Assessment should be carried out on notified new substances or on priority existing substances, according to the present EU concept (Commission Directive 93/67/EEC and Council Regulation 1488/94/EEC). This risk assessment should proceed in the following sequence [1]:

- Hazard identification
- Dose (concentration) - response (effect) assessment
- Exposure assessment
- Risk characterisation

The risk assessment can be carried out for all three environmental compartments, i.e.

aquatic environment, terrestrial environment and air.
In essence, the procedure for the Environmental Risk Assessment of a substance consists of comparing the concentration in the environmental compartments with the concentration at which no effects on organisms or ecological systems are expected to occur. In principle, human beings as well as ecosystems (aquatic, terrestrial or air compartments) are to be protected.
Environmental Risk Assessment focuses attention on the quantification of the risk by comparing the Predicted Environmental Concentration (PEC) with the Predicted No Effect Concentration (PNEC) (Figure 1). The environmental risk can be characterised quantitatively by the Risk Quotient. The Risk Quotient equals to the PEC/PNEC ratio, what indicates whether a substance (inorganic or organic) presents a risk to organisms in the environment or not, as well as helping to elaborate appropriate decisions for sustainable soil/water/air management.
The Predicted Environmental Concentration (PEC) can be derived from available monitoring data (unidentified pollution source), exposure assessment (using contaminant loads), model calculations, etc. The Predicted No Effect Concentration (PNEC) is the outcome of the dose (concentration) - response (effect) assessment. PNEC is determined on the basis of the biological effects, usually taking into account the results from monospecies laboratory tests (ecotoxicological tests, bioassays) or in a few cases established concentrations from model ecosystem tests. Adequate safety factors are also taken into account. The PNEC is regarded as a level below which the probability suggests that an unacceptable effect will not occur.

Risk Quotient. The quantification of the Environmental Risk Assessment for soils can be demonstrated with the examples of heavy metals (Table 5). Predicted Environmental Concentrations can be evaluated by using the data of the Soil Information and Monitoring System. Average heavy metal contents of 1196 soil samples represent the total soil cover of Hungary while the maximum heavy metal contents characterise the most contaminated sites of the monitoring system. Target values can be used to characterise Predicted No Effect Concentrations. The reason for this is that target values are considered as the environmental quality levels and concentrations at which the risks of adverse effects (risks to ecosystems, functional properties of the environmental compartments, etc.) are negligible. The target values can be accepted as safe environmental concentrations of predicted no effect.
The comprehensive Dutch concept [2, 3] is applied to calculate the target values for soils. In the Netherlands the background levels in soils of relatively uncontaminated areas were used as a basis to derive target values for contaminants in the case of a so-called standard soil with 10 % organic matter and 25 % clay content. This approach is unique because the organic matter and clay content can be taken into consideration when target values of non-standard soils are determined. As a consequence of this, the target values of Hungarian soils were also calculated by taking into account their measured humus and clay contents. Average or site-specific organic matter and clay contents were used in the case of the average or the maximum heavy metal contents, respectively.

It can be seen in Table 5 that the Risk Quotients of the average soil are less than 1, so it is concluded that the soil cover of Hungary is not contaminated with Cd, Pb and Zn. There is no need for further testing or risk reduction measures according to the upper part of Figure 1.
However, the Risk Quotients are very high in the case of the three contaminated soil samples. Further tests or further information should be collected to revise these Risk Quotients (see the lower part of Figure 1). More detailed studies and site specific Environmental Risk Assessment should be carried out by reliable experts.

TABLE 5. The Environmental Risk Assessment of heavy metal contents in soils

	Cd	Pb	Zn
Average heavy metal content (n=1196)			
PEC average	0.5	21	51
PNEC average	0.6	77	126
Risk Quotient	*0.8*	*0.3*	*0.4*
Maximum heavy metal content			
PEC maximum	13.4	2372	618
PNEC maximum	0.6	74	167
Risk Quotient	*22*	*32*	*4*

3. Comparison of Environmental Monitoring and Environmental Risk Assessment

The processes occurring in our natural environment can be followed and/or predicted either by Environmental Monitoring (EM) or by Environmental Risk Assessment (ERA). It is worth comparing the main characteristics of these two concepts.
The traditional, routine monitoring is compared with the present EU risk assessment approach for new and existing substances. The EM and ERA are compared on the basis of their dominant features (Table 6).

Time Relationships: Past - Present - Future. Environmental Risk Assessment (ERA) concentrates on the probable effects of the environmental exposure in the future. ERA tries to forecast the relationship between the present and the future. Environmental Monitoring (EM) characterises the present state of the environment, giving information about the pollutants that have been accumulated up today. EM connects the past with the present.
Generalisation. ERA can be generally used since Predicted Environmental Concentrations can be calculated for any large areas. Both generic ERA and site specific ERA can be carried out. EM characterises the sampled area and the possible extension of the results is often limited.

Exposure. ERA takes into account the source of the pollution and estimates the magnitude of the pollution via the determination of the Predicted Environmental Concentrations. EM characterises, among other things, the potentially dangerous pollutants but the possible source of pollution is not identified, and unnecessary properties may be monitored in non-polluted areas as well.
Effect. ERA calculates the extent of the exposure and the results are interpreted by taking into account the most frequently ecotoxicological effects as well. EM indicates the concentration of pollutants, but the effects of pollution are usually not taken into consideration.

TABLE 6. Comparison of traditional Environmental Monitoring and the Environmental Risk Assessment

	Environmental Monitoring	Environmental Risk Assessment
Time relationships	*past & present*	*present & future*
Generalisation	*sampled area*	*any large area*
Exposure	*unknown source*	*identified source and load (PEC)*
Effect	*not evaluated*	*evaluated*
Risk Quantification	*not quantified*	*quantified (Risk Quotient)*
Measured parameters	*many parameters*	*key parameters*
Costs	*very expensive*	*less expensive*
Time delay	*controlled present*	*controlled in future*
Natural processes	*followed*	*oversimplified*
Territorial extension	*sampled area*	*space variability*
Interpretation	*by measurements*	*by exposure & effects*
Cost efficiency	*low (analyses)*	*favourable (calculations)*
Risk modification	*not possible*	*risk reduction*
Environmental protection	*not conservative*	*realistic conservative*

Risk Quantification. ERA quantifies the magnitude of the risk via the Risk Quotient. EM gives information about the concentration of the pollutants. The evaluation of the result is done by comparing the measured concentration with the accepted limit values. The risk is not quantified.
Measured parameters. ERA focuses attention on those parameters that may have a risk to the environment. When ERA models are being built the potentially hazardous parameters are selected and evaluated. As a consequence of this, only the key parameters are important. Environmental Monitoring also follows the parameters of environmental importance but as our knowledge grows, more and more parameters are being followed. This is often the case in waters and soils. Many parameters are followed as a function of time without revision.
Costs. ERA is less expensive since selected key parameters are analysed. EM is rather expensive because regular measurements of many parameters are needed.
Time delay. ERA forecasts future risk. Due to the time delay, the error(s) of the developed ERA model may only be realised after a long time period. There is a danger that

FIGURE 1. General procedure for Environmental Risk Assessment of new and existing substances

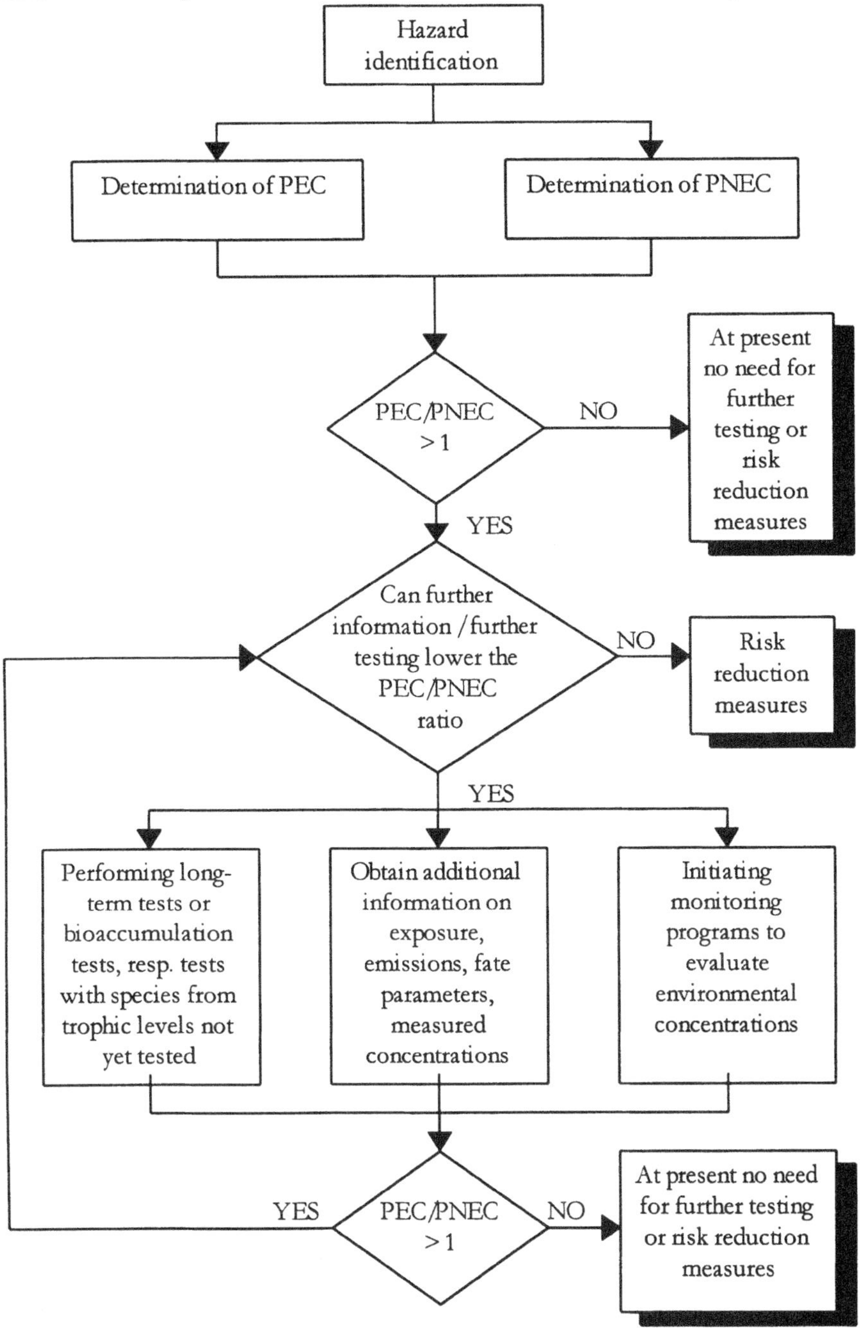

harmful consequences may occur when the model does not characterise the environmental situation properly. EM gives safer results because of the regular analyses.
Natural processes. Natural processes are very complex in the environmental compartments, especially in terrestrial ecosystems. They are very difficult to describe and characterise in models or schemes. So ERA is obliged to simplify (in most cases oversimplify) the real natural processes. The necessary simplifications lead to less accurate estimations of future events. EM follows the actual changes occurring in the environment (air, waters, sediments, soils).
Territorial extension. The regional ERA uses Predicted Environmental Concentration. The average value does not characterise the total concentration distributions because of the natural heterogeneity of our environment. So the existing 'hot spots' of a large area may be hidden Site specific ERA is more reliable in this respect. EM either collects average samples or characterises the 'hot spot' itself.
Interpretation. ERA takes into account both the exposure of the loaded area and the expected effects as well. It gives a complex interpretation. Studying the causes and the results at the same time gives a comprehensive overview about the possible risks. EM registers what has occurred in the area under study up to the day of sampling.
Cost efficiency. ERA usually integrates the existing knowledge into the given model, so that more calculations rather than more measurements are needed. The costs of the EM are much more expensive because of the regular measurements of (usually many) parameters. As a consequence, the cost efficiency of ERA is more favourable (higher) than that of the EM.
Risk modification. ERA makes it possible to reduce the risk, since the source and the pathways of the pollution are identified. EM usually does not define the origin of the pollution and hence risk reduction is rarely possible.
Environmental protection. When appropriate Environmental Risk Assessment models are being built, realistically conservative predictions are always kept in mind in order to protect the environment as much as possible. Environmental Monitoring only follows the changes; it is not conservative.

The detailed comparison of the Environmental Risk Assessment and Environmental Monitoring reflects the limitations and/or advantages of the two concepts. It is easy to realise and to conclude that their appropriate integration is the best solution to characterise the state of our environment. This integrated concept can reduce or eliminate the weak points of the two approaches. It can be expected that the integration of Environmental Risk Assessment and Environmental Monitoring will be the focus of environmental researches in the near future. Such an integrated environmental characterisation seems to be a rational tool for describing the state of our environment and to predict the trends of its changes. However, it can never substitute our efforts and duty to study our environment in detail and to understand better the natural biogeochemical processes occurring in aquatic and terrestrial ecosystems.

4. Long-term Effects - Chemical Time Bombs

In the long-term, however, the importance of the accumulation of potentially toxic pollutants in soils keeps growing. Due to immobilisation, a large reservoir of metals will be built up, which may turn out to become a so called 'Chemical Time Bomb'. The Chemical Time Bomb (CTB) was defined by Stigliani [17] as "an unforeseen chain of events resulting in the delayed and sudden occurrence of harmful effects due to the mobilisation or chemical transformation of chemicals stored in soils and sediments in response to saturation or alteration in certain environmental conditions".
Mineral fertilisers can also be the reason for contaminant accumulation in agricultural soils. Several studies have reported an increase in the Cd content of plants and soils following the application of high rates of phosphates. In a study of the status of Cd, Pb, Co and Se in soils and plants of thirty countries (including Hungary) Sillanpaa and Jansson found that the relatively firm correlation left no doubt as to the increasing effect of phosphorus fertilisation on the Cd contents of both plants and soils [14]. In Hungary the pH of the soils of the high Cd group, with two exceptions, was on the acidic side (varying from 5.0 to 6.7), while those of the low Cd group were all alkaline (pH_{CaCl2} 7.4 or higher). The effect of pH on the availability of Cd, if taken into account as a pH correction, improved the plant-soil Cd correlation.

Polluting concentration. The extent of the chemical contamination is particularly important when the accumulation of pollutants are studied. This can be characterised by the Polluting Concentration what is defined as [9]:

Polluting Concentration = Measured Concentration - PNEC.

The Polluting Concentration gives the excess (or deficit) of a pollutant over the PNEC. The calculated target value, at which adverse effects cannot be expected, can be applied as PNEC. If Polluting Concentration is positive harmful effects can arise in soils. The Polluting Concentration of Cd, Pb and Zn in the three contaminated soils is 12.8 mg/kg, 2298 mg/kg and 451 mg/kg, respectively (Table 5). The Polluting Concentration reflects the actual extent of pollution and determines the excess of the contaminant over the acceptable level. The Polluting Concentration can be used for developing remediation plans or for monitoring the process of contamination or remediation, etc.
However, the extent of pollution can be characterised not only by this absolute concentration but also by a relative value. Toxic Unit [12] can be applied for this purpose.

Toxic Units. The Toxic Unit of organic and inorganic contaminants is defined as:

Toxic Unit = Measured Concentration / PNEC.

Toxic Unit represents a proportional value (always a positive number) indicating how much the measured concentration of the pollutant is higher or lower than the PNEC, that is the acceptable limit. If Toxic Unit is higher than one, the pollutant may cause toxic effects. The Toxic Unit of Cd, Pb and Zn in the three contaminated soils is 22.3, 32.0 and 3.7, respectively (Table 5). The toxicity of Pb as well as Cd is very high. The advantage of Toxic Unit is that it is an additive property. If these pollutants were in the same soil the Sum of Toxic Units would be 58, extremely high.

Heavy metal mobilisation. The concept of Salomons can serve as a basis for studying Chemical Time Bomb - related problems, including heavy metal mobilisation [13]. The risks of delayed and non-linear responses of heavy metal pollution can be assessed by understanding the biogeodynamics of pollutants in soils and sediments. For short-term risk assessment (5-10 years) it is sufficient to understand how the Capacity Controlling Parameters (CCPs: pH, redox, organic matter, exchange capacity, microbial activity) control the balance between retention and mobility, and hence their leaching and bioavailability. Much research is needed to evaluate the effect of CCPs on the behaviour of heavy metals in soils and sediments.
Over the long-term, however, changes in the CCPs can cause mobilisation of vast stores of pollutants. One example of this is the extensive acidification of the environment, leading to large-scale leaching of heavy metals from acid impacted areas. For example, in Scandinavia, Hg levels in freshwater fish are increasing even though Hg loads have been decreasing for several decades. Hg locked in watershed soils decades ago is now mobilised by soil acidification. Soil acidification of the root environment can also mobilise the contaminants, permitting them to enter into the food chain [8].

5. Long-term Soil Quality

The continuous pollution of our soils by chemicals forces us to review the factors / processes determining Soil Quality. As has been shown Environmental Monitoring focuses attention on the chemical concentrations in soils. It is a chemically oriented concept. The Environmental Risk Assessment predicts the risks by taking into consideration the ecotoxicological effects, so that it is based on biological responses. The Chemical Time Bomb concept emphasises the importance of time in the long run. These concepts are also relevant when long-term Soil Quality is characterised (Figure 2).

Chemical Soil Quality. The concept of the Chemical Time Bomb focuses our attention on the capacity factors of soils. The capacity factors can be considered as fundamental soil chemical properties. They are represented by the solid phase composition (including ion-, compound-, mineral contents) and characterise the total quantity of the chemical components. They play an important role in essential soil functions like fertility (nutrient reservoir) and resilience (long-term buffering).
The capacity factors are in equilibrium with intensity parameters. The intensity parameters are represented by the soil solution composition and characterise the available fraction of the chemical components. The solid phase can dissolve and precipitate in soils, controlling the composition of the soil solution. Solubility equilibria are important because they determine the mobility of chemical elements in soils and affect their availability to plants. While the capacity factors (total quantities) are stable, the intensity parameters (soluble quantities) are unstable, changeable properties. Soluble concentrations

are easily modified by environmental conditions (pH, redox conditions, etc.). Lindsay bridged the gap between soil science and chemistry and showed that most reactions taking place in soils can be understood and predicted from basic chemical relationships [6]. Sparks focused our attention on time dependence of chemical reactions [15]. Most of the chemical processes that occur in soils and sediments are dynamic and a knowledge of the mechanisms and kinetics of these reactions is fundamental. One must study kinetics to properly understand the fate of applied fertilisers, pesticides and inorganic/organic contaminants in soils with time in order to control their bioavailability and the quality of soils and groundwater.

Soil Quality can be described by a soil chemical approach, where solid phase composition is fundamental and this together with the solution composition and kinetics determine the chemical behaviour of soils, including their contaminants. This soil chemical approach does not take into consideration however, that the soils are alive.

Biological Soil Quality. Soil Quality can be characterised by a soil biological approach. In this case the community composition (bacteria, actinomycetes, etc.) can be considered as the foundation for the biological processes occurring in soils. The composition of the community can be regarded as the capacity factor of soils in a microbiological sense. The microorganisms play a central role in soil functions. On the one hand soil fertility is closely related to organic matter formation/decomposition, on the other hand the microorganisms can serve as the biological dimension of soil resilience [7].

The community composition of soil biota is determined by the microorganisms (Table 7). Both the number of individuals and the living mass of the microflora is much larger than that of the fauna. Soil life is dominated by soil microorganisms.

It is also clearly demonstrated how important the living conditions are. Under optimum condition, the number of microorganism is increased by 100 - 1000 times and the living mass of microorganisms is 5 - 10 times higher. The fungi are the most sensitive in this respect. The living conditions of soil biota have been rather neglected for a long time. This will probably change as the importance of organic farming increases. These considerations suggest that the utilisation of beneficial and effective microorganisms for sustainable agriculture and the environment is a viable concept [5].

As in the chemical approach, capacity factors are also in equilibrium with the intensity parameters represented by the activities of microorganisms. This determines the microbiological processes in soils. For instance, nitrifying bacteria form a part of the microbial community but their activity is relevant in the case of ammonium fertilisation when they transform the ammonium to nitrate and determine the nitrate availability to plants. The activity of microorganisms is highly sensitive to the environmental conditions.

FIGURE 2. Soil Quality concept based on natural sciences

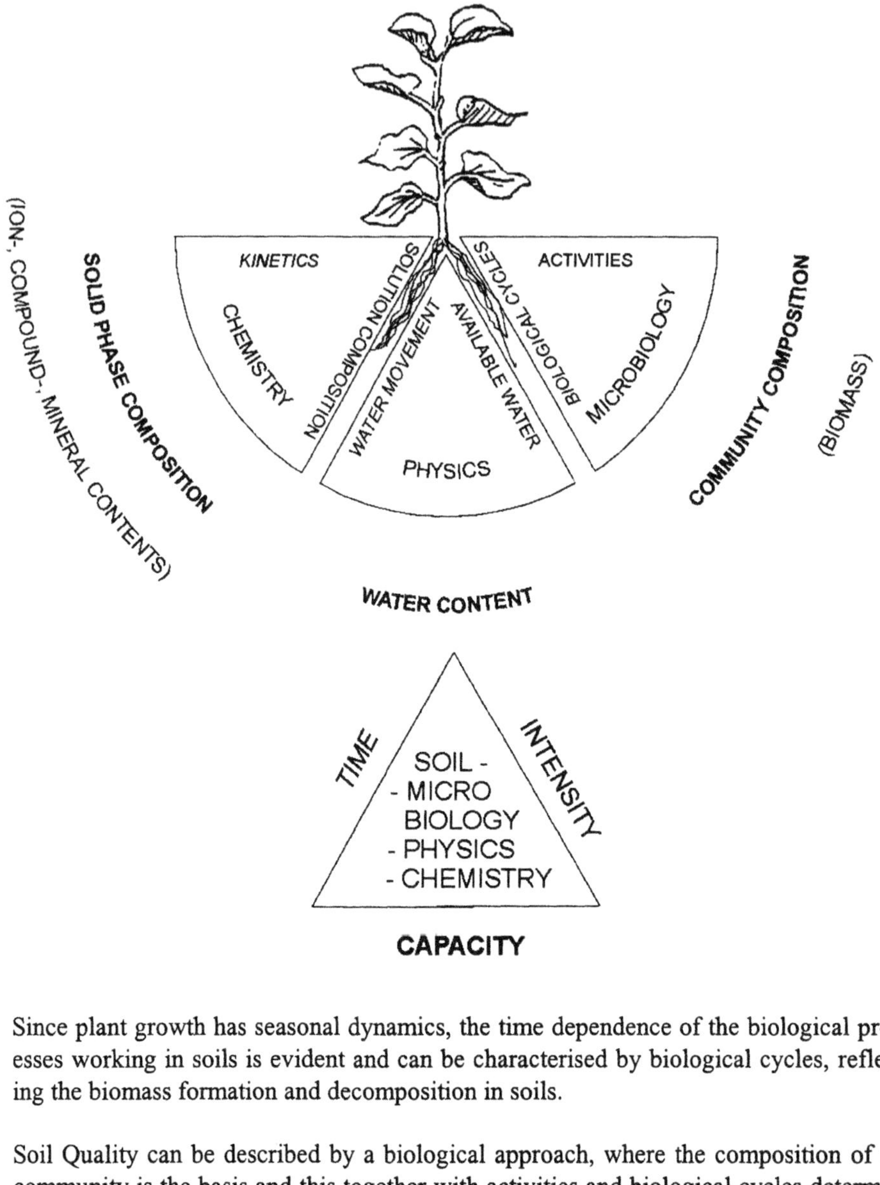

Since plant growth has seasonal dynamics, the time dependence of the biological processes working in soils is evident and can be characterised by biological cycles, reflecting the biomass formation and decomposition in soils.

Soil Quality can be described by a biological approach, where the composition of the community is the basis and this together with activities and biological cycles determine the microbiological behaviour of soils.

TABLE 7. The number of individuals and the living mass of soil organisms in temperate soils under average and optimum living condition [4]

	Individuals / m^2			Living mass, g / m^2		
	average	optimum	avr./opt.	average	optimum	avr./opt.
Microflora						
bacteria	10^{+14}	10^{+16}	0.010	100	700	0.14
actinomycetes	10^{+13}	10^{+15}	0.010	100	500	0.20
fungi	10^{+11}	10^{+14}	0.001	100	1 000	0.10
algea	10^{+8}	10^{+11}	0.001	20	150	0.13
sum	$1.1*10^{+14}$	$1.1*10^{+16}$	0.010	320	2 350	0.14
Fauna						
microfauna	$1.1*10^{+8}$	$2.0*10^{+10}$	0.006	5	150	0.033
mesofauna	$1.1*10^{+6}$	$1.0*10^{+8}$	0.011	6	59	0.104
macrofauna	$3.1*10^{+4}$	$3.2*10^{+5}$	0.096	45	346	0.129
megafauna	10^{-2}	10^{-1}	0.100	0.1	10	0.010
sum	$1.1*10^{+8}$	$2.2*10^{+10}$	0.006	56	564	0.099

It can be concluded that soil chemistry and soil microbiology determine Soil Quality together and that their integrated use is an appropriate key for the characterisation of the state of our soils in the long run.

6. Conclusions

This paper attempts to summarise the concepts that are relevant for characterisation of the environmental status of our soils at present as well as in the long-term. Environmental Monitoring detects changes in soils and characterises their quality. The Soil Fertility Monitoring System of Hungary focuses attention on changeable soil characteristics and this monitoring program clearly proved that soil acidification is one of the most significant degradation processes in Hungary. Moreover, the degradation of soil quality by acidification was clearly related to the decline of the soil fertility function, that is the yield as well as the NPK fertiliser efficiency decreased. The Soil Information and Monitoring System characterises the chemical, physical and biological status of soils in Hungary. Studying the heavy metal contents of this complex database it can be concluded that the soil cover of Hungary is not polluted by heavy metals, but a small minority of soil samples do represent polluted sites.

Environmental Risk Assessment takes into consideration the effects of pollution loads . The environmental risk is quantified by the Risk Quotient, which is equal to the ratio of the Predicted Environmental Concentration (PEC) and the Predicted No Effect Concentration (PNEC). Large Risk Quotients were determined in contaminated soils.

Environmental Risk Assessment and Environmental Monitoring were compared in detail. Based on their advantages and limitations it was concluded that their appropriate integration is the best solution for characterising the state of our environment and predicting trends and changes.
The accumulation of toxic contaminants in soils may become a Chemical Time Bomb in the long run. The extent of the chemical contamination can be characterised either by the Polluting Concentration (= Measured Concentration - PNEC) or by the Toxic Unit (= Measured Concentration / PNEC). The Capacity Controlling Parameters determine the behaviour of accumulated contaminants in the short-term as well as in the long-term.
The factors and processes determining Soil Quality were reviewed because of the continuous pollution of soils. Soil Quality can be characterised by a soil chemical approach, where solid phase composition is fundamental and this together with the solution composition and kinetics determine the chemical behaviour of soils, including that of contaminants. Soil Quality can also be described by a soil biological approach, where the composition of the community is the basis and this together with activities and biological cycles determine the microbiological behaviour of our soils. It was concluded that soil chemistry and soil microbiology determine the Soil Quality together and their integrated use is an appropriate key for characterising the state of our soils in the long-term.

7. References

1. ECB (1996) Technical Guidance Document on risk assessment for new notified substances and for existing substances, Manuscript, Ispra, Italy, 19th April 1996.
2. DGEP (1994) Intervention values and target values - soil quality standards. Ministry of Housing Physical Planning and Environment, Hague, the Netherlands.
3. DGEP (1991) Environmental quality standards for soil and water. Ministry of Housing Physical Planning and Environment, Hague, the Netherlands.
4. Dunger, W. (1983) *Tiere im Boden*, A. Ziemsen Verlag, Wittenberg Lutherstadt (in German).
5. Higa, T. and Wididana, G.N. (1991) The concept and theories of effective microorganisms, in J.P. Parr, S.B. Hornick and C.E. Whitman (eds.) *Proc. First Int. Conf. on Kyusei Nature Farming*, U.S. Dep. Agric., Washington, D.C., USA, pp. 118-124.
6. Lindsay, W. L. (1979) *Chemical Equilibria in Soils*, John Wiley & Sons, New York.
7. Lynch, J.M. (1994) The biological dimension of soil resilience: the impact of molecular biology, in D.J. Greenland and I. Szabolcs (eds.), *Soil Resilience and Sustainable Land Use*, CAB International, Wallingford, pp. 69-76.
8. Murányi, A. et al. (1993) Acidification in the rhizosphere of rape seedlings and in bulk soil by nitrification and ammonium uptake, *Z. Pflanzenernahr. Bodenk.*, **157**, 61-65.
9. Murányi, A. (1999) The extent of pollution and its ecotoxicological effects, in A. Kettrup and K.-W. Schramm (eds.) *Proc. SECOTOX 99 Fifth Eu. Conf. on Ecotoxicology and Environmental Safety*, GSF Inst. Ökologische Chemie, Munich, Germany.

10. Németh, T. et al. (1997) The environmental background values of soils in Hungary. Project report, KTM KEV-2631/96, Budapest (in Hungarian).
11. Ódor, L. et al. (1998) The environmental effects, the accumulation and the transformation of pollutants. Project report, KVM 310/F, Budapest (in Hungarian).
12. Pedersen, F. et al. 1997. Characterization of sediments from Copenhagen Harbour by use of biotests, in *Int. Conf. Contaminated Sediments*, 7-11 Sept. 1997, Rotterdam, The Netherlands. Preprints Vol. I. 67-74.
13. Salomons, W. (1995) Long-term strategies for handling contaminated sites and large-scale areas, in W. Salomons and W.M Stigliani (eds.) *Biogeodynamics of Pollutants in Soils and Sediments*, Springer, Berlin, pp. 1-30.
14. Sillanpaa, M. and Jansson, H. (1992) *Status of cadmium, lead, cobalt and selenium in soils and plants of thirty countries,* FAO Soils Bulletin 65. FAO. Rome.
15. Sparks, D. L. (1989) *Kinetics of Soil Chemical Processes*, Academic Press, San Diego.
16. Stefanovits, P. (1964) Soil erosion in Hungary; OMMI Genetikus Talajtérképek Kiadványai, Budapest, (in Hungarian).
17. Stigliani, W.M. (1993) Overview of the Chemical Time Bomb problem in Europe, in G.R.B. ter Meulen et al. (eds.) *Chemical Time Bombs*, Foundation of Ecodevelopment, Hoofddorp, the Netherlands, pp. 13-29.
18. Szabó, P. (1989) Changes of soil reaction and fertility in Hungary, in I. Szabolcs (ed.), *Ecological Impact of Acidification,* Hungarian Academy of Sciences, Budapest, Hungary, pp. 95-102. (Proc. Joint Symp. "Environmental threats to forest and other natural ecosystems". Oulu, Finland, 1-4 Nov. 1988.)
19. Várallyay, G., Rédly, M., and Murányi, A. (1989) Map of the susceptibility of soils to acidification in Hungary, in I. Szabolcs (ed.), *Ecological impact of acidification,* Hungarian Academy of Sciences, Budapest, Hungary, pp. 79-94. (Proc. Joint Symp. "Environmental threats to forest and other natural ecosystems". Oulu, Finland, 1-4 Nov. 1988.)
20. Várallyay, G. et al. (1997) Soil vulnerability assessment in Hungary, in N.H. Batjes and E.M. Bridges (eds.) *Implementation of a soil degradation and vulnerability database for Central and Eastern Europe,* ISRIC, Wageningen, pp. 43-50.
21. Villars, M.T. (1995) *Monitoring Water Quality in the Future*, Executive Summary, Delft Hydraulics, Delft, The Netherlands, May 1995.

Acknowledgement

The author wants to thank the EU INCO COPERNICUS Programme No. ERB IC15-CT98-0133, the PHARE Regional Environmental Programme No. EU/AR/101/91, the EU PHARE TDQM Programme No. H9305-02/1111 and the OTKA Programme No. T026084 for the financial support of the presented researches.

CONTAMINATION AND DEGRADATION OF SOILS IN THE CZECH REPUBLIC - CONTEMPORARY AND FUTURE STATE

E. PODLEŠÁKOVÁ[1], J. NEMECEK[2]
[1]*Research Institute for Soil and Water Conservation, Prague*
Zabovreská 250, 15627 Prague 5, Czech Republic
[2]*Czech Agricultural University, Dept. Soil Science and Geology*
Kamýcká 127, 16521 Prague 6, Czech Republic

1. Introduction

Great attention has been focused on soil quality assessment from the viewpoint of sustainable development of agriculture and its environment. The term soil quality is used in soil survey interpretations for a certain functional set of soil properties and characteristics like soil erodibility, soil pollution etc. or in general for satisfying the demand of multifunctionality on some kind of soil. Sustainable development in agriculture in the Czech Republic (CR) emphasizes especially the following aspects:

- development of agriculture without soil degradation and pollution and without contributing to pollution of waters and biological resources,
- introduction of agricultural systems making possible the production of healthy and reasonably priced products by making use of input regulations, but at the same time maintaining economic effectiveness,
- adaptation of agricultural production to adverse natural and anthropogenic factors along with limitation of the latter where necessary,
- landscape conservation both in regions of intensive use and restricted agricultural production.

The assessment of soil qualities concerning soil degradation and soil pollution can be done in two different ways:

- by an evaluation of degradation status and its risks, or
- by an evaluation of soil vulnerability.

Both mentioned approaches are mutually complementary. Soil vulnerability may be thought of as a buffering potential that prevents some kind of degradation or pollution.

2. Previous studies of land degradation in the CR

Such studies have been aimed at assessing the situation in the CR with respect to soil degradation resulting from erosion, acidification, loss of organic matter, pollution by heavy metals and organic compounds etc. Each of these threats will be briefly discussed in turn

Erosion. Erosion maps have been compiled at a scale 1:25.000 in some areas and at smaller scale for the whole territory of the republic. They show vulnerability to erosion, evaluated on the basis of climate, soil (factor K), slope and land use parameters of the USLE. It is estimated that water erosion is causing the deterioration of 17% percent of the country's land base and affects approximately 50% of its lands, while wind erosion affects 10,4% of all the arable land in the Czech Republic. Attempts have been made to map the consequences of erosion and accumulation, as reflected in reduction of soil profile depth and deposition of erosion sediments (soil maps at 1:10.000 scale of the systematic soil survey).

M.J. Wilson and B. Maliszewska-Kordybach (eds.),
Soil Quality, Sustainable Agriculture and Environmental Security in Central and Eastern Europe, 79–86.

Soil Acidification. Soil acidification due to severe acid deposition is an important degradation process that relates to dying forests in some areas of the CR. In addition, it could cause in future mobilisation of hazardous trace elements (especially Cd) in arable soils. But the pH data from dystric Cambisols and Podzols of forest sites do not show convincingly that soil acidification has occurred. In arable soils it was found that even strong liming did not change the base saturation in the B horizons, so that this feature was permanent enough to be used in general soil taxonomy.

Loss of organic matter. Soil dehumification trends have been investigated by means of retrospective monitoring [1]. It has been found that dehumification took place during the last 30-40 years only in regions of dystric Cambisols which were affected by liming, erosion, and deep ploughing and in drained Stagnosols and Gleysols.

Soil Pollution. Great attention has been paid to the evaluation of pollution trends during the last 30-40 years of industrial and agricultural activities, which neglected to take environmental aspects into account. However, retrospective monitoring [2], which made use of archived samples from the systematic soil survey, carried out at the beginning of the period where severe impacts on soils could be anticipated (1960-1970), found that there was no significant non-point increase of hazardous trace elements during this time, even in the most impacted regions. On sites where increased contents of trace elements were found this status had been reached even 30-40 years ago, due to the long-term industrial emissions. Recent retrospective monitoring [3] of PAHs reveals a decline of most PAHs except for B(ghi)perylene, probably as a result of more perfect combustion technologies, despite the increase in traffic. In Fluvisols, however, the accumulation of PAHs, especially B(a)A, N and A has continued. With regard to other possibly contaminated inputs, such as sewage sludge and other waste materials, these were rarely applied to the soils of the CR. Even in the period when Togo-phosphate was imported, the Cd inputs into soils were kept within the range of 2-3 $kg.ha^{-1}$ y^{-1}. However, the use of DDT in the past can be detected, along with its metabolites, especially in soils of the former state farms and in forestry.

3. Recent developments in methodology for investigating polluted soils

In recent years more attention has been paid to soil pollution and a systematic inventory of Cd, Pb, Cr (in 2M HNO_3) and Hg has been conducted by the Central Institute for Supervising and Testing in Agriculture (CISTA). The results have been presented in a cartographic form [4]. On selected sites 10 trace elements (aqua regia) and some pesticides are monitored, as well as inputs from atmospheric deposition. The Research Institute for Soil and Water Conservation (RISWC) has performed an inventory of the most severely impacted regions in comparison with areas which are free of such impacts. All regions are characterised for total contents of 12 potentially hazardous trace elements and principal groups of persistent organic pollutants - POPs (5 MAHs, 12 PAHs, PCBs, HCB, DDT and metabolites, petroleum hydrocarbons) of the Dutch list.

A three-level system of reference values has been proposed in the RISWC.

First level contamination limits reflect the upper boundary of variability range of natural and/or anthropogenic diffuse background of contaminants [2, 5, 10].Second level reference values are empirical official maximum permissible levels. New recommended criteria are based on trace element mobilities and their transfers into crops, leading to the concept of critical crop loads.

The third level is represented by intervention-action values for health requirements developed by the Ministry of Environment.

The contamination assessment method developed in RISWC is based on total contents of trace elements in top soils of agriculturally used soils and in their profiles. The soils include the main taxonomic units and parent material types, representative of the CR and of the main critical ecological regions. Contamination limits are differentiated in accordance with topsoil mineralogical and textural features into 9 classes. They are determined on the basis of 90% percentiles and $GM.GD^2$ of logarithmic transformed values of the uncontaminated "normal" soils where the trace elements are mostly of "geogenic" origin. Exceedance of these limits (Table 1) indicates increased anthropogenic inputs into the soil.

TABLE 1. Contamination limits (mg /kg) of trace elements (totals) in soils and lowest values of geogenic anomalies (in bold).

	Principal Groups of Parent Materials and Weathered Materials				
Trace Elements	**Sediments (Clays and Sands)**	**Granites, Gneisses, Schists**	**Mafic, Ultra-Mafic**	**Metal-Rich Zones**	**Calcar-eous**
Hg	0.5	0.5	0.5	0.5	0.5
As	25	30	30	**>80**	30
Cd	0.4-0.5	0.5	0.5	0.5	**>1.0**
Be	3.5-5.0	5.0	5.0	**>7.0**	5.0
Pb	70-80	90	80	**>100**	80
Zn	120-150	150-160	**>200**	**>200**	160
Cu	50-70	50-70	**>100**	**>100**	70
Mn	1100-1300	1100-1400	**>2000**	1100	1100
Co	25-30	30-40	**>80**	40	**>50**
Ni	50-70	50-70	**>80**	70	**>80**
Cr	100-130	100-180	**>200**	180	**>200**
V	150-180	150-190	**>300**	190	**>200**

For the assessment of geogenic extremes, profile analyses are indispensable. The impacted and non-impacted regions were characterised by the percentage of samples exceeding the contamination limits and in this way the contribution from anthropogenic sources as opposed to lithogenic/geogenic sources were differentiated [6].

More recently, RISWC has paid attention to the principal groups of POPs. Investigations have been made both in polluted and relatively unpolluted regions. POP limits have been based [7] on the same principles as the limits for trace elements (Table. 2).

4. Distribution of contaminated soils in the CR

Trace elements and POPs were determined in soils as well as in fodder plants , [7, 3]. The most severely POP-contaminated soils in the CR are Fluvisols on the lower terraces downstream from industrial cities (Table 3). The pollution which is caused by waste waters, often surpasses the contamination limits shown in Table 2. PAHs have particularly high average values, especially in B(a)A, N and A. These soils are also contaminated by Cd, Zn, Hg, Pb, Cu and Cr. The metropolitan area of Prague is particularly contaminated. Here, PAHs attain maximum values in all soils, as do some other substances. From a qualitative point of view a progressive increase of B(a)P and a relative increase of B(b)F and B(ghi)Pe, indicating the traffic impacts, may be noted. Contamination by POPs is accompanied by high levels of Cd, Pb, Cu, Zn and Hg. Another highly impacted area is the North Moravian region, characterised mainly by metallurgical industry, black coal mining and (former) cokeries. Soil contamination by POPs is due to PAHs and petroleum hydrocarbons. B(a)P shifts to high values and Cd, Zn and Pb dominate among the trace element contaminants. In the North and Northwest Bohemian region the centre of contamination by PAHs is the Sokolov district and the emissions have impacted the Karlovy Vary district. The identification of MAHs indicates the influence of the chemical industry. The maximum contents of PAHs in the North Bohemian region are found in the Most, Teplice and Ústí n. L districts. The predominant contamination with PAHs and As, Cd, Be indicates an origin through the combustion of brown coals. MAHs occur in the soils of the Most and Teplice districts due to the petrochemical industry. PCBs also have maximum values in the districts with chemical industry, such as Ústí, Litomerice, Decín. Contamination with DDT and its metabolites is found on the territory of the former big state farms. The contamination of the soils of all these regions differs markedly from the relatively unpolluted soils in Šumava and the Bohemo- Moravian highlands. Maps show that the transmission of PAHs follows the direction of the dominant winds. This assessment of soil contamination in the CR is confirmed by other data concerning the amount of PAHs participating in the contamination and by the relative toxicity against B(a)P. It is also supported by the systematic inventory of trace elements, completed by the State Institute for Agricultural Supervision and Testing. It was found that the contamination with Cd, Cr, Hg and Pb occurs in 6.9, 2.4, 0.8 and 2.1% respectively of the samples tested. Official maximum permissible limits of these elements are exceeded in 1.3, 0.5, 0.2 and 0.6% respectively of the samples tested. The limits of Ni were exceeded in 0.7% of samples and of Hg, Be, Co, Cu, V and Zn in less than 0.4% of samples [4, 8]

5. Impact on agricultural soils

Comparison of soil and fodder plants loads indicated that emissions are responsible for fodder plant contamination. The vulnerability of agricultural soils to pollution by trace elements is also being assessed on the basis of the ability of the soil to reduce

TABLE 2. Proposal of soil contamination limits of persistent organic xenobiotic substances, and geometric means of their background values

Substance	μg.kg^{-1}	
	CL	**GM**
Monoaromatic Hydrocarbons (MAH)		
Benzene (B)	**30**	**5**
Toluene (T)	**30**	**5**
Xylene (X)	**30**	**5**
Ethylbenzene (Eb)	**40**	**5**
Polyaromatic Hydrocarbons (PAH)		
Fluoranthene (Fl)	**300**	**60**
Pyrene (P)	**200**	**50**
Phenanthrene (Ph)	**150**	**20**
Benzo(b)fluoranthene (B(b)F)	**100**	**15**
Benzo(a)anthracene (B(a)A)	**100**	**10**
Anthracene (A)	**80**	**10**
Indeno(cd)pyrene (I(cd)P)	**100**	**5**
Benzo(a)pyrene (B(a)P)	**75**	**10**
Benzo(k)fluoranthene (B(k)F)	**50**	**10**
Benzo(ghi)perylene (B(ghi)P)	**60**	**10**
Chrysene (Ch)	**50**	**10**
Naphtalene (N)	**30**	**5**
Chlorinated Hydrocarbons		
PCB	**20**	**2**
HCB	**20**	**2**
DDT	**15**	**1**
DDE, DDD	**10**	**1**
HCH	**10**	**1**
Others		
Styrene (S)	**30**	**3**
Petroleum hydrocarbons (Pe.H.) (mg.kg^{-1})	**100**	**30**

CL ... contamination limit
GM ... geometric mean

TABLE 3. Anthropogenic (A) and geogenic (G) loads of soils in the most critical areas in the Czech Republic

A) POTENTIALLY HAZARDOUS ELEMENTS

District, Site	N	in % of samples exceeding the contamination limits of geogenic normal soils																					
		As		Be		Cd		Co		Cr		Cu		Hg		Mn		Ni		Pb		Zn	
		A	G	A	G	A	G	A	G	A	G	A	G	A	G	A	G	A	G	A	G	A	G
Prachatice [I]	48	0	6	0	12	0	2	0	0	0	8	0	0	0	0	0	0	0	8	0	2	0	4
Most [II]	63	25	19	22	5	13	0	0	5	0	14	0	11	0	0	0	3	0	19	0	21	0	8
Teplice [III]	75	18	23	26	20	17	4	0	15	0	21	0	11	1	0	0	11	0	17	0	27	0	13
Sokolov [IV]	54	4	54	5	33	24	0	0	2	0	7	0	5	2	0	0	7	0	0	0	7	0	22
Ostrava-Karviná [V]	33	6	0	0	0	25	5	0	0	6	0	3	0	3	0	15	0	0	0	6	0	36	0
Prague [VI]	39	0	0	0	0	36	0	0	0	0	0	33	0	38	0	0	0	0	0	33	0	28	0
Fluvisols [VII]	33	7	0	29	0	93	0	7	0	31	0	56	0	88	0	33	0	10	0	53	0	88	0

B) ORGANIC XENOBIOTIC SUBSTANCES

District, Site	N	in % of samples exceeding the contamination limits of POPs																		
		PAH												MAH				ClH		PeH
		Fl	P	Ph	B(b)F	B(a)A	A	B(a)P	I(cd)P	B(k)F	B(ghi)P	Ch	N	B	T	X	Eb	PCB	HCB	PeH
Prachatice [I]	32	9	12	3	3	6	6	6	3	3	3	3	9	0	0	3	3	3	0	12
Most [II]	39	23	8	13	13	5	5	8	8	8	10	13	13	8	18	15	15	2	0	15
Teplice [III]	48	23	19	8	4	4	6	6	8	6	10	8	10	2	4	10	15	0	0	6
Sokolov [IV]	33	24	21	36	12	45	12	9	9	6	12	38	51	0	0	0	0	21	6	0
Ostrava-Karviná [V]	25	20	24	36	20	36	12	16	8	20	12	36	0	4	0	0	0	4	0	12
Prague [VI]	63	67	10	10	62	40	33	60	13	62	52	41	0	0	0	0	0	43	0	5
Fluvisols [VII]	17	100	41	47	94	94	47	94	53	94	88	6	100	0	0	12	23	82	35	71

District, site	Contamination sources region
I	Emission free region
II, III	North Bohemian region, power plants, mining (brown coal), metallurgy chemical industry
IV	West Bohemian region, mining (brown coal), chemical industry
V	North Moravian region, mining (black coal), metallurgy, cockeries
VI	Traffic Emission
VII	Flooding - waste waters, downstream industrial cities

bioavailability and the feasibility of regulating this process. The assessment is based on the following experimental data.

Sequential analyses of trace elements in Ap horizons of representative soils [9]

Sequential analyses after artificial pollution [9]

Experiments with the mobility and transfer of trace elements from soils to crops after artificial pollution [10]

A second level of reference values [11] will rest upon our own studies of trace element contents and mobilities, both potential and effective, in the principal soils of the republic in relation to their transfer to plants. The investigations are being carried out in both field conditions and in greenhouse pot experiments with naturally contaminated soils (airborne, fluvial, geogenic loads). For mobile elements (Cd, Zn, Ni, Mn Co) with high transfer factors, reference values will be derived by means of regression analyses in relation to fodder and food crop load studies. For other elements characterised by low mobilities and transfers, critical reference values will be based on studies of crop uptakes from anthropogenic and extreme geogenic loads.

6. Conclusion

The concept of sustainable agriculture supports the introduction of a Soil Protection Act, involving the protection of the soil resources of the CR against all kinds of land contamination and degradation, along with the elimination of highly contaminated soils and establishment of a professional soil protection service, devoted to these activities. The protective measures will be more effective because they will be matched to the category of the soil resource, which reflects the degree of need and specificity of soil conservation and protection of other parts of the ecosystem related to the soil. Each category of soil will stress specific soil functions and specific limitations of use and management. Protection and conservation of the various soil categories will need legislation norms, as well as economic and administrative tools.

Such preventive measures will help to relieve the decline in soil quality and help improve the current status of the soils in the Czech Republic. Further development of soil protection and conservation will be faced with problems of:

- the lack of an overall strategic concept,
- neglect of the role of agriculture and soil resources,
- clarification of the responsibilities of the Ministries of Agriculture and the Environment.

7. References

1. Podlešáková, E., Nemecek, J. (1983) Changes in Humus Content of Agriculturally Used Soils, *Proc. 7th Int. Symp. Humus et Plants*, pp. 140-144.

2. Podlešáková, E., Nemecek, J. (1995) Retrospective Monitoring and Inventory of Soil Contamination in Relation to Systematic Monitoring, *In: Environmental Monitoring and Assessment*, Kluwer Academic Publishers, Netherlands, pp. 121-125.

3. Nemecek, J., Podlešáková, E., Pastuszková, M. (1997) Some Specific Features of the Contamination of Soils of the Czech Republic with Persistent Organic Xenobiotic Substances (in Czech), *Rostlinná Výroba* **43 (8)**, 365-370.

4. Mazanec, O., Nerad, J. et al. (1994, 1996) *Inventory of the Content of Hazardous Trace Elements in Soil of the Czech Republic I., II* (in Czech). ÚKZÚZ, Prague.

5. Podlešáková, E., Nemecek, J. (1996) Soil Contamination and Pollution Criteria (in Czech), *Rostlinná Výroba* **42 (8)**, 357-364.

6. Nemecek, J., Podlešáková, E., Vácha, R. (1996) Geogenic and Anthropogenic Soil Loads (in Czech), *Rostlinná Výroba* **42 (12)**, 535-541.

7. Podlešáková, E., Nemecek, J., Vácha, R., Pastuszková, M. (1998) Contamination of Soils with Persistent Organic Xenobiotic Substances in the Czech Republic, *Toxicological and Environmental Chemistry* **66**, 91-103.

8. ÚKZÚZ (1996) Basic Monitoring of Agricultural Soils (in Czech), in Ministry of Agriculture (eds.), *Results of the Investigations of Pollutants in 1996*, pp. 36-40.

9. Nemecek, J., Podlešáková, E., Pastuszková, M. (1998) Use of the Sequential Analysis for the Assessment of Trace Elements Bonds in Principal Soils (in Czech), *Rostlinná Výroba* **44 (5)**, 203-207.

10. Podlešáková, E., Nemecek, J. (1998) Criteria of Soil Contamination of Organic and Inorganic Pollutants in the Czech Republic, *Advances in GeoEcology* **31**, 735-743.

11. Podlešáková, E., Nemecek, J. (1998) Mobility of Trace Elements in Soils and their Transfer into Crops, *Contaminated Soil'98*, Thomas Telford Publishing, London 961-962.

THE PRESENT STATUS OF THE SOIL ENVIRONMENT IN RUSSIA

S.Andronikov
V.V.Dokuchaev Soil Institute, Pijevsky per.7
Moscow, 109017, Russia

1. Introduction

Russia is the largest country in the World. However, if the average amount of land is calculated per head of population, it is found that Russia is in third place (11.6 ha per person) after Australia and Canada. Moreover, the distribution of population in Russia is rather unequal. If the amount of land per head of population is calculated for different regions, it is found that in the Moscow region it is 0.3 ha per person, in the Central Economic Region it is 1.59 ha per person and in the European part of Russia, where 80% of population live, it is 3.7 ha per person. The vast territory of Russia (1709.6 mln.ha) therefore gives a wrong impression that there are an infinite amount of soil resources suitable for agricultural and natural development in the country. In fact, in Russia land resources suitable for agriculture are rather limited due to the influence of unfavourable bioclimatic conditions and increasing pressures related to anthropogenic activities

With respect to the bioclimatic situation of Russian lands, about 60% are in permafrost conditions (1100 mln.ha.) and large areas are also situated in desert and semi-desert dry zones. Thus, agricultural lands occupy 222.3 mln.ha or only about 13% of the total territory. The amount of arable lands is about 132 mln.ha or 7.7% of the total territory. These lands are more or less suitable for agriculture and have to produce 98% of the population nutrition requirements. If the biological productivity of Russian arable lands is assessed according to an energetic coefficient based on water and heat provision, it is found that Russia is in seventh place in the World, corresponding to 0.17 ha per head of population (Table 1).

As the total amount of arable land per person in Russia is 0.89 ha, in the Moscow region 0.08 ha, in the Central Economic Region 0.4 ha and in the European part of Russia is 0.86 ha (1), it is clear that arable land resources in relation to the size of the population is much less than in some other countries.

During the last fifty years the total amount of agricultural land has decreased by 26 mln.ha, but the amount of arable land has slightly increased by 12 mln.ha due to ploughing soils that are partly suitable for agriculture. These are situated in the northern part of European Russia. Nevertheless, it can only be concluded that there is a lack of natural resources to increase the total amount of arable lands.

2. Anthropogenic influence on soil degradation

The lack of juridical and socio-economic rules and laws in agriculture, the population migration from rural to urban areas and the lack of infrastructure development in rural areas has led to an increasing negative anthropogenic influence on the soils and lands

M.J. Wilson and B. Maliszewska-Kordybach (eds.),
Soil Quality, Sustainable Agriculture and Environmental Security in Central and Eastern Europe, 87–95.

TABLE 1 Total amount of arable lands in different countries.

Country	Energetic Coefficient of Bioproductivity	Arable Lands (ha per head of population)
Australia	0.58	1.65
Argentina	0.80	0.46
USA	0.42	0.46
Brazil	0.85	0.29
Canada	0.16	0.28
Spain	0.36	0.22
Russia	0.19	0.17
France	0.32	0.10
United Kingdom	0.23	0.03
China	0.43	0.03

on the soils and lands of Russia In addition, negative processes occurring within the soils have led to a deterioration of major soil properties. As a result, many soils have become much more sensitive to anthropogenic influences and less stable to degradation processes. Thus, the qualitative status of many arable soils is rather inadequate and soil degradation processes have become more intense over a large area. The different degradation processes will be discussed in turn.

2.1 SOIL EROSION

According to agricultural statistics more than 53 mln.ha of agricultural lands in Russia are eroded (Fig. 1 and coloured plate).The area of gullies is 2.4 mln. ha and is increasing annually by 1-0.2 mln.ha. During the last 50 years the speed of erosion has increased by about 30 times and crop production on such soils has been reduced by 50 to 60%.

2.2 DECLINE IN SOIL FERTILITY

During recent years the fertility of many soils in Russia has decreased (1). The current input of organic fertilisers is about 2.6 tons per hectare, a decrease during the last three years of 2.5 times. In 1986 -1990 the input of lime, organic and mineral fertilisers to soils was 60 to 70 % of scientifically based standards. In 1991 this amount was reduced by one third and in 1992 by one half. As a result organic matter content has decreased in many soils due to intensive agricultural use and lack of fertilisers. Soil biota have declined and the soils are less resistant to degradation processes. The calcium in many soils has become rather low, leading to soil acidification, and has to be immediately restored (2).

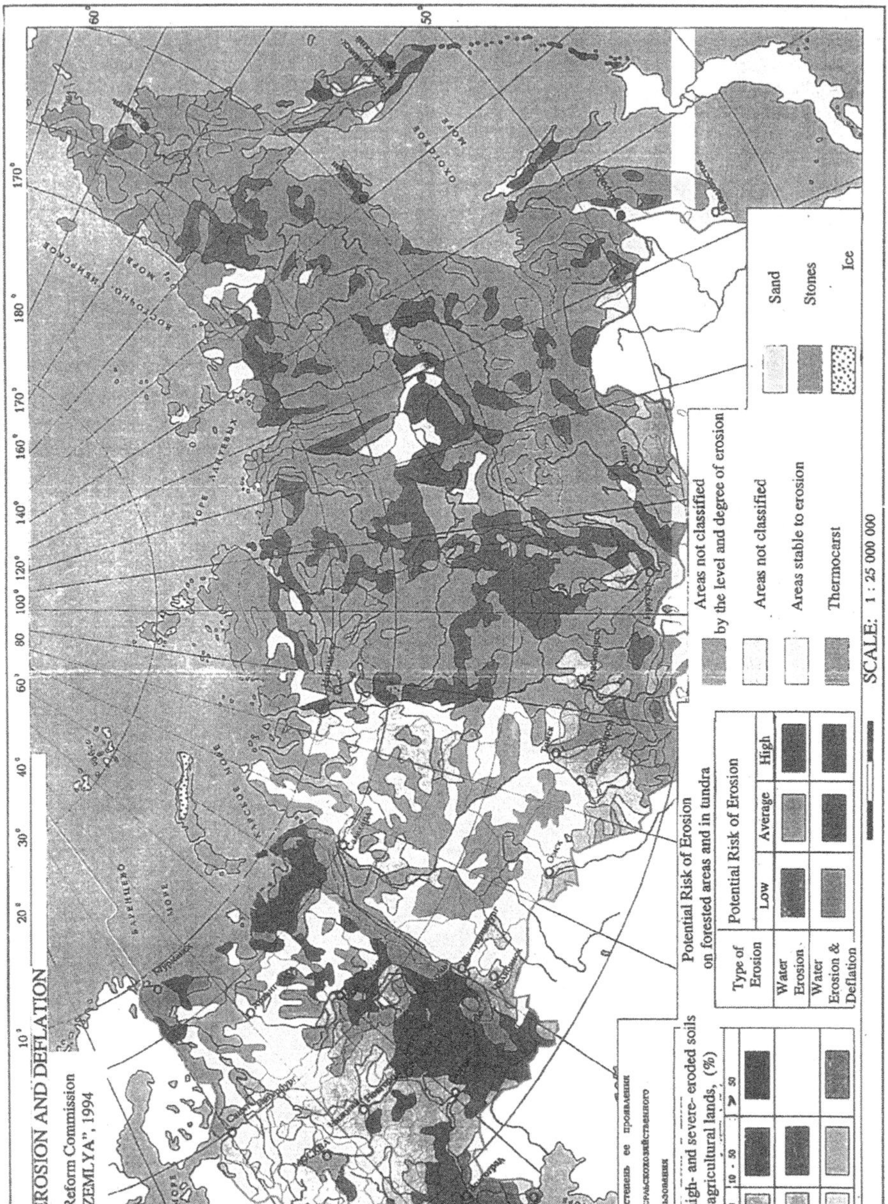

(see also coloured plate at end of book)

2.3 SALINISATION AND WATER-LOGGING.

Soil irrigation and melioration policy has helped to increase the amount of favourable arable lands in Russia. However, in some areas a high groundwater level, because of the breakdown of drainage construction, combined with ineffective irrigation equipment, has led to increasing swampy and waterlogged areas and secondary salinisation in the soils. Gleyic, anaerobic conditions in topsoil are common in many irrigated soils. Again, the amount of peat soils has decreased rapidly due to peat cutting and existing irrigation policy (3).

2.4 OVERGRAZING AND DESERTIFICATION.

Intensive uncontrolled agricultural land use and overgrazing in dry regions leads to the development of deflation and desertification processes in soils. In 15-20 years the total open sandy area will reach 1 mln.ha. Fig.2 (and coloured plate) demonstrates the degradation levels of the natural fodder lands of Russia. The highest level is in the southern part of European Russia, where the desertification effect is very strong. Moreover, overgrazing and uncontrolled land use in the northern part of Siberia and the Far-East has caused high degradation levels of reindeer pastures.

2.5 SOIL POLLUTION

A specific problem is the effect of soil pollution by industrial and agrochemical contaminants (4). The main pollutant components in soils are pesticides, herbicides, heavy metals, oil products and radionuclides. Fig. 3 (and coloured plate) shows the total index of pesticide loading on arable lands in Russia as well as the risk level of pollution by pesticides. The highest contamination risk level is near the border with Ukraine where the arable lands developed on chernozems are under intensive anthropogenic pressure. Fig.4 demonstrates anthropogenic/technogenic loadings and the risk of pollution by heavy metals on arable lands. In general, the most polluted areas are situated in the vicinity of the main industrial centres as well as around big cities. The highest risk of contamination of arable lands is in the region of heavy industry development in the central and southern part of Russia (Fig.4 and coloured plate). Finally, more than 4.9 mln.ha of soils are polluted by radionuclides to varying extents as a result of the Chernobyl accident.

In addition to the degradation pressures described above, socio-economic problems in rural areas of Russia have led to an uncontrolled migration of rural populations. As a result, in last 25 years the total cultivated area has decreased by 33 mln.ha and is still continuing to decrease. There is clearly an urgent need in Russia to improve the quality of agricultural soils over large areas.

3. Recent Developments

Recently, with the development of land reform in Russia the situation has begun to change for the better. Several research institutes have been involved in analysing and

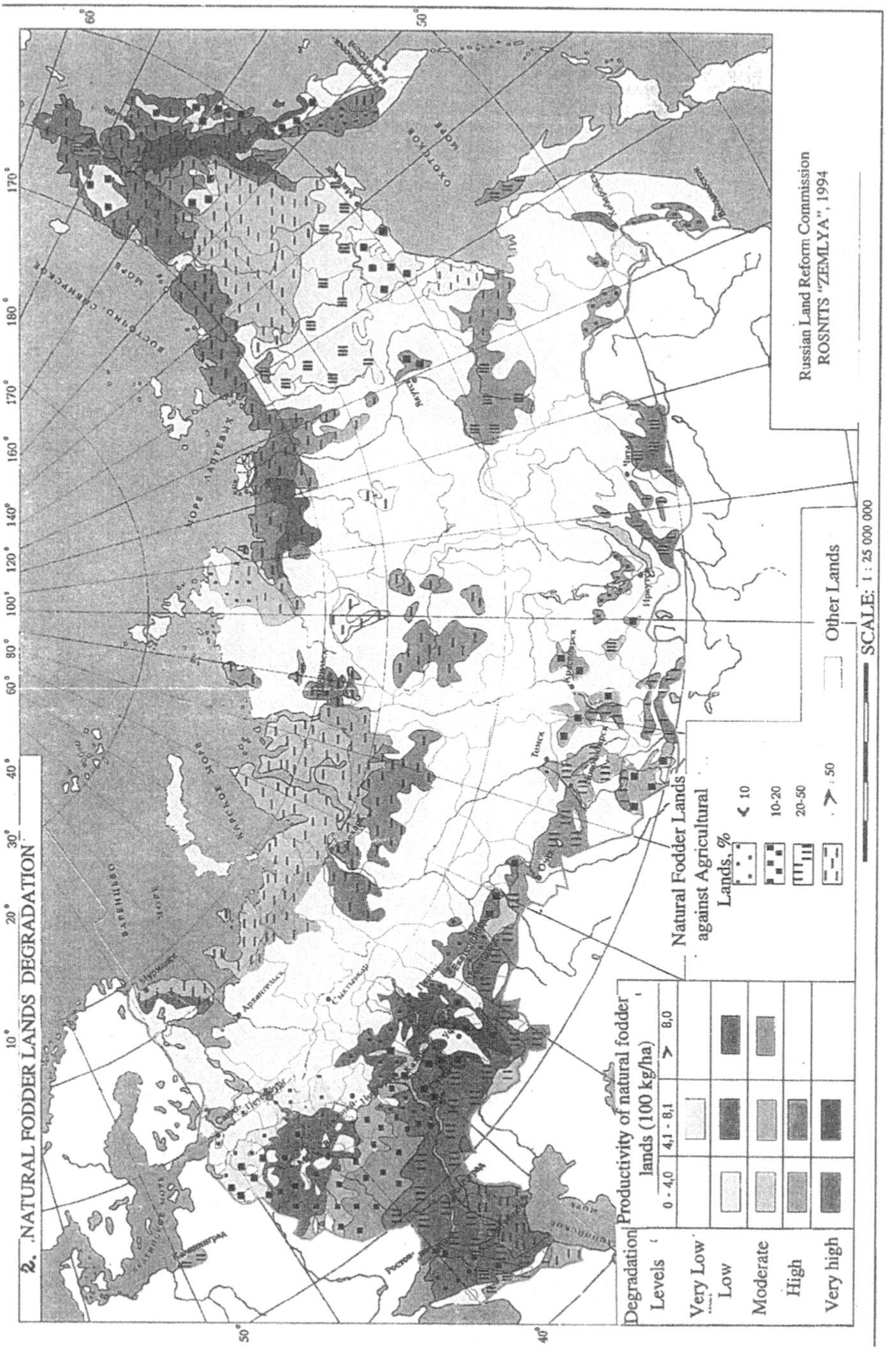

(see also coloured plate at end of book)

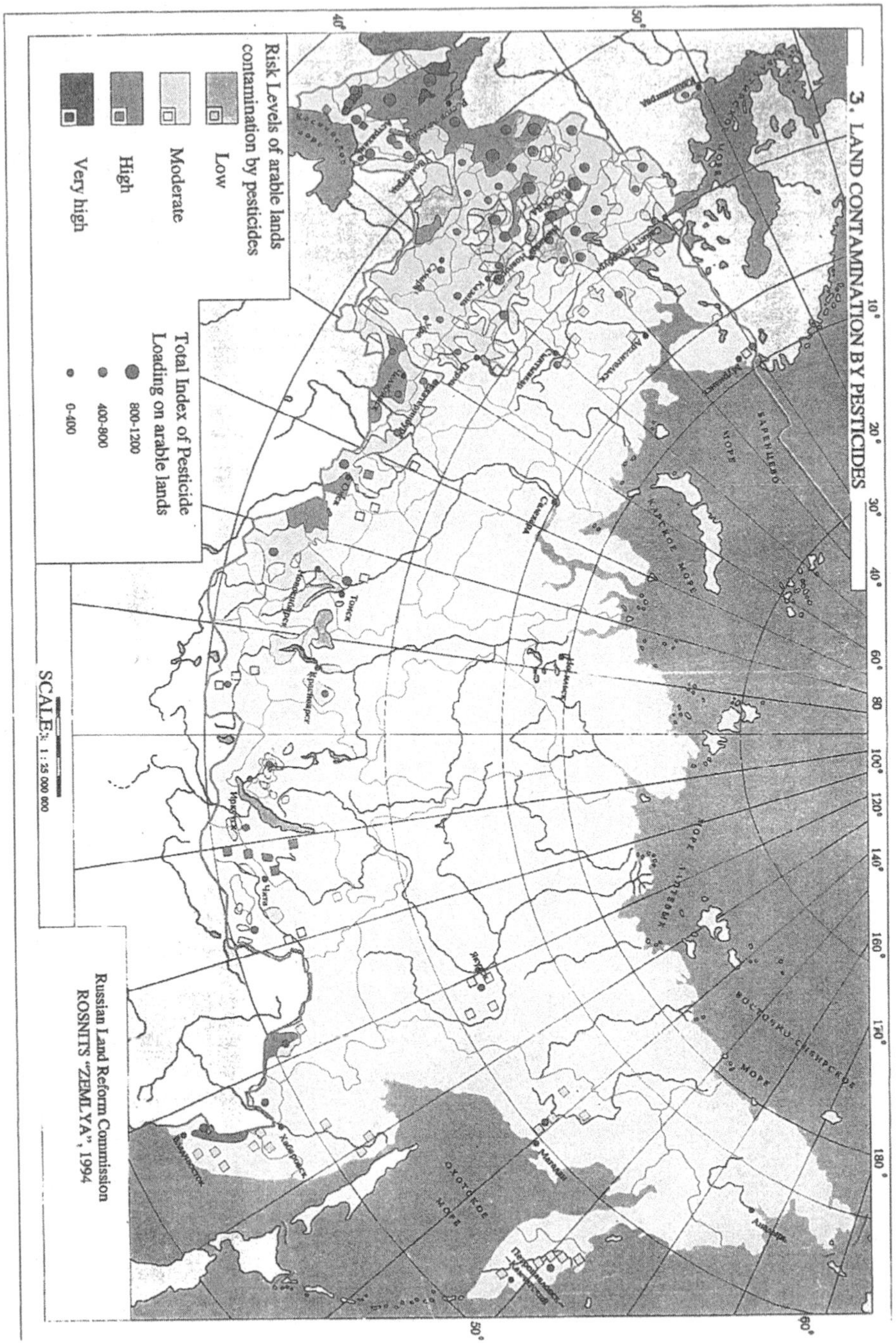

(see also coloured plate at end of book)

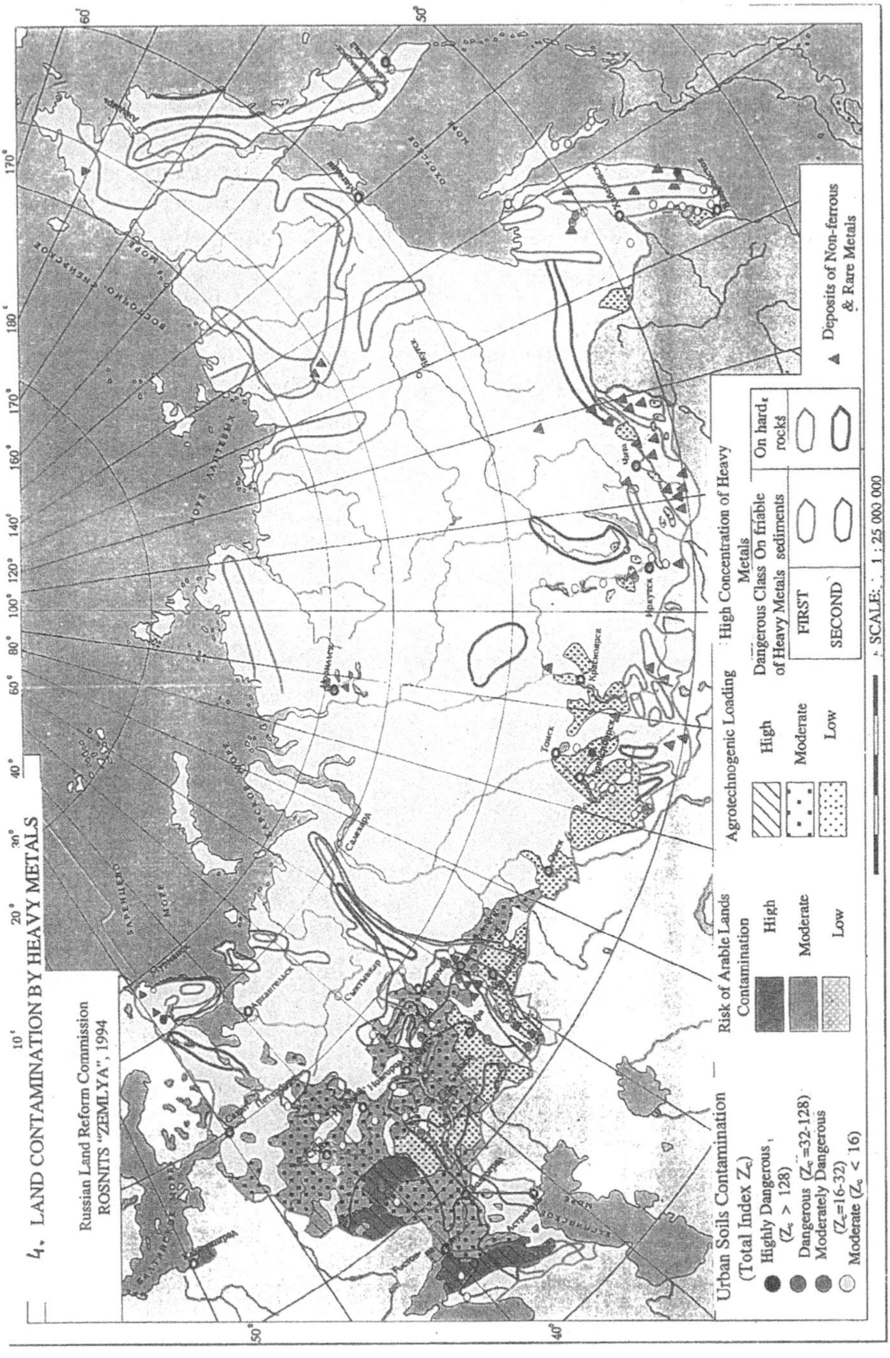

(see also coloured plate at end of book)

improving the situation (2) and an international conference on problems of anthropogenic soils, held in 1996, in Moscow, at the V.V.Dokuchaev Soil Institute, has marked the latest achievements in soil quality assessment. As an example, the transformation of chernozems through unwise agricultural use has been assessed and analysed and a strategy for reducing degradation and restoring fertility has been proposed. Again, water and salt regimes in irrigated soils have been thoroughly examined. Salt in drainage water is now twice as high as in rivers and lakes and a system of agroecological monitoring of irrigated soils has been proposed to improve the current situation. A methodology to assess the level of salinity and alkalinity in soils has been presented with the aim of establishing a correlation with crop biological activity.

Scientists at the V.V.Dokuchaev Soil Institute have been involved in several important areas of research relating to soil quality. Anthropogenic changes in soils of the main agricultural regions of Russia have been described and show how soil quality has changed in the past and will change in the future, and suggest ways of stabilising and increasing soil fertility. A strategy for land use changes according to regional conditions has been proposed (5) and a conceptual monograph has been prepared to describe the scientific and practical basis for creating highly fertile soils in different parts of the country. The main part of this monograph is a theory of organo-mineral interactions and their influence on soil properties and this would help to develop practical principles of agrochemical and agrophysical soil evaluation so as to obtain high yields (5). Again, a model of agricultural landscapes stable to erosion in intensively eroded regions has been created. This should help to increase the yield of arable crops in these areas by up to 60%.

A special focus of Institute research activity is directed to the problems of pollution and contamination. Using remote sensing images of different resolution the level of contamination by heavy metals has been determined for different soil types and crop species. To make a thorough study of the pollutant influence by radionuclides and to analyse the fluxes of radionuclides in agroecosystems a pilot test area near Bryansk has been created. Ultimately, research such as that described above, should help to improve soil quality in different regions of Russia.

4. Conclusions

The current situation is that a critical qualitative condition of soil and land resources in Russia still exists (1). The main reasons for the continued existence of this situation are :

(a), the lack of juridical rights and laws to promote environmentally safe use of various soil types as well as the failure of farms to use land in a sustainable way, mainly for socio-economic reasons;

(b), the unbalanced nature of agricultural investment policy, as well as decision making in agriculture from the top - down;

(c), the lack of a privatisation mechanism in agriculture, and the low development of land cadastral systems;

(d), the lack of coordination between research results and their practical implementation, as well as the low level of soil quality assessment.

5. References

1. The State of the Soils in Russia (1994). A Report of Land Reform Committee, Moscow, 6pp.

2. Measures to Reduce Soil Degradation in Russia. (1994). A Report of Agricultural Working Group, Moscow, 18pp.

3. Kreshtapova, V. (1993). Trace Elements in Peat Soils and Peat Landscapes, Rossel'khosizdat, Moscow, 148pp.

4. Bolshakov, V.A. Krasnova, N.M.,Borisochkina, T. I. Sorochkin, S.E. and Grakovscki, V.G. (1993). Aerotechogenic pollution of soils by heavy metals. Dokuchaev Soil Institute, Moscow, 92pp.

5. The Annual Report of the V.V. Dokuchaev Soil Institute. (1996). Moscow, 35pp.

PART III

STUDIES EMPHASIZING SOME WESTERN EXAMPLES OF MANAGEMENT PRACTICES AND STRATEGIES RELATING TO AGRICULTURAL SUSTAINABILITY AND ENVIRONMENTAL SECURITY

Synopsis and Overview

This section discusses some examples of management practices and strategies in agriculture, mainly resulting from work in the west, that seek to maintain or enhance soil fertility while at the same time minimising environmental impacts and enabling the production of high quality foodstuffs. Amongst the practices considered are the disposal of industrial and urban wastes to agricultural lands, precision farming, organic farming, management to enhance soil physical quality and minimum tillage technology. The application and relevance of such practices for central and eastern European agriculture will be discussed in the next section.

As pointed out by *Haneklaus et al,* sustainable agriculture ideally should use as fertilisers, available materials that save natural resources and do not adversely affect the environment, either now or in the future. It is in this context that they examined recycling of plant nutrients from industrial and urban wastes. The particular wastes examined included flue gas desulphurisation (FGD) and spray dry absorption (SDA) products as sources of sulphur and sewage sludge ashes (SSA) as a source of phosphorus. Both FGD and SDA products were effective as sulphur fertilisers, but their heavy metal contents meant that an objective risk assessment was required. It was established that, generally, heavy metal inputs from these products were no higher than those from NPK fertilisers or pig slurry, although with continued application there will be an accumulation of metals in the soil over the long term. Incineration of sewage sludges may be necessary to cope with the problem of potentially harmful organic xenobiotics, but this procedure naturally concentrates their heavy metal contents. SSA may be as effective as conventional phosphate fertilisers and it contains much less cadmium, although all the other heavy metals are enriched. It follows that application rates of SSA must be limited on the basis of the heavy metal content of the original sludge. *Haneklaus et al,* also consider the applicability of "Precision Agriculture" to inputs of secondary raw materials such as those described, and conclude that the technique will become increasingly important in the future, given its potential to combine economic profitability with protection of the environment. Although such developments may not be of immediate concern to the central and eastern European farmer, they may well be in the not too distant future.

Alcock and Jones consider the long-term fate and behaviour of polychlorinated biphenyls (PCBs) in soils from experimental plots to which sewage sludge was applied several decades ago. The results demonstrate that atmospheric inputs now

M.J. Wilson and B. Maliszewska-Kordybach (eds.),
Soil Quality, Sustainable Agriculture and Environmental Security in Central and Eastern Europe, 97–100.

dominate the PCB content of the soils, even where large applications of sludges were made. In sludge-amended plots established in 1968, PCB concentrations reached control plot concentrations in the late 1980s. Volatilisation was thought to be account for the losses observed and this possibility was confirmed by short-term experiments.

Organic farming has become increasingly popular in the West and *Bachinger and Stein-Bachinger* describe various management strategies that they have used in east Germany and which may well have applications to countries in central and eastern Europe. The strategies are aimed at minimising inputs from outside the farm as well as losses from within the farming system. In this way, crop yields and quality are optimized but environmental problems such as nitrate leaching are avoided. Crop production is essentially nitrogen limited and inputs rely principally upon fodder and grain legumes in the crop rotation. With careful management techniques involving quantification of nutrient fluxes, *Bachinger and Stein-Bachinger* found that there was a low potential for nitrate loss over the whole crop rotation, such losses being several times lower than in conventional systems. Other environmental benefits include a positive humus balance in the soil, a decreased potential for soil erosion, no contamination of ground and surface water by pesticides and increased biodiversity of flora and fauna. The organic farming strategies described by *Bachinger and Stein-Bachinger* were applied to large farms, formerly state-owned in East Germany, on sandy soils and showed that, in principle such farms could be successfully converted to organic farming, despite difficulties arising from farm structure and complexity. Nevertheless, it is recognized that the viability of organic farming in the CEE counties does depend on the development of specific high price markets for organic produce and/or economic rewards for the positive environmental effects.

Conventional tillage (CT) is used for the establishment of most crops in western Europe, involving mouldboard ploughing to a depth of 200-300cm, secondary cultivation to produce a seedbed, sowing the crop and finally rolling the sown seedbed. The system optimises the soil conditions for crop growth but can lead to erosion under certain conditions. Evidently, systems based upon minimum tillage would tend to reduce soil erosion, although as *Ball* points out, this advantage may be nullified by other deleterious effects resulting from slugs, weeds, residues of former crops and soil compaction. Nevertheless, given a high standard of management these problems can be overcome and the potential environmental benefits to be obtained from minimum tillage systems are such that further consideration needs to be given to its use. *Ball* compares the physical, chemical and biological qualities of soils after CT and after minimum tillage, involving reduced tillage (RT) and no-tillage (NT), particularly from the environmental point of view. It was found that under both RT and NT there was an increase in soil bulk density, soil strength, and aggregate stability and a decrease in surface cracking and hydraulic conductivity. In addition, there was an accumulation of organic material and nutrients near the soil surface and an increase in microbial and enzyme activity. The main environmental benefit of RT and NT systems is a reduced risk of erosion and decreased losses of phosphorus through mobile particulates. *Ball* concludes that minimum tillage provides most benefit to soil quality in warm, dry areas such as Spain, where it is used widely. Whilst recognizing the problems involved in adopting RT and NT systems in CEE countries, it is suggested that such systems may be used to advantage in vulnerable areas where is an erosion risk.

The physical properties of soil in relation to overall soil quality are examined in further detail by *Dexter and Czyz,* in particular soil strength, soil stability, hydraulic properties and soil aeration. It is suggested that the key to good physical soil quality is a stable structure. Stability of soil structure can be promoted by the application of inorganic compounds, such as lime or gypsum, to the soil, and by building up the soil organic matter content. A stable soil structure can be maintained through the use of appropriate tillage systems and careful control of wheel traffic. *Dexter and Czyz* conclude by pointing out that there are many soils in the CEE countries that are physically degraded and that need to be improved so that their proper functioning can be maintained in both agricultural and environmental contexts. Such improvement could be readily brought about by the adoption of the management techniques proposed. The conservative outlook of farmers and the lack of resources in the agricultural sector remain as formidable barriers.

Over the last 50 years, the use of herbicides has come to play an essential role in crop production, particularly in the West, and the practice is recognized in the main as being beneficial. Thus, herbicide applications reduce both the amount of labour and the mechanical cultivation required and generally increase crop productivity, harvest efficiency and the grazing capacity of pastures. However, environmental concerns remain and *Clay* identifies the two major ones as being the impact upon ground and surface water quality through offsite movement and the development of herbicide resistant weeds. *Clay* reviews the practical management practices that have been shown to be effective in reducing herbicide environmental impacts and discusses the principles and mechanisms that underpin these practices. To a large extent, these practices work because they reduce or eliminate herbicide movement to non-target areas through sorption and retention in the soil, a process controlled by the physical and chemical properties of both soil and herbicide. Soil management through tillage and fertilisation also influence the sorption, leaching and degradation of herbicides, as does placement, the planting of filter strips and other practices. *Clay* also describes a variety of best management practices aimed at reducing the development of herbicide resistant weeds. It is concluded that with the implementation of such practices as well as continuing research to develop better management solutions, herbicides may continue to be used with minimal offsite environmental effects.

In farming systems involving intensive animal production, losses of N to ground and surface waters and accumulation of P in the soil may attain a level that is unacceptable from the environmental point of view. In Denmark, for example, incidents of severe oxygen deficiency in the Kategat and a number of Danish fjords appeared to be directly attributable to N losses from agriculture and prompted Parliament to institute stringent rules for the use of manure and chemical fertilisers. These rules are outlined by *Pedersen* and involve harmonized regulation, on a national scale, of storage, application rates and timing of application of fertilisers and manures, as well as the return to the authorities of fertiliser balances at the end of the growing season. The effect has been that in the last 10 years there has been an overall reduction of the surplus of nitrogen from Danish agriculture by 25% and further reductions could be achieved if necessary. *Pedersen* concludes that as far as nitrogen is concerned, Danish agriculture can continue to produce large amounts of animal products for the world market in a sustainable way.

Finally, *Paterson et al,* show how systematically collected soils data may be used to predict environmental sensitivity to pollution through examples using the Soils Database in Scotland. The first example describes the use of the critical loads concept to acidic inputs to soils and illustrates how a decision framework involving mineralogical, chemical and biological reactions can be spatially represented in the form of a soil mapping unit. However, the use of a similar approach to illustrate point data concerning potentially toxic elements (PTE) encounters some difficulties. One problem is that the degree of risk associated with determination of the "total" PTE content of the soil is highly variable, suggesting that a more refined decision framework needs to be developed before soil mapping units are used as carriers of spatial information. This point is illustrated by the use of process-based indicators for the assessment of sensitivity to PTE risk in Scottish soils, as opposed to the more conventional compositionally-based indicators that are usually used for this purpose. *Paterson et al* conclude that the use of soil data bases and maps will be important in the assessment of the sustainablity of land use systems.

AGRO-ENVIRONMENTAL ASPECTS OF LAND DISPOSAL OF INDUSTRIAL WASTES

S. HANEKLAUS, J. FLECKENSTEIN AND E. SCHNUG
Institute of Plant Nutrition and Soil Science, Federal Agricultural Research Centre, Bundesallee 50, 38116 Braunschweig, Germany

1 Introduction

Processing, recycling and use of plant nutrients from industrial processes are central concepts in waste avoidance. However, pressure to use agriculture as a way of disposing of industrial wastes may conflict with requirements to maintain the rural environment and guarantee food quality. Sustainable agriculture should use any resource in such a way that does not compromise the way in which the present and future human needs for food or other agricultural goods are satisfied, and ensures that the quality of the environment and natural resources remains preserved.
Recycling of plant nutrients from industrial wastes can be a contribution to saving of natural resources if both the demands of the soil and environmental protection are fully satisfied. Where natural resources are limited, as in the case of phosphate deposits, an efficient recycling of P-containing secondary compounds will be of increasing importance in future. This together with variable rate application techniques, (better known under the term 'Precision Agriculture'), where a spatial distribution of applied nutrients according to the demand of the crop can be realised, can offer major agro-environmental advantages.

However, a number of secondary raw materials often contain not only essential plant nutrients but also varying amounts of inorganic xenobiotics. Wastes which contain known, but also possibly unknown organic xenobiotics such as sewage sludge should be excluded from disposal on agricultural farmland as no objective risk assessment is possible. Thus in this case there is a conflict with the purpose of agricultural production (which is to produce high quality foodstuffs) and protection in terms of preservation of soil fertility and minimisation of environmental burdens. For this reason, it is essential that procedures and criteria for a comprehensive agro-environmental evaluation of the residues in industrial wastes need to be established in order to guarantee their safe disposal on soils. This question is one of some concern to all EU countries and will inevitably become more important in CEE countries with the globalisation of markets including those related to agriculture and harmonisation of measures to protect the environment.
In this contribution an agro-environmental evaluation of the use of sulphur containing secondary raw materials from the desulphurisation process and sewage sludge ashes is

M.J. Wilson and B. Maliszewska-Kordybach (eds.),
Soil Quality, Sustainable Agriculture and Environmental Security in Central and Eastern Europe, 101–113.

assessed and potential for the implementation of new fertiliser technologies are discussed.

2 General Comments on the Use of Industrial Wastes

The steadily increasing number and quantity of industrial wastes is intensifying the social pressure on agriculture to dispose of them. Their acceptance by farmers will depend firstly on the safety of the products in terms of foodstuff quality and soil protection, secondly on the nutrient efficiency of the product itself and thirdly on the economic benefit.

The application of secondary raw materials may lead to an accumulation of environmentally important elements in soil which is a more or less irreversible process. It would take only 115 years to increase the lead content in the top soil from an average background value of 50 $\mu g\ g^{-1}$ to 100 $\mu g\ g^{-1}$ (which corresponds with the threshold defined in the German sewage sludge regulation [1]) by applying a maximum permitted load of 15 g Pb yr^{-1}. But it would take more than 246,000 years to half the content again by crop removal and leaching. All efforts to grow crops which accumulate heavy metals have failed [2] so that there is virtually no method to extract them from the soil again.

The question of whether for fertilisers and secondary raw materials there should be a threshold system involving zero accumulation of heavy metals in soils is discussed extensively by Witter [3].

The conclusion is that input of heavy metals, whether from common organic and mineral fertilisers or from secondary raw materials, should be minimised as far as possible as a precautionary measure in order to protect soil fertility and the overall environment, as well as animal and human health.

Sources of recycled plant nutrients from industrial processes are manifold and their number is increasing due to the 'Waste Avoidance, Recovery and Disposal Act' [4]. Legal and administrative aspects on the utilisation of secondary raw materials in agriculture are discussed extensively by Oswald [5] and a comprehensive overview of sources of major plant nutrients in industrial wastes and amounts produced is given by Werner [6].

The application of secondary raw materials in Germany requires that contamination by the product should be lower than the values defined for sewage sludges [1] but legal thresholds have not been defined so far. Other than this, restrictions concerning the total heavy metal loads applied by secondary raw materials are still under discussion. Table 1 shows the legal thresholds for heavy metals in sewage sludges and treated soils, in comparison with average background values in uncontaminated soils.

Table 1. Thresholds for contents of environmental elements in sewage sludges and soils according to the German Sewage Sludge Regulation [1] and EEC legislation.

Element	Sewage sludge (dm)	Soil (D)	Soil (EEC)	Agricultural soils (Background values)
	[1]	[1]	EEC 86/278/EWG	[4]
Cd	10	1.5	1 - 3	<0.3 - 1.1
Cr	900	100	---	---
Cu	800	60	50 --140	13 - 71
Hg	8	1	1 --1.5	0.08 - 0.35
Ni	200	50	30 --75	<3 - 339
Pb	900	100	50 - 300	32 - 75
Zn	2,500	200	150 - 300	51 - 168

note: maximum load of sewage sludge: 5 tonnes/3 yr. (dm); background values in soil reflect ranges for different soil types

2.1 AGRO-ENVIRONMENTAL ASPECTS OF THE USE OF FLUE GAS DESULPHURISATION (FGD) PRODUCTS: SULPHUR

As an example an agro-environmental evaluation of different flue gas desulphurisation (FGD) products will be given because of the expanding market for sulphur-containing fertilisers in Europe. The desulphurisation of flue gases from coal-fired power plants has reduced SO_2-emissions drastically and thus greatly reduced the sulphur (S) supply to agricultural crops [7, 8]. Severe S-deficiency is meanwhile the most widespread nutrient disorder in northern Europe and both losses in yield and quality can only be avoided by crop specific S-fertilisation [9, 10]. The use of FGD products as S-fertilisers in agriculture could cover the sulphur supply of crops and would simultaneously contribute to an appropriate recycling of these residues. Gypsum from German coal-fired power plants is totally recycled [11] while spray dry absorption (SDA) products are only partly re-utilised.

Different FGD products were tested in field experiments on four sites in northern Germany with regard to assessing their suitability as a S-source and their environmental impacts. The highest S-content (17.5% S) was found in SDA-product A and the lowest (6.6% S) in SDA-product D (table 2). SDA-products with fly ash contents <8% may contain up to 68% calcium sulphite, while in products with fly ash contents between 20 and 85%, calcium sulphite does not exceed 47% [12]. The required time for a complete oxidation of sulphite is about two weeks [13].

Bertelsen and Gissel-Nielsen [14] and Ritchey et al. [15] found a negative influence of the varying amounts of sulphite applied by SDA-products on the growth of oilseed rape or cereals which was, however, not observed by Paulsen et al. [16] on the different test sites. These results are confirmed by earlier experiments of Mortensen and Nielsen [13], [17] and [4]. Negative impacts of sulphite on crops following in the rotation did not occur either [12].

Oilseed rape is a crop with a high demand for S because of the high protein content in its seeds and also because of the characteristic presence of S-containing glucosinolates. S-fertilisation with SDA-products and FGD-gypsum improved the S-supply of oilseed rape significantly (figure 1); differences between the products were not statistically significant.

Table 2. Mean total concentrations of sulphur, calcium and environmentally important elements in SDA-products and FGD-gypsum [16].

Element	LLD[1]	FGD-gypsum	SDA-product			
			A	B	C	D
			(<8% FA[2])	(<8% FA)	(20% FA)	(60% FA)
SO_3 (%)	180	44.16	45.19	32.23	28.24	16.63
CaO (%)	2.400	35.37	43.30	48.15	34.58	20.16
Cu (μg g^{-1})	5.3	<LLD	<LLD	38.7	79.0	139.0
Zn (μg g^{-1})	5.7	<LLD	8.7	41.7	42.7	174.7
Cd (μg g^{-1})	0.8	<LLD	<LLD	<LLD	<LLD	<LLD
Ni (μg g^{-1})	5.3	<LLD	10.0	7.0	36.3	149.3
Cr (μg g^{-1})	7.5	12.7	16.3	25.3	49.0	126.7
Pb (μg g^{-1})	6.1	22.3	33.3	172.0	64.7	172.0
Hg (μg g^{-1})	0.1	<LLD	<LLD	<LLD	<LLD	<LLD
Co (μg g^{-1})	3.3	<LLD	5.0	6.7	19.3	61.3
Se (μg g^{-1})	0.9	5.0	13.0	17.0	10.0	10.3
As (μg g^{-1})	5.5	<LLD	8.0	30.7	12.0	29.3

note: [1]Lower Limit of Detection in μg g^{-1} (practical detection limit); [2]Fly ash

The S-supply of the oilseed rape crop in the control plots was in the range of severe S-deficiency with total S contents of <3.5 mg g^{-1} S at the beginning of stem elongation. Application of 40 kg S ha^{-1} was sufficient to improve the S-supply above the critical threshold of 6.5 mg S g^{-1} in order to achieve maximum yields. As expected S-fertilisation increased the total S content in straw and seeds significantly. S-fertilisation with SDA-products led to a significant yield increase of oilseed rape at all sites.

In oilseed rape and also in cereals the total S-content in the leaf tissue at stem elongation (growth stage GS 32; [18]) and straw at harvest increased significantly with the S-dose, although differences between the products were not statistically significant. The results clearly reveal that FGD products were suitable for satisfying the S-demand of agricultural crops.

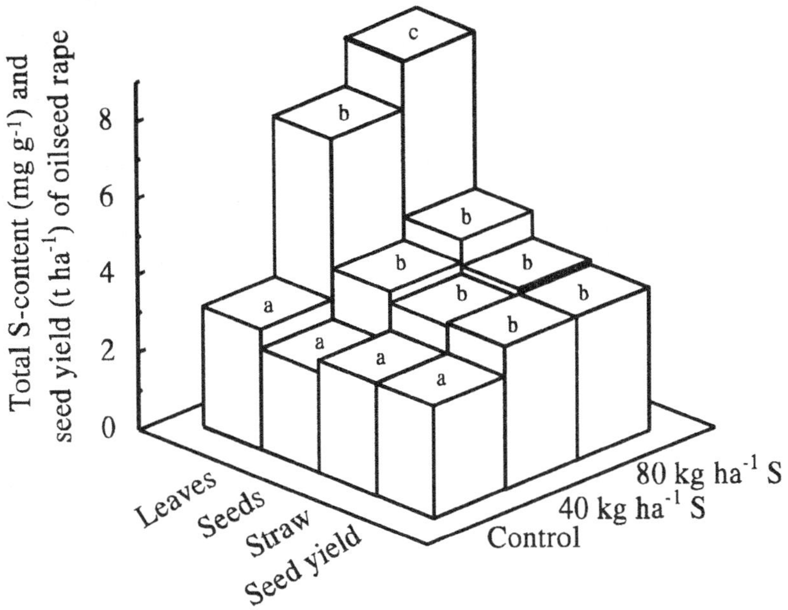

Figure 1. Influence of S-fertilisation by SDA-product on the total S-content in vegetative and generative plant parts of oilseed rape [16] (note: different characters designate statistically significant differences of means at 5% level in the t-test).

2.2 LIME/CALCIUM

SDA-products contain considerable amounts of Ca bound to SO_3, SO_4, Cl, F, CO_3, O, or OH. The liming potential of the products depends on the basic reaction of Mg and Ca carbonate, oxide and hydroxide. Ca-sulphite and Ca-sulphate do not affect soil pH but increase the calcium saturation of clay minerals in soil and thus soil aggregation wihout increasing soil pH. This may be an important factor on heavy soils where a high Ca saturation of clay minerals is required for flocculation of soil aggregates and for avoiding an immobilisation of micro nutrients induced by too high pH values [19].
In general, average liming rates over three years vary between 500 kg ha^{-1} CaO on sandy soils and 1,500 kg ha^{-1} CaO on loamy soils. SDA-products used as S-fertilisers supplied 130 - 480 kg ha^{-1} Ca but only between 46 and 357 kg ha^{-1} CaO cumulatively in three years [16]. This means that basic compounds in SDA-products and additional Ca will positively influence soil structure but do not produce a significant pH effect. This was verified by pH measurements on all test sites [16].

3 Environmental Aspects of Utilisation of SDA-products in Agriculture

Undesired, environmentally relevant elements may accumulate in soils and enhance uptake by plants when SDA-products are applied, so that an objective risk assessment is required. The heavy metal content of SDA-products is proportional to their fly ash

content but is in all products distinctly below the critical thresholds defined in the sewage sludge regulation [1] (see table 2). The Se content of SDA-products, however, (table 2) prevents 'a priori' their use on grassland because of the German Foodstuff Law [20].
The loads of undesired elements applied to the soil in proportion to 50 kg S ha^{-1} in the SDA-product are shown in table 3. This input is compared with the average heavy metal input by common annual application rates of pig slurry and NPK-fertilisers.

The results reveal that significantly higher Cu and Zn loads are applied with pig slurry than with SDA-products. Cd was below the detection limit in all FGD-products. SDA-products with fly ash contents <20% added Ni and Cr in amounts comparable to those in common NPK-fertilisers and pig slurry. Pb loads of SDA-products with >8% fly ash were distinctly higher than those of slurry or NPK fertiliser.
The application of all products will lead to an accumulation of environmentally relevant elements in soil on a long term basis though the time period needed for an accumulation of 1 $\mu g\ g^{-1}$ is at least 27 years in the case of Pb and Zn.

Table 3. Average loads of environmentally relevant elements applied with FGD-products together with 50 kg ha^{-1} S in comparison with other sources [16].

Element	FGD-gypsum	SDAP (<8% fly ash)	SDAP (60% fly ash)	NPK fertilisers[1]	Pig slurry (30 m^3)	Crop removal[2] (g ha^{-1})
As	< LLD[3]	2	22	3	2	0.1
Cd	<LLD	<LLD	<LLD	4	<1	0.64
Cr	4	5	95	53	15	1.6
Co	<LLD	1	46	-	-	-
Cu	<LLD	<LLD	101	6	372	36
Hg	<LLD	<LLD	<LLD	-	---	0.021
Ni	<LLD	2	112	9	39	2.5
Pb	6	9	130	8	12	1.9
Se	1	4	8	-	-	-
Zn	<LLD	2	132	91	1.170	213

note: [1] Average application rates to oilseed rape: 700 kg ha^{-1} calcium-ammonium-nitrate, 130 kg ha^{-1} Triplephosphate, 200 kg ha^{-1} potassium chloride acc. to Boysen [21]; [2]Annual crop removal for a sugar beet-oilseed rape-cereals crop rotation acc. to Andersson [22]; [3] LLD = Lower Limit of Detection

4 Agro-environmental Aspects of the Use of Sewage Sludge Ashes

For the inorganic elements Cd, Cu, Cr, Hg, Ni, Pb and Zn thresholds in soils and sewage sludge are legally defined. In comparison, 21 different groups of organic compounds have been identified [23], while an ecotoxicological evaluation together with critical thresholds and tolerance values for these substances is often missing or relies on single data so that a risk assessment for the whole ecosystem is not possible. Furthermore new compounds may be identified in future as potentially harmful. This unnecessary and unquantifiable risk related to application of sewage sludges to soils could effectively be avoided by their incineration. First experiments on using such sewage sludge ashes (SSA) show that they could be a valuable source for phosphorous in future, particularly as the natural resources of P are limited and estimated to satisfy agricultural demands for only another 100 years [24]. Therefore the P-efficiency of these materials is of primary interest.

The P-efficiency of sewage sludge ashes (SSA) on yield parameters of maize was tested in a pot trial with two soils differing in pH and compared to the effect of commercial fertilisers. Values for the solubility of the products tested in different extraction solutions and their mean total heavy metal content are given in table 4.

Table 4. Composition of P-fertilisers and P-containing secondary raw materials.

Product	Solubility	P_2O_5 (%)	Cd	Cr	Cu	Ni (mg kg^{-1})	Pb	Zn
Triple phosphate[4]	100% in water	45	30.6	300	23	31	< 1	448
Hyperphos[5]	80% in FA[1] 20% in conc. acids	29	46.3	245	---	15	---	400
SSA (A)	2.1% in AC[2] 8.2% in AC + CA[3]	15	4.1	330	911	132	325	2,513
SSA (B)	3.3% in AC 9% in AC+CA	12	3.1	177	1,306	69	168	2,695

note: [1]FA = Formic Acid; [2]AC = Ammonium Citrate; [3]CA = Citric Acid; [4][21]; [5][25]

SSAs are soluble to about 9% in a mixture of ammonium citrate and citric acid. In comparison triple phosphate is water soluble to 100%, while Hyperphos, a fine ground raw phosphate, dissolves to 80% in formic acid.

The cadmium content in SSAs is only one half to one fifth of that of triple phosphate and Hyperphos, respectively due to volatile losses during incineration. All other elements are enriched in the SSA. It becomes clear that it is necessary to limit application rates on the basis of the original sludge. The maximum permitted loads need to be defined, not only for SSAs but also for other secondary raw materials, in addition to the maximum permitted heavy metal content.

Differences in yield were significant between the two soils and also between the different products and application rates (figure 2). On both soils, increasing amounts of P were related to increased grain yield of maize.

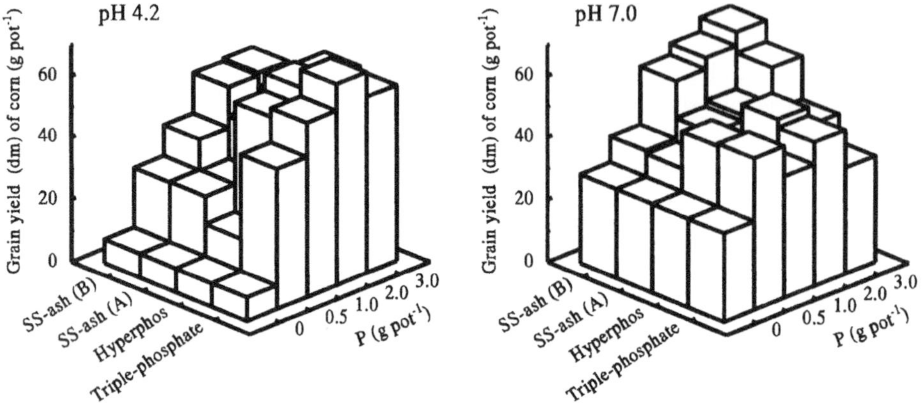

Figure 2. Influence of P-applications on straw yield of corn in dependence on soil pH and P-source.

SSA, type B yielded best results on the soil with pH 7 and on the soil with pH 4.2 it was equivalent to Hyperphos. The results show that SSAs were equivalent to commercial products in obtaining maximum yield, especially when soil pH is in the neutral range. Differences in straw yield were not significant between the two soils. The highest straw yield was achieved by the highest application rates for all products (figure 3). SSA, type A resulted in the same yields as Hyperphos while those of SSA, type B were significantly lower.

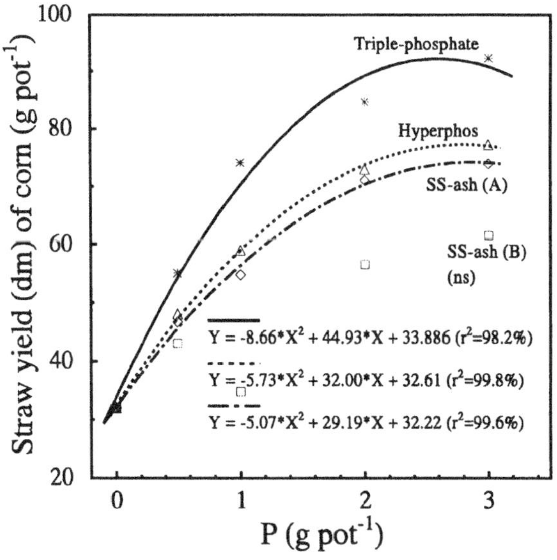

Figure 3. Influence of P-applications on straw yield of maize in dependence the P-source.

5 Applicability of New Fertiliser Technologies for Secondary Raw Materials

Faced with limited resources, as in the case of P, or increased environmental burdens caused by an under- and over-supply with nutrients in the field as in case of N and S [26], it is more than ever important to use new fertiliser technologies so that the variable nutrient supply in the field matches the local demand of the crop [27]. Man has probably been aware of the spatial variability of soil parameters, resulting in different patches of plant growth and yield in the field, ever since cropping began in ~10.000 BC [28] but until the end of this millennium no suitable technique was available to tackle the problems and opportunities that this variability presented. With the development of geostatistics by Krige [29] and Matheron [30], the realization of satellite aided positioning (GPS, Global Positioning System) in the late 70's [31] and the availability of affordable fast computers with ample storage capacities since the late 80's this ancient problem could finally be addressed in land use. Today, the application of GPS, GIS (Geographical Information Systems) and variable rate application technology in agriculture are described under the term 'Precision Agriculture' [32;33]. But the conceptual idea behind all these technical aids is that of managing resources locally in order to match inputs correctly to soil features (Local Resource Management [34; 35; 36; 37]. The concept of LRM opens challenges in the control and spatial variation of agricultural inputs such as fertilisers or secondary raw materials, and provides the best opportunity for sustainable plant production where the economic profitability of agriculture and ecological demands are harmonised.

Spatially variable nutrient management has been successfully implemeted [35, 36, 37, 38] and could be extended to secondary raw materials incorporating aspects relating to both nutrient and contaminants. An example showing variable application rates of P on the basis of spatial variability of plant available P in the soil is given in figure 4. This highlights different zones with different P-demands as determined by soil analysis. A uniform fertilisation of the crop would result in an undesirable under- or over-supply while the spatially variable application shown would mean a sustainable use of the nutrient. More sophisticated strategies relying on local calibration of soil and plant data against yield have also been developed [41].

Waste management techniques could be developed which use, for example, spatial data on the variability of heavy metals at a field scale. In a nested sampling area of 50 meter square Wopereis et al. [42] found that there was no spatial correlation of different total heavy metal contents of a forest soil. In comparison, on a large scale basis, available zinc and copper contents in top soil samples collected in northern Germany showed ranges in their variogrammes between 50 to 335 and 45 to 480 meter, respectively [43].

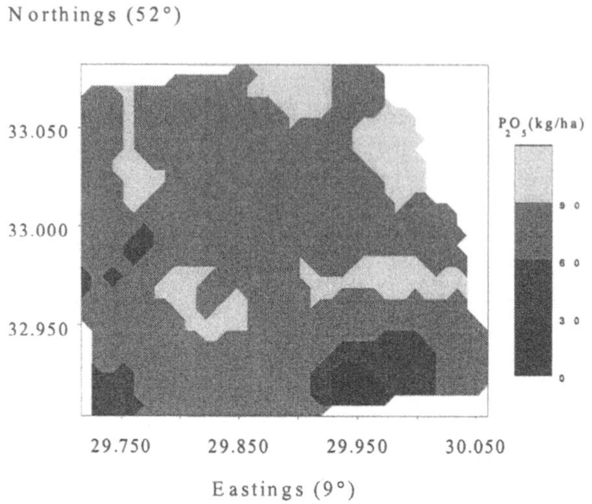

Figure 4. Variable P-rates according to the spatial variability of plant available P-contents in soil.

Knowledge of spatial correlations of total heavy metal contents in soil, followed by transformation of the data into digital agro-resource maps [44], will help to distinguish different levels of contamination and exclude 'hot spots' from application of 'contaminated' products.

As yet no information is available about the influence of anthropogenic factors, such as application of slurry or sewage sludges on the spatial distribution of heavy metals or their different chemical species. The knowledge of such relationships is, however, essential if refined strategies for application are to be developed.

6 Conclusions

Sources of recycled plant nutrients are plentiful and are available on regional or national scale [6]. The acceptance of those secondary raw materials by farmers will depend on the safety of the products and their efficiency and competitiveness compared with commercial fertilisers. Some of that raw materials contain xenobiotics which may be considered as potentially harmful, and as no objective risk assessment for such wastes is possible at present, they should be further processed e.g. by incineration. Incinerated ashed sewage sludge can be a useful phosphorus source, as was proved in a greenhouse experiment.

The application of FGD products, and also P-fertilisers and slurry will lead to an accumulation of environmental elements in soils in the long term and this should be avoided or restricted by reducing the permitted thresholds of such elements in fertilisers and secondary raw materials. Sustainable agricultural production may ultimately require new farm management practices such as 'Precision Farming', so that the luxury use of nutrients/resources can be avoided and inorganic xenobiotics/industrial wastes can be disposed of safely at the farm level. Such developments, although not of immediate concern to CEE farmers at present, will inevitably become more important in the not too distant future.

7 Acknowledgements

The authors wish to express their gratitude towards Dr. Kerr C. Walker (SAC, Aberdeen) for the linguistic revising of this paper.

8 References

1. Anon (1992) Klärschlammverordnung (AbfKlärV), *Bundesgesetzblatt*, Jahrgang 1992, Teil I.
2. Padeken, K. (1998) Schwermetallaufnahme verschiedener Pflanzenarten unter besonderer Berücksichtigung der N- und P-Ernährung. *Landbauforschung Völkenrode,* Sonderheft No 182.
3. Witter, E. (1995/96) Towards zero accumulation of heavy metals in soils: an imperative or a fad? *Proc. Int. Symposium 'Fertilisers and Environment'* Salamanca (Spain), 413-421.
4. Anon (1994) Gesetz zur Vermeidung, Verwertung und Beseitigung von Abfällen, Krw-/AbfG. *Bundesgestzblatt* v. 6.10.1994.
5. Oswald, P. Legal and administrative aspects of recycled plant nutrients. *Proc. 10th Int. Symposium of CIEC, Braunschweig,* Dec. 9-11, 1996, 17-29.
6. Werner, W. (1997) Sources of recycled plant nutrients from industrial processes. *Proc. 10th Int. Symposium of CIEC, Braunschweig,* Dec. 9-11, 1996, 1-16.
7. Schnug, E. and Haneklaus, S. (1994) Sulphur Deficiency in *Brassica napus*- Biochemistry- Symptomatology- Morphogenesis. *Landbauforschung Voelkenrode,* Sonderheft 144.
8. SCHNUG, E. (1991): Sulphur nutritional status of European crops and consequences for agriculture. *Sulphur in Agriculture,* 15, 7-12.
9. Schnug, E. and Haneklaus, S. (1998) Diagnosis of sulphur nutrition. In: Schnug, E. and Beringer, H. (ed.): *Sulphur in Agro-Ecosystems*. Vol. 2 of the series 'Mineral Nutrition in Ecosystems', Kluwer Academic Publ. Dordrecht, 1-38.
10. Schnug, E. 1997. Significance of sulphur for the nutritional and technological quality of domesticated plants. In: Cram, W. J., De Kok, L. J., Stulen, I., Brunold, C. and Rennenberg, H. "*Sulfur Metabolismus in Higher Plants - molecular, ecophysiological and nutritional aspects*". Backhuys Publishers, Leiden, The Netherlands, 109-130.
11. Puch, K.-H., W. vom Berg (1997) Nebenprodukte aus Kohlebefeuerten Kraftwerken. *VGB Kraftwerkstechnik* **11 (7)**, 604-610.
12. Paulsen, H. M. (1998) Produktionstechnische und ökologische Bewertung der landwirtschaftlichen Verwertung von Schwefel aus industriellen Prozessen. Sonderheft Landbauforschung Voelkenrode (in press).

13. Mortensen, J. and Nielsen, J. D. (1995) Use of a sulphite containing desulphurization product as sulphur fertiliser. *Z. Pflanzenernährung & Bodenkunde* **158**, 117-119.
14. Bertelsen, F. and G. Gissel-Nielsen (1987) Toxicity of root applied sulphite in Zea mays. Enironm. *Geochemistry and Health* **9**, 12-16.
15. Ritchey, K. D., T. B. Kinraide, R. R. Wendell, R. B. Clark and V. C. Baligar (1994) Strategies for overcoming temporary phytotoxic effects of calcium sulfite applied to agricultural soils. In: S. H. Chiang (ed) *Proc. 11th Ann. Int. Pittsburgh Coal Conf., Pittsburgh, P. A.*, 457-462.
16. Paulsen, H. M., Haneklaus, S. and Schnug, E. (1998) Suitability of SDA-products for supplying the sulphur demand of agricultural crops. VGB/BVK-Konferenz "Nebenprodukte aus Kohlekraftwerken", 11-12. Februar 1998, *VGB-Tagungsband* **233**.
17. Anon (1993) Oversigt over Landsforsϕgene. Forsog og undersϕgelser i de landϕkonomiske foreniger. *Samlet og udarbejdet af Landsudvalget for Planteavl* (ed. C.A. Pedersen), Arhus.
18. Zadoks, J. C., T. T. Chang and C. F. Konzak, (1974) A decimal code for the growth stages of cereals. *Weed Research* **14**, 415-421.
19. Finck, A. (1979) *Dünger und Düngung*. Verlag Chemie, Weinheim, New York.
20. Haneklaus, S., H. M. Paulsen and E. Schnug (1995) Field studies on the use of desulphurisation products as sulphur fertilizers. *VGB/BVK-Konferenz "Nebenprodukte aus Kohlekraftwerken"*, 1-13 (1995).
21. Boysen, P. (1992) Texte Umweltbundesamt 55/92, research report 10701016/01, UBA-FB 92-104.
22. Andersson (1993): Trace elements in agricultural soils. Fluxes, balances and background values. SNV 4077-DE93-778470.
23. Anon (1993) *Untersuchung der Möglichkeiten zur Reduzierung der Belastung von Klärschlämmen mit organischen Schadstoffen durch Maßnahmen bei privaten Haushaltungen und im gewerblichen Bereich*. Batelle-Institut e.V..
24. Anon (1996) USGS Mineral Commodity Summaries, 1996.
25. Heiland, K. (1986) Spurenelementuntersuchungen von Rohphosphaten verschiedener Lagerstaetten und industriell bearbeiteten Rohphosphaten mittels der instrumentellen Neutronenktivierungsanalyse. Diplomarbeit, Fachbereich Chemie, Hamburg.
26. Schnug, E. and Haneklaus, S. (1994) The ecological importance of sulphur. *Norwegian Journal of Agricultural Sciences* , **15**, 149-156.
27. Haneklaus, S., Schroeder, D. and Schnug, E. (1996) Strategies for fertilizer recommendations based on digital agro resource maps. *Proc. of the 3rd Int. Conf. on Precision Agriculture*, June 23-26, 1996, , Minneapolis, MI, P. C. Robert, R. H. Rust and W. E. Larson (eds.), 361-368.
28. Graichen, G. (1990) *Das Kultplatzbuch*. Hoffmann und Campe, Hamburg.
29. Krige, D. G. (1951) A statistical approach to some basic mine valuation problems on the Witwatersrand. *J. Chem. Metall. Min. Soc. S. Africa* **52**, 119-139.
30. Matheron, G. (1971) *The theory of regionalized variables and its applications*. Les Cahiers du Centre Morphologie Mathèmatique de Fontainebleau No. 5.

31. Bauer, M. (1994) *Vermessung und Ortung mit Satelliten*. Wichmann Verlag Heidelberg.
32. Anon (1997) *Precision Agriculture in the 21st century: Geospatial and Information Technologies in Crop Management.* Board on Agriculture, National Research Council, National Academic Press, Washington, D. C. 1997.
33. Johnson, R. C. (1996) *Target farming - a practical guide to precision agricul*ture. Published by J. Johnson, Rosetown, Saskatoon, Canada.
34. Schnug, E. (1996) Lokales Ressourcen-Management landwirtschaftlicher Böden. *Landbauforschung-Völkenrode* **46**, 111-113.
35. Schnug, E. and Haneklaus, S. (1996) Application of local resource management (LRM) of agricultural soils to fertilisation. Seminar on Site Specific Farming, Arhus, Denmark, 20.-21.03.1995, *SP-Report* **26**, 115-122.
36. Haneklaus. S. and Schnug, E. (1998) Impacts of precision agriculture technologies on fertilization. Proc. 11th Int. Symposium of CIEC. "*Codes of good fertilizer practice and balanced fertilization*", eds E. Schnug and M. Fotyma, 95-107.
37. Schnug, E. and Haneklaus, S. (1999) Self-Surveying and its Utilisation for Variable Rate Nitrogen Fertilisation. *Actes du colloque*, Gestion du phosphore et des herbicides, 8-9 mars 1999 au Saint-Jean-sur-Richelieu, 17-26.
38. Haneklaus. S., Schroeder. D. and Schnug, E. (1998) Decision Making Strategies for Fertilizer Use in Precision Agriculture. Proc. of the 4th Int. Conf. on Precision Agriculture, Minneapolis (in press).
39. Schroeder, D., Haneklaus, S. and Schnug, E. (1997) Information management in Precision Agriculture with LORIS. Proc. 1*st European Conf. on Precision Agriculture*, Warwick, Vol. II, 821 - 826, J. V. Stafford (Ed.), BIOS Scientific Publishers Ltd., UK.
40. Schnug, E., Murphy, D., Evans, E. and Haneklaus, S. (1993) *Local Resource-Management in computer aided farming: A new approach for sustainable agriculture.* In: Fragoso, M. A. C. and Beusichem, M. L. van (eds.), Optimization of Plant Nutrition, 657-663, Kluwer Acad. Publ. Dordrecht.
41. Schroeder, D., Haneklaus, S. and Schnug, E. (1997) Information management in precision agriculture with LORIS. *Proc. 1st European Conf. on Prec. Agric., Warwick*, **2**, 821-826.
42. Schnug, E., Heym, J. and Murphy, D.P.L. (1995) Boundary line determination technique (BOLIDES). Site-Specific Management for Agricultural Systems, ASA-CSSA-SSA, 899-908.
43. Wopereis, M. C., Gascuel-Odoux, C., Bourrie, G. and Soignet, G. (1988) Spatial variability of heavy metals in soil on a one-hectare scale. *Soil Sci.* **146 (2)**, 113-118.
44. Schroeder, D. (1998) Investigations on including the spatial variability of soil fertility parameters into nutrient management strategies for Precision Agriculture. PhD thesis, TU-Braunschweig (in preparation).

LONG-TERM FIELD EXPERIMENTS AND IMPLICATIONS FOR SOIL QUALITY ASSESSMENT WITH RESPECT TO ORGANIC CONTAMINANTS

The significance of sewage sludge applied to agricultural land.

R. E. ALCOCK and K.C. JONES
Institute of Environmental and Natural Science
Lancaster University, Lancaster, LA1 4YQ, UK

1. Introduction

Over 1 million tonnes of sewage sludge is produced in the UK annually of which about 50% is applied to agricultural land (both arable and grassland). However, only around 1% of all UK agricultural land receives sludge in any given year. Its use as a soil conditioner and fertiliser is attractive economically, while also recycling nutrients and generally improving soil structure. Substantial quantities of sewage and sludges currently disposed of to sea in Europe will require an alternative method of disposal by the end of 1998, due to changes being implemented under the Urban Waste Water Treatment Directive. Also, as specified in the Directive COM 911271 on wastewater treatment, the installation of treatment systems for all populations exceeding 2000 inhabitants must be installed before the end of the year 2005. As a result, UK sludge production is predicted to rise by 30-50% over the next decade, thus doubling the demand for land-based disposal options. An increase of ~185% in sludge production (about 1 million ton/year) has been estimated recently for the year 2000 in Spain. [1]

Sewage sludges are contaminated with a wide range of trace organic contaminants from diverse sources, both natural and anthropogenic [2]. Many organic components measured in sludge are thought to affect the sustainable use of soil as a result of their persistent nature. Current management objectives in Europe for the land disposal of sludge are based on providing for the nutrient requirement of a crop without causing pollution of water courses and without exceeding the soil loading limits for a number of heavy metals. The focus on organic contaminants in sewage sludge creates a difficult management problem as there are no generally accepted soil concentrations limits for organic compounds in sludges. To date, only Germany has established limit concentrations for dioxins and furans and PCBs in sludge used for agricultural purposes, and the European Union is currently studying the issue of limit concentrations for organic compounds in sewage sludge.

This paper will provide an overview of soil quality issues in terms of sewage sludge application to agricultural land and its impact on organic contaminant concentrations in soil over time. The value of controlled long-term experiments to our understanding of the fate and potential accumulation of a single group of trace organics, polychlorinated biphenyls (PCBs) is evaluated. There is a paucity of reliable field data concerning the long-term degradation of organic chemicals in soil, their ultimate fate in agricultural systems, food-chain biomagnification and possible effects on biota and humans.

M.J. Wilson and B. Maliszewska-Kordybach (eds.),
Soil Quality, Sustainable Agriculture and Environmental Security in Central and Eastern Europe, 115–124.

PCB fate and behaviour has been investigated using two long-term controlled sludge experiments, namely, the Woburn Market Garden Experiment and the Luddington and Lee Valley Experimental sites. Long-term soil collections are able to place sewage sludge inputs to agricultural soils in context with 'background' PCB concentrations in surface soils by analysis of a set of archived soil samples from long term agricultural plots. These samples have been the subject of several studies on long term trends of heavy metals and other trace organic compounds in agroecosystems and their integrity has been well characterised [3;4].

Analysis agricultural soils.of archived soil and sludge samples from these experiments, along with unsludged control plot soils, has provided a unique field - based insight into the long-term behaviour of individual PCB congeners in both sludge-amended and typical background UK agricultural soils.

2. Polychlorinated biphenyls

PCBs form a class of 209 individual compounds, in which between 1 and 10 chlorine atoms are substituted on a biphenyl nucleus. Atmospheric deposition is the major source of PCBs to the terrestrial environment and has resulted in trace levels of PCBs being present in all soils. Their commercial production started in the USA in 1929, with world production reaching a maximum, estimated to be in the range of 2 x 10 9 kg [5] in the early 1960s. Because of elevated levels found in a range of environmental samples, their industrial use in much of Europe and America has been restricted since the mid 1970s. Despite these legislative controls, contemporary PCB concentrations in marine predators are still high enough to be linked to adverse ecotoxicological effects [51. For these reasons, PCBs remain compounds of significant ecological concern.

Transport and fate of PCBs and their partitioning in different compartments of the environment depend to a large degree on sorption reactions. PCBs have very low vapour pressures which like their solubility in water, decreases with increased degree of chlorination. Generally, sorption increases with increasing chlorine content of the biphenyl ring and with the surface area and organic carbon content of the sorbent. They are ubiquitous, hydrophobic compounds that concentrate onto organic-rich particles during wastewater treatment. Concentrations in sewage sludge are generally in the range of 1 - 10 mg kg-1 [2]. Contemporary concentrations in Western Europe are thought to be generally less than 1 mg kg-1 (see Table 1). No published data are available for PCB concentrations in Central European sewage sludges or soils. The wide range of concentrations observed in Table 1 arises from a number of factors including variability in the nature of the catchment, point sources in the waste water collection system, sampling differences and various analytical protocols used. These factors are superimposed on temporal variations in background PCB levels in the environment. It is apparent from Table 1 that contemporary values of total PCB measured in the early 1990s throughout Europe are significantly lower than those reported during the 1980s.

PCBs are present in all soils as a result of atmospheric deposition. However, application of sludge to agricultural land will result in an increase in the PCB content of a given soil. Recently, concern has focused on the pathways and transfers of these compounds in soil and has highlighted the potential of PCBs to transfer to livestock grazing sludge-treated pasture and ultimately the human food chain.

3. The significance of long-term experiments in the UK

Because of the uncertainties surrounding the fate and persistence of PCBs and similar compounds (e.g. PCDD/Fs), it is important to investigate the long-term field (rather than laboratory) behaviour of these compounds within the soil environment. Reliable quantitative time trend data are important in order to establish how environmental levels have changed over time, the significance of various anthropogenic emissions as a source and to evaluate the progress of measures designed to limit the release of chemicals into the environment. Good quality temporal trend data_are largely unavailable for this group of compounds. It is important to appreciate that little reliable environmental analytical data existed for PCBs prior to the 1970s. It is only relatively recently that sufficiently sensitive and reliable procedures have become available to determine them routinely. As a result, the only satisfactory way of obtaining information on these issues is through the retrospective analysis of archived soils from known locations, and comparison with contemporary material.

4. The Woburn Market Garden Experiment, Rothamsted

Between 1843 and 1856 Lawes and Gilbert began a series of field scale experiments at Rothamsted Experimental Station, to the north of London. Eight of these have continued more or less as originally planned to the present day: these are the Rothamsted Classical Experiments'. Their main objective was to measure the effects on crop yields of inorganic fertilizer additions and various organic manures. The Woburn Market Garden Experiment was initiated in 1942 by Rothamsted, originally to test the manurial value of different bulky organic manures. Full details of the experiment, the soils and crops have been reported previously [8]. In summary, the soils at Woburn are free draining sandy loams with a 9% clay content and a pH maintained at 6.5-7.0.

A plot was selected for the purposes of this study that had received 24 additions of sludge between 1942 and 1961 at a maximum rate of ca 75t (fresh weight) ha-1 year-1. In 1942, sludge was applied at a rate of 20 t ha-1 whereas from 1943 to 1961 the application rate was 75 t ha-1 yr -1. In total 1520 t ha-1 (fresh wt., equivalent to 834 t ha-1 dry wt.) was applied between these years. All sewage sludges were produced at the west Middlesex Drainage Works at Isleworth, West London. Following each sludge addition the soil was ploughed to a depth of 23 cm. Since 1961, the plots have received only inorganic fertilisers. A control plot was also chosen that had received only inorganic fertilisers throughout the whole of the experimental period. This was to enable a comparison to be made between the sludge-amended plot and a plot which had only received PCB inputs via atmospheric deposition. Soils from both these plots (0-23cm depth) were collected using a semi-cylinder auger in 1942, 1951, 1960, 1967, 1972, 1980, 1984 and 1992. Following collection they were air dried, ground, bulked and sieved prior to storage in sealed glass containers. Sample extraction and analysis have been detailed previously [9].

TABLE 1. PCB concentrations in sewage sludge reported in the literature (mg/kg). Table adapted from Alcock and Jones [6]

Country	**Range**	**Mean**	**Congener**					
	PCB	**PCB**	**28**	**52**	**101**	**138**	**153**	**180**
Mumma et al, 1984 *USA*	0.15 - 3.6	1.19	-	-	-	-	-	-
McIntyre & Lester 1982 *UK*	0.02 - 0.46	0.16	-	-	-	-	-	-
Tarradellas et al. 1985 *Switzerland*	0.5 - 8.0	-	-	-	-	-	-	-
Webber et al. 1983 *Canada*	0.13 – 1.63	0.55	-	-	-	-	-	-
Fieggen 1985 *Holland*	0.6 – 6.6	0.95	-	-	-	-	-	-
Wierner 1987 *Germany*	0.52 – 15.2	4.1	-	-	-	-	-	-
Burgermeister et al. 1981 *Switzerland*	0.36 – 7.6	2.13	-	-	-	-	-	-
Witte et al. 1988 *Germany*		-	0.054	0.022	0.061	0.093	0.1	0.064
Kedding et al. 1989 *Germany*	1.38 – 6.65	1.69	0.041	0.028	0.052	0.082	0.084	0.053
Alcock & Jones 1993 *UK*	0.12 – 0.71	0.20	0.009	0.009	0.011	0.012	0.011	0.013
Berset & Holzer 1995 *Switzerland*	0.29 – 0.54	0.41	0.021	0.039	0.077	0.089	0.085	0.042
Eljattat et al. 1997 *Spain*	0.19 – 0.53	0.34	0.006	0.010	0.009	0.015	0.014	0.010

PCB concentrations in the applied sludges varied by more than an order of magnitude over the 20 years (1942 – 1962). From 1942 to 1952, concentrations varied between 0.4 and 0.8mg/kg, rising to over an order of magnitude to reach a maximum value of 4.3mg/kg in the final sludge applied in 1961. From 1942 to 1961, about 1 kg of PCB would have been applied per hectare in sludge.

Total PCB concentrations measured in the control and sludged plots at Woburn from 1942 to 1992 are shown in Figure 1. PCB kg-1. In 1942 prior to sludge addition, the soil contained ~63 g PCB kg-1. The PCB burden of the control plot increased steadily from that time, as a result of cumulative atmospheric inputs, to reach a maximum in the 1972 sample of 600 g kg-1 PCB. This increase coincides with trends in the industrial manufacture of these chemicals both in the UK and on a global scale. PCB levels have declined substantially since 1972, to such an extent that the contemporary burden is lower than that found in the soil from 1942. This trend mirrors those observed in archived soil samples from other long-term UK agricultural experiments analysed recently [10] and time trends observed throughout Europe through the analysis of lake sediment cores [11]. It reflects a change in the balance between atmospheric deposition inputs and losses from soil via volatilisation, biodegradation etc.

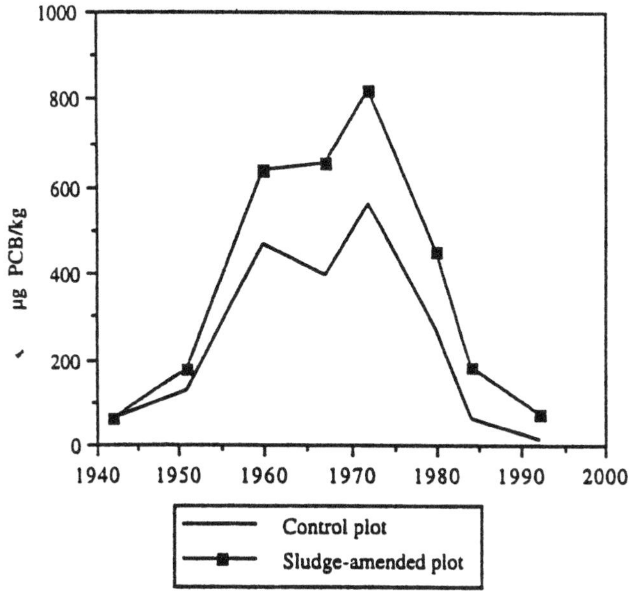

FIGURE 1. Total PCB concentrations on the control plot and the sludge-amended plot at Woburn.

Total PCB trends in the sludge-amended plot closely follow that of the control plot, with concentrations reaching a maximum in 1972, even though sludge was applied only between 1942 and 1961. Cessation of sludge inputs did not result in an immediate gradual decline in PCB concentrations. Concentrations remained relatively constant between 1960 and 1980, then fell sharply to present day levels (70 g PCB kg-1).

By 1992, 31 years after the last sludge was applied, the sludge-amended plot contained over 5 times more PCB than the control plot. In 1984, the control plot-contained 0.96% organic carbon, whilst the sludge-amended plot contained 1.99%. It is assumed that PCB concentrations became elevated for a time following each successive application of sewage sludge (between 1942-1961), then declined. Such a pattern would have been repeated throughout the experimental period after each sludge addition, so that by 1960 an accumulated fraction had built-up, with an increased proportion of the more recalcitrant PCB congeners.

5. The long-term experimental sites at Luddington and Lee Valley

In 1968, the UK Agricultural Development Advisory Service (ADAS) and the Macaulay Institute for Soil Research (now the Macaulay Land Use Research Institute) set up two field experiments to investigate the uptake of selected metals by crops on sewage sludge amended soils. The two field stations are located at Lee Valley, in Cambridgeshire, and Luddington, near Stratford-upon-Avon in Warwickshire. The soil at Luddington Experimental Station, is described by Berrow and Burridge [12] as a sandy loam, containing 79% sand, 6% silt and 15% clay with slightly imperfect drainage. The soil cation exchange capacity is 11.5 meq/100g with a base saturation of 74%, an organic matter content of ~2% and a pH in water of 5.8. At Lee Valley Experimental Station the soil is a stony silty loam of higher pH (6.5), overlying a silty/clay loam subsoil. It has an organic matter content of ~5% and a carbon content of ~3. 1 %. The cation exchange capacity at Lee Valley is higher than at Luddington, while the bulk density is lower.

The experiments at both sites were identical and involved the application of different metal contaminated sewage sludges to soil plots, indigenously contaminated with either Cr, Cu, Ni or Zn and applied at a rate of 125 tonnes (dry weight) per hectare. Details of these experiments can be found in Berrow and Burridge [13]. Importantly, corresponding untreated control plots were also available for analysis at each site. Following application, the soil was dug or rotivated to a depth of 15 cm. Soil samples were collected before any sludge applications in 1968, and periodically up to the late 1980s. Samples were collected from the Lee Valley site in 1968, 1972, 1977, 1981 and 1988, while at Luddington samples were obtained in 1968, 1972, 1976, 1981, 1985, 1988 and 1990. Following collection they were air dried, ground and sieved prior to storage in sealed glass containers. Samples of the original sewage sludge were also archived.

For the purpose of this paper, data from the Zn-rich sludge plot of both Luddington and Lee Valley are reported. [14]. The Zn-rich sludge contained about 1.0 mg PCB kg -1. Using both sludge PCB concentrations and the known dry weight of sludge applied, the total amount of PCBs originally added to the soil at each site can be calculated. About 0. 1 kg PCB/ ha was added to the Zn-rich plot.

Figures 2 and 3 illustrate trends in PCB concentrations on both the control and Zn-rich sludge-treated plots from 1968 to the present day at Luddington and Lee Valley.

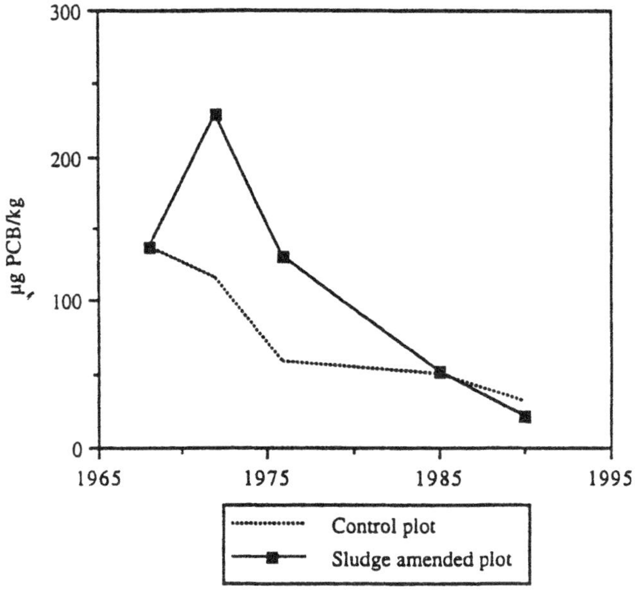

FIGURE 2. Total PCB concentrations on the control plot and the sludge amended plot at Luddington

Luddington Prior to sludge additions in 1968, the soil at Luddington contained ~137 g PCB kg -1, this being dominated by the. low molecular weight congeners (tri- and tetra-chlorinated biphenyls, particularly 28 and 52). PCB levels show a steady decline with time from ~137 to ~32 g PCB kg-1 (see Figure 2).

Lee Valley At Lee Valley, control plot PCB levels show a marked decline with time from ~299 g PCB kg-1 to ~57 g PCB kg-1 (see Figure 3). Initial PCB concentrations are approximately double those found at Luddington, probably because soil characteristics at Lee Valley favour the retention of organic contaminants (notably they contain higher amounts of clay and organic matter than at Luddington) and/or greater atmospheric deposition of PCBs has occurred at the Lee Valley site.

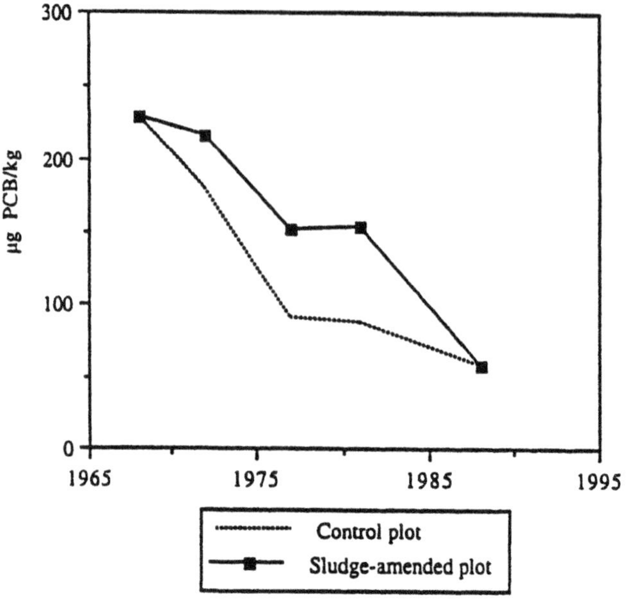

FIGURE 3. Total PCB concentrations on the control plot and the sludge amended plot at LeeValley.

These control plot trends correspond well with results from Woburn described earlier and match the production, use likely atmospheric releases of PCBs. The substantial decline of PCB concentrations at both sites suggests that there has been a substantial reduction in the potential sources of PC13s to the environment over the last 20 years. In summary, the trends in control plot soil concentrations reflect a decline in composition since peak atmospheric deposition inputs to the sites in the 1960s [101. Clearly, trends in the sludge-amended plots could not be reliably interpreted without this information.

Through timc thcrc is a clear substantial decrease in the PCB burden of each sludge-amended plot. Initially, application of sewage sludge caused significant increases in the soil PCB in each plot at Luddington and Lee Valley. A similar temporal trend is observed in each of these plots, notably an initial high level following sludge application followed by a steady decline to low contemporary concentrations. As in the control plots, temporal changes in the relative proportions of individual congeners generally occurred in the sludge-amended plots. With time, the proportion of tri- and tetra-chlorinated congeners generally declined, while higher molecular weight compounds i.e. hexa-and hepta- biphenyls, generally increased.

It is hypothesised that the effect of a single, large sewage sludge application has effectively raised the retention capacity of the soil for both sludge-derived and aerially-derived PCBs (compared to that of the control soil). This will presumably affect both the efficiency of retention of aerially-derived PCBs and their availablity for volatilisation and biodegradation.

Furthermore, the loss curves generated seem likely to be a reflection of a long-term equilibrium process between the soil and the air (similar to that on the control plot), governed by kinetically constrained air-soil partitioning, due to a declining input of PC13s to the environment over the last 20 years (Alcock et al., 1993).

6. Discussion

One fascinating feature of the time trend data is that PCB concentrations at these sites have now declined to levels close to those prior to sludge inputs, albeit with a somewhat different congener composition, with the heavier homologues constituting a greater proportion of the PCB content. Clearly, manufactured 'Aroclor' inputs now exert a far less significant influence on PCB levels at these sites than during the years of peak use. Atmospheric inputs appear to dominate the PCB content of soils at Woburn, Luddington and Lee Valley, even though large applications of sludge were applied. The addition of sewage sludge simply added to this atmospheric burden. Contemporary PCB concentrations in the sludge amended plots are between 2 to 5 times higher than the respective control, up to 30 years after sludge addition.

Control plot soils provide a record of long-term temporal trends in background soil quality, clearly showing the timing and magnitude of the peak PCB concentrations and subsequent decline for sites in rural southern England. Importantly, they allow contemporary levels to be set in context and allow the impact of sludge inputs to be evaluated. These results suggest that the contribution of PCB inputs to the current UK environmental burden has been substantially reduced in the last 20 years. Clearly, given the dramatic changes in background soil PCB concentrations at each site, it would be impossible to meaningfully interpret the changes in the PCB concentrations of the sludge-amended soils without data for the archived control soil samples These studies provide valuable field-based evidence on how PCBs behave in soil and an insight into how concentrations have changed over time. This type of approach also indicates that there may be discrepancies between PCB residence times or half-lives estimated for soils under artificial laboratory conditions and those actually observed in the field.

Using archived samples from long-term experiments has provided data to help assess the implications of organic compounds applied in sewage sludge to agricultural land. Anthropogenic influences inputs can be placed in context with processes occurring in the agricultural system as a whole. Applying large amounts of relatively corntaminated sludge (1 – 5 mgkg-1) at Woburn, Luddington and Lee Valley has simply raised PCB concentrations from that of typical rural soils, to that of urban levels. It should be notedthat the sludge applied to the experiments contained far higher PCB concentrations than contemporary concentrations in European sewage sludge (see Table 1) and the application rates were far higher than those used for current operational practice on the majority of agricultural land.

Current environmental concerns over the presence of trace organic contaminants applied to agricultural soils in sewage sludge extend to the potential for them to transfer to humans via ground waters (used for drinking water), crop plant uptake, and livestock ingestion. Given the very low aqueous solubilities of PCBs, the first pathway is unlikely to be a limiting factor controlling the rates at which sludge is applied to land. Plant uptake of soil-borne PCBs is generally very inefficient, although enhanced levels in some root crops may be expected 15]. In some countries, sludge is not applied to grazing land. This practice is continued in the UK, with a compulsory 3 week "no grazing" ban after sludge spraying. Direct ingestion of contaminated surface soils may be important under some circumstances [16, 17] Future studies need to consider whether PCBs present in agricultural soil following sludge addition are bioavailable and whether they transfer into animal tissue efficiently when ingested. Again, archive samples from UK long-term experiments represent an invaluable long-term resource currently being used to explore these issues.

7. Acknowledgement

We are grateful to the Water Research Centre, the NERC, the AFRC and MAFF for financial support for our research on PC13s in the environment.

8. References

1. Eljarrat, E., Caixach, J and Rivera, J. (1997) Effects of sewage sludges contaminated with PCDD/Fs and PCBs on agricultural soils, *Environ. Sci. Technol 31,* 2765-2771.

2. Jacobs, L.W, O'Connor, G.A., Overcash, M.A., Zabik, MJ. and Rygiewicz, P. (1987) Effects of trace organics in sewage sludge on soil-plant systems and assessing their risk to humans, In *Land Application of Sludge*, ed. A.LPage, T.G. Logan and LA. Ryan. Lewis Pub Co., Michigan, USA, pp. 101-143.

3. Jones, K.C., Sanders, G, Wild, S.R., Burnett, V. and Johnston, A.E. (1992) Evidence for a decline of PCBs and PAHs in rural vegetation and air in the United Kingdom, *Nature 356,* 134-137.

4. Kjeller, L-O., Jones, K.C., Johnston, A.E. and Rappe, C. (1991) Increases in the PCDD/F content of soils and vegetation since the 1840's. *Environ. Sci. Technol 25,* 1619-1627.

5. Tanabe, S. (1988) PCB problems in the future: foresight from current knowledge, *Environ. Pollut. 50,* 5-28.

6. Alcock, R.A. and Jones, K.C. (1993) Polychlorinated biphenyls in digested UK sewage sludge,*Chemosphere 26,* 2199-2207.

7.Berset, J.1. and Holzer, R. (1995) Organic micropollutants in Swiss agriculture: distribution of PAHs and PCBs in soil, liquid manure, sewage sludge and compost samples, *Intern. J. Environ. Anal, Chem. 59,* 124165

8. McGrath, S. (1987) Long-term studies of metal transfers following application of sewage sludge. In *Pollutant Transport and Fate in Ecosystems* ed. P.L Coughtrey, M.H. Martin and M.H. Unsworth. British Ecological Societ Special Publication No 6. Blackwell, Oxford, UK, pp. 301-317.

9.Alcock, R.A., S.P. McGrath and Jones, K.C. (1995) The influence of multiple sewage sludge amendments on the PCB content of an agricultural soil over time, *Environ. Toxicol and Chem. 14,* 553-560

10. Alcock, R. A., Johnston, A.E., McGrath, S.P., Berrow, M.L. and Jones, K.C. (1993) Long-term changes in the PCB content of United Kingdom soils. *Environ. Sci- Technol 27,* 1918-1923.

11. Beurskens, J.E.M., Mol, G.A.L, Barreveld, H., Van Munster, B and Winkels, H. (1995) Geochronology of priority pollutants in a sedimentation area of the River Rhine, *Environ. Toxicol. Chem, 12,* 1549-1566.

12. Berrow, M.L. and Burridge, LC. (1977) *Inorganic Pollution and Agriculture.* MAFF Reference Book 326. HMSO, London, UK, pp. 159-183

13. Berrow, M.L. and Burridge, LC. (1990) Persistence of metal residues in sewage treated soils after seventeen years, *Int. J. Environ. Anal. Chem. 39,* 173-177.

14. Alcock, R.A., Bacon, L, Bardget, R.D., Beck, XL, Haygarth, P.M., Lee, R.G.M., Parker, C.A. and Jones, K.C. (1996) Persistence and fate of PCBs in sewage sludge-amended agricultural *soils, Environ. Poll. 93,* 83

15. Wild, S.R. and Jones, K.C. (1992) Organic chemicals entering agricultural soils in sewage sludge: screening for their potential to transfer to crop plants and livestock, *Sci. Total Environ. 119,* 85-119.

16. Fries, G.F. (1982) Potential PCB residues in animal products from application of contaminated sewage sludge to land, J. *Environ. Qual. 11,* 14-20.

17. Fries, G.F. (1996) Ingestion of sludge applied organic chemicals by animals, *Sci. Total Environ. 119,* 85-119.

ORGANIC FARMING ON LARGE FARMS WITH SPECIAL REFERENCE TO EASTERN GERMANY

Management strategies, environmental effects and economic aspects

BACHINGER, J. und K. STEIN-BACHINGER
Center for Agricultural Landscape and Land Use Research (ZALF); Institute for Land Use Systems and Landscape Ecology, Eberswalder Str. 84, D-15374 Müncheberg, Germany
e-mail: jbachinger@zalf.de and kstein@zalf.de

1. Introduction

Organic agriculture aims at the organisation of a complete farm as an organism, orientated on the rules of natural ecosystems. Therefore a strong renunciation of specific external control mechanisms and a low consumption of non-renewable natural resources is required. According to these aims there are three main rules [2, 15] (Table 1):

Firstly, no mineral nitrogen fertiliser is allowed. Hence, the N-demand of crop production relies principally upon the nitrogen accumulated during the ley period, meaning that fodder and grain legumes are the main sources of N-input.

Secondly, there is a limitation of additional brought-in feed which means that livestock units per hectare have to be based on the fodder sources of the farm. Hence, the amount of farmyard manure is limited and its application must be calculated over the whole crop rotation cycle in order to avoid nutrient losses and to improve yield and quality of the crops [30]. According to the IFOAM-standards [15] a maximum of 2 livestock units per ha can be kept. In this context improving animal welfare is also an aspiration.

Thirdly, no synthetic pesticides are allowed which means that the plant protection mainly has to be prophylactic, e.g. by selecting site-adapted varieties, crop rotations and cropping systems based on biological methods.

The standards of organic agriculture and food processing [2] allow a distinction with other agricultural systems in juridical terms. Within the framework of the EC-regulation 2092/91 [10] the farms must be controlled each year by a certification organisation. Moreover, every farm in Germany that belongs to one of the associations which are members of the AGÖL [2] has to be controlled in addition.

M.J. Wilson and B. Maliszewska-Kordybach (eds.),
Soil Quality, Sustainable Agriculture and Environmental Security in Central and Eastern Europe, 125–138.

TABLE 1. Characteristics of organic agriculture

renunciation of external control mechanisms (non inherent to the biological system)		
no mineral N-fertiliser	limitation of brought-in feed	no synthetic pesticides
⇓	⇓	⇓
N-autarky of the farming system	site-adapted livestock units per ha (based on the fodder sources of the farm)	prophylactic and biological plant protection
∇	∇	∇
crucial: ratio of legumes in the crop rotation and their nitrogen-fixation capacity	crucial: productivity (quantity and quality) of forage production	crucial: enhancement of natural pest control mechanisms, site-adapted crop diversity and crop rotations

2. Situation in East Germany

Since the reunion of West and East Germany there has been a far-reaching restructuring of the agricultural sector in the eastern part of the country. According to analyses of Stolze [32, 33] the farms in East Germany were confronted with special problems, such as the distribution of the assets of the collective farms to their former members, residual redemptions, unsettled property conflicts and capital procurement. Beside a strong decrease in total farm sizes and animal livestock (down to 0.5 livestock units per ha), nearly 3 % of the total arable land of the five New States has been converted to organic farming according to the AGÖL standards [2] at the present time (Table 2). The average farm size in East Germany is more than six times that in West Germany, mainly for historical reasons going back to the Middle Ages. Some of the large organic farms cultivate up to 2000 ha.

Most of the organic farms in West Germany were converted from mainly traditional family dairy farms with a low degree of specialisation. Therefore, the changes within the farm management were not very far reaching. Accordingly, there was only a little knowledge and experience on how to convert and restructure the large former state farms in East Germany. In contrast to most of the farms in West Germany the structure of the large farms in East Germany is characterised by a higher complexity, mainly blue collar workers, low livestock units per hectare and large, often heterogenous fields with mainly sandy soils. After the reunion there were doubts about a successful conversion to organic farming under these conditions.

TABLE 2. Numbers and sizes of organic farms (AGÖL) in Germany (as at January 1st, 1997) [14]

	number of farms	agricultural land use area (ha)	% of the total agricultural land use area	average farm size (ha)
Germany (total)	6 465	327 329	190	51
West Germany	5 616	168 868	145	30
East Germany	849	158 461	285	187

3. Management strategies concerning nutrient fluxes in organic farming systems

Especially, the conversion of large farms to organic agriculture needs a professional management of the system because of the special structural conditions as mentioned above. There was a strong demand for developing and introducing planning strategies and instruments in order to optimise the management of production factors, e.g. nutrients.

In contrast to conventional farming, organic farming systems are mostly nitrogen-limited with two different consequences. On the one hand, there is a lower potential for nutrient leaching (see chapter 4), on the other (especially in the case of non system- and site-adapted nutrient management) there can be a risk of nitrate leaching in connection with the cultivation of legumes or fallow crops as well as of strong yield losses.

One of the main goals of management strategies concerning nutrient fluxes in organic agriculture is the prediction and optimisation of potential nutrient input and the prevention of deficiencies and losses. With regard to this fact it is necessary to take long-term and short-term planning periods and management strategies into account simultaneously.

3.1. LONG-TERM MANAGEMENT STRATEGIES

3.1.1. *Humus Balance*

A well-balanced humus budget supports soil organisms and soil structure and provides a continuous nutrient supply, especially nitrogen, for the plants. Therefore, it is necessary to pay attention to the reproduction of humus, especially in organic farming due to its aims and regulations (Table 1). Even profound changes in the humus content of the soil within a few years cannot be identified by analysing the content of organic carbon because of seasonal and rotational fluctuations [18]. Therefore, indirect methods such as humus balancing can be used for characterising potential humus surpluses or deficits [4, 9, 20, 22]. The evaluation of long-term experiments with regard to the demand of organic matter by cultivation of e.g. fallow crops as well as the amount of humus reproduction of e.g. legumes and organic fertiliser led to coefficients of crops corresponding to their humus-demand and -formation. In order to take the special situation in organic farms into account modifications of these coefficients are suggested [21, 23]. With these methods only long-term trends due to the humus supply of the arable fields of the farms can be calculated. Moreover, it is important to note that use of the coefficients does not take into account yields or special soil cultivation methods, which are the main sources of uncertainty in interpreting the results.

Our investigations of large organic farms (sizes up to 1200 ha, 0.3 - 0.6 LU·ha^{-1}) in the state of Brandenburg show equilibrium or positive humus balances, mainly caused by the portion of legume-grass used as perennial fodder crops and rotational fallow (30-40 % in the crop rotation). The calculations were carried out with different methods according to the above mentioned authors [4, 9, 20, 22, 23]. All calculated results indicate that under the present crop ratio it can be predicted that there is a sustainable development concerning the humus content of the arable land of these farms.

3.1.2. *Nitrogen Balance*

Nutrient balances are used increasingly for the analysis and assessment of parts as well as of total farms, in order to identify weak points within the farming system and to estimate potential environmental pollution. In comparison with the humus balances, N-balances can show more detailed aspects of the internal nitrogen fluxes but the calculations are more difficult and time-consuming. When balancing the nitrogen fluxes of the whole farm estimates are necessary in many cases, because direct measurements are impossible. Especially the estimation of the N_2-fixation is necessary in organic farming systems because it is the most important nitrogen source of the whole system and the portion of legumes can range between 20-50 % in different crop rotations.

Table 3 shows calculations of nitrogen balances of two large organic farms in North-East Germany. Minimum and maximum values result from estimation with different assumptions concerning the N_2-fixation capacities or N-losses within the farm [31]. When maximum output was subtracted from minimum input and vice versa the nitrogen balance, described as change in soil nitrogen pool [26], as shown in Table 3, resulted. On the average it can be concluded that the nitrogen balances of these farms under the current yield levels were nearly equalised, but low N-input from legumes and high losses within the farm due to non site- and situation-adapted cultural practices can cause negative N-budgets.

Therefore, it is of great importance to improve the N-input from legumes as well as estimation methods for nitrogen fixation, and to reduce losses in stables and fields. The improvement of special weak points concerning cultivation measures have to be done with short-term management strategies.

TABLE 3. Estimated overall nitrogen balances [kg N·ha^{-1}·a^{-1}] of large organic farms in North-East Brandenburg (farm 1: 1200 ha, 0.6 LU·ha^{-1}, farm 2: 800 ha, 0.5 LU·ha^{-1}, means of 3 years)

			farm 1	farm 2
	Input			
purchase with fodder, seeds			1	0.2
rainfall			10	10
N_2-fixation	symb.	min.	30	20
		max.	68	45
	asymb.		5	5
total		min.	46	35.2
		max.	84	60.2
	Output			
sale with products			10	16.2
balance within the farm				
stable		min.	23	12
		max.	31	16
fields			22	17
		min.	55	45.2
total		max.	63	49.2
Balance = change in soil nitrogen pool			- 17 to + 29	- 14 to + 15

3.2. SHORT-TERM MANAGEMENT STRATEGIES

Short-term management strategies aid site- and situation-adapted decision making in crop cultivation systems to optimise the nitrogen uptake and to minimise nitrate leaching.

Under the conditions of North-East Germany (mean precipitation ~ 550 mm, cold winter periods, sandy soils, similar to many regions in Poland and the Baltic States) pulses are a substantial source for nitrogen supply of the non-legume plants within crop rotations of organic farming systems. Because of the fast mineralisation of their organic nitrogen residues after harvesting and the low field capacity of the sandy soils, pulses can be one of the main reasons for nitrate leaching within the whole rotation. This is a problem especially when pulses are used for green manure. In our investigations high nitrate contents (up to 100 kg $N \cdot ha^{-1} \cdot a^{-1}$) could be found in the soil layer 0-90 cm after the end of the growing season (Table 4).

TABLE 4. Effects of undersown grass on the nitrate content in the soil layer 0-90 cm at the end of the growing season (mid November 1995) under different species of pulses and management methods (kg $N \cdot ha^{-1} \cdot a^{-1}$) [5]

		without	with
		undersown grass	
combine harvest	yellow lupin *(Lupinus luteus)*	55.9 a*)	17.5 b
	blue lupin *(Lupinus angustif.)*	21.3 a	17.5 a
	pea *(Pisum sativum)*	551	**)
rotational fallow (mulching)	yellow lupin *(Lupinus luteus)*	55.1 a	15.3 b
	blue lupin *(Lupinus angustif.)*	89.5 a	14.0 b
	pea *(Pisum sativum)*	101.0	**)

*) values followed by the same letter in the same line do not differ significantly at the 5 % LSD-level (t-test), **) undersown grass was totally suppressed

Two different strategies to reduce the risk of nitrate leaching can be considered into addition to intercrops such as mustard (*Brassica nigra* L.) or radish (R*aphanus sativus* L. var. *oleiformis* PERS.).

The first one is the undersowing of grass as a catch crop, under the specific site conditions mentioned before. In most years grass can be successfully undersown in pulses [27] and the nitrate content in the soil can be strongly reduced by this measure (Table 4). Because of the low precipitation in August and September, as well as sandy soils, intercrops can be cultivated less successfully after harvest of the main crop (Table 6). Non winter hardy intercrops are least suitable to preserve the received nitrogen until the spring under these climatic conditions (Table 6), because high quantities of nitrogen (up to 50 kg $N \cdot ha^{-1}$) in the frozen dead organic matter can be mineralised and leached in mild winter periods [7].

The second successful strategy can be the sowing of winter rye about 3-5 weeks earlier than usual (at the end of September until beginning of October) (Table 6). Through plant uptake the nitrate amount in the soil which could be lost through leaching can be reduced significantly (Table 5). The synergetic effects of increasing yields of winter rye and reducing nitrate leaching as shown in Table 5 were also found in other field trials. Early sown winter rye did not show an increase in plant diseases nor a reduction of winter hardiness due to the climatic situation in North-East Germany. This practice leads to the second best evaluation result (Table 6). On large farms the period for the seeding of winter cereals lasts for more than three weeks, so it is recommended to begin with the fields with the highest potential for nitrate leaching which mainly depends on the kind of fore-crop and soil texture.

TABLE 5. Effects of sowing date of winter rye on the nitrate content in the soil of winter rye crop, the nitrogen content in the shoot biomass at the end of the growing season (mid November) and the grain and straw yield, 1996/97

	sowing date		
	6. Sept.	26. Sept.	16. Oct.
nitrate content in the soil layer 0-90 cm, mid November ($kg\ N{\cdot}ha^{-1}$)	43.3 a*)	72.8 b	91.5 b
nitrogen content in the shoot biomass mid November ($kg\ N{\cdot}ha^{-1}$)	57.5 a	35.8 b	**)
grain ($t{\cdot}ha^{-1}$)	4.9 a	4.8 a	3.9 b
straw ($t{\cdot}ha^{-1}$)	6.0 a	5.4 a	4.5 b

*) values followed by the same letter in the same line do not differ significantly at the 5 % LSD-level (t-test), **) not analyzed due to very low shoot biomass

TABLE 6. Evaluation of different strategies to reduce nitrogen losses through leaching by the cultivation of pulses in different situations under the site conditions of North-East Germany (increasing numbers of '+' = decreasing risk of nitrate leaching, *) with peas: no risk of delayed maturity) [5]

<table>
<tr><th colspan="2">decision making</th><th rowspan="2">strategies</th><th rowspan="2">evaluation result</th></tr>
<tr><th>production</th><th>follow up crop</th></tr>
<tr><td rowspan="4">rotational fallow (mulching)</td><td rowspan="3">spring crop (corn)</td><td>intercrops (not winter hardy)</td><td>+</td></tr>
<tr><td>intercrops (winter hardy)</td><td>++</td></tr>
<tr><td>undersown grass</td><td>++++</td></tr>
<tr><td>winter crop</td><td>early sowing date of winter rye (from the end of August)</td><td>+++</td></tr>
<tr><td rowspan="4">combine harvest</td><td rowspan="3">spring crop (corn)</td><td>intercrops (not winter hardy)</td><td>+*)/-</td></tr>
<tr><td>intercrops (winter hardy)</td><td>++*)/+</td></tr>
<tr><td>undersown grass</td><td>+++</td></tr>
<tr><td>winter crop</td><td>early sowing date of winter rye (from the beginning of Sept.)</td><td>++</td></tr>
</table>

4. Environmental effects

4.1. ABIOTIC ENVIRONMENTAL EFFECTS

Calculations for selected farms (average farm sizes approx. 35 ha) on the basis of the agricultural report of the BML (Federal Ministry of Food, Agriculture and Forestry) with regard to the area-related CO_2-emission from fossil energy sources showed different results for conventional and organic farming systems [13]. Under the conditions of West Germany significantly lower values resulted for organic farms (Table 7). The difference is mainly caused by the renunciation of N-fertilisier and brought-in feed.

TABLE 7. CO_2-emission from fossil fuels in organic and conventional farming systems in Germany 1991/92 [13]

	CO_2-emission (kg $CO_2 \cdot ha^{-1}$)	
	organic	conventional
fuels etc.	256.5	241.6
drying, flame weeding	20.0	41.3
pesticides	-	28.7
seeds	23.3	25.9
N-fertilisier	-	371.6
P-, K-, Ca-fertilisier	62.3	171.4
machinery	92.1	68.1
brought-in feed	48.9	247.8
total	503.2	1196.4

Depending on the cultivated crop the yields in organic farming systems are lower by 30-40 %. Nevertheless, Haas *et al.* [13] calculated for winter wheat an organically produced energy yield of 1522 $MJ \cdot t^{-1}$ as opposed to a conventionally produced yield of 2652 $MJ \cdot t^{-1}$. In Germany, organic farming can contribute to a reduction of the emission of the climate-relevant trace gas carbon dioxide from agricultural production. Under the conditions of North-East Germany (large farms, more extensive conventional farming, lower livestock units) this differentiation may be lower.

But in this context it is important to take into account that it is not possible to produce the same quantity of meat, especially pork meat, with organic agriculture than with conventional on the same size of arable land [13].

As shown in Table 3, there were relatively low potentials of nitrate losses on the field scale over the whole crop rotation of the two farms (average calculated losses of 17 to 22 kg $N \cdot ha^{-1} \cdot a^{-1}$), which includes the potential of denitrification and leaching. The amount of nitrogen lost by leaching and denitrification cannot be calculated exactly, but with regard to the site and the average excess it can be predicted that the annual nitrate losses concerning ground water protection could be about 20 kg N per hectare per year. Nolte and Werner [26] calculated N-losses in the fields of an organic farm with 40 kg $N \cdot ha^{-1} \cdot a^{-1}$, which is composed of a mean leaching potential of 25 kg $N \cdot ha^{-1} \cdot a^{-1}$ and a denitrification potential of 10 to 20 kg $N \cdot ha^{-1} \cdot a^{-1}$. Many investigations of the ground- and drainage-water confirm that the nitrate loss in organically managed fields was significant lower than in integrated and conventionally managed fields [6, 8, 28, 34]. Other investigations of organic farms in West Germany confirm the finding that there are low calculated N-surpluses concerning the whole farming system [17, 26]. Due to the renunciation of pesticides no contamination of the ground water with pesticide residues is possible through organic farming.

Because of the described positive effects for the ground water, in particular, many of water authorities in Germany prefer and especially promote organic farming [3] in their main collection areas. For this reason, they sometimes give financial support during the time of conversion or even longer. They realised that this is economically sensible because the costly nitrate- and pesticide-elimination in the drinking-water production would then not be necessary [3].

4.2. BIOTIC ENVIRONMENTAL EFFECTS

According to the basic standards of organic agriculture [2] demands for the protection and promotion of biotic aspects are additional goals of this land use system. In order to support the self-regulation of the system, e.g. the construction of and attention to hedges and other semi-natural habitats are also of importance for beneficial organisms.

Many investigations have been carried out to examine the effects of organic farming on flora and fauna. Investigations in 1960 showed (Table 8) that in 81 % of organic fields and 61 % of conventional fields endangered weed species could be found. When the investigations were repeated in 1987, there was a strong reduction of the weed species only in conventional fields. Other investigations on organic fields, which were converted to organic farming in the last 10 years showed similiar effects.

Frieben and Köpke [12] evaluated 18 studies which compared organically and conventionally managed arable fields. In organically managed fields the number of weed species was 1.2 to 6 times higher. These investigations were carried out mainly on small farms. At the moment there are no published data available concerning large farms in East Germany. First results of our own investigations on large organic and conventional farms show similiar effects with regard to the numbers of weed species already five years after conversion.

Investigations on animals (insects, spiders, millipedes, crustacea) in conventional and organic farms in West Germany showed that there were significantly higher numbers of species in organically managed farms [16, 22].

TABLE 8. Percentage of organic and conventional fields with at least one endangered weed species [11]

endangered weed species	organically managed fields	conventionally managed fields
investigations 1960	81 %	61 %
investigations repeated 1987	79 % [1)]	29 %
other investigations 1987	50 % [2)]	16 %

[1)] beginning of organic agriculture before 1962
[2)] beginning of organic agriculture after 1977

One goal of the government of the State of Brandenburg is to put 10 % of the land to a nature reserve and 30 % as an area of outstanding natural beauty [25]. In some nature conservation areas in Brandenburg (e.g. biosphere reserve Schorfheide-Chorin) the portion of the total area managed organically reaches 15 % [29]. This is due to the fact that in these sensitive areas ecologically orientated land use systems are preferred to abandoning agriculture, because the majority of the endangered plant and animal species are dependent on the man-made landscape. Organic agriculture is a preferred land use system for these areas because of its desirable features, e.g. the renunciation of synthetic pesticides and a reduction of nutrient inputs. This means, on the one hand, that some conflicts between agriculture and nature conservation do not occur, and on the other that control of special requests by the administration of the nature conservation areas is not necessary in organic farms. Moreover, crop rotations in organic agriculture should be as varied as possible and should include legumes or temporary pastures which include legumes, green manure and deep rooting plants [15]. This has positive effects on many biotic aspects as shown above as well as on the structuring of the landscape.

5. Economic viability

Organic farming products have higher costs due to the observance of strict ethical principles. Therefore, it needs either specific markets with higher prices or subsidies [35], if it is to be practised on a larger scale.

Stolze [32, 33] conducted economic analyses on 32 large organic farms in East Germany in 1994 and 1995. Cash crops showed a high gross margin mainly caused by subsidies and premium prices. In contrast to this, 94 % of the organically produced milk was sold to dairies for the regular non-premium price, because the marketing of organically produced milk has still to be build up. High fodder and labour costs are the reasons why the milk production is still not profitable for most large organic farms.

The strategy of some European countries includes financial support schemes to assist farmers converting to organic production. As a consequence, the organic sector has grown rapidly in recent years. From 1994 to 1997 the certified and policy-supported organic and in-conversion land area has increased, e.g. in Sweden from 1.5 % up to 8.9 % and in Austria from 2.4 % up to 8.6 % [19] but there are still low percentages in France (0.3%) and Great Britain (0.3%).

The situation of organic agriculture in Central and Eastern Europe is charactericed by the fact that only few countries have significant percentages of land area with organic farms, e.g. the Czech Republic, Hungary and Estonia have 0.2 - 0.5 % organic and in-conversion land area [19]. The main differences for the future development of organic agriculture in these countries will be caused by their possibility for their joining the EC and using the increased budgetary resources for agri-environmental measures e.g. organic farming, as proposed in “AGENDA 2000" agriculture [1]. Favourable for the conversion to organic farming in the CEE countries is the fact, that there are low wage costs and low input agricultural systems established due to the recent economic developments. To reach economic viability of organic farming in these countries the building up of special markets and regional processing is essential.

6. Conclusions

Nearly 3 % of the total arable land of the five new States of Germany have been converted to organic farming according to the AGÖL standards [2]. The average farm size reaches about 200 ha, while the average size of organic farms in West Germany amounts to 30 ha. The structure of the large farms in East Germany contrasts with the organic farms in West Germany and are mainly characterised by a higher complexity of organisation, involving mainly blue collar workers and low livestock units.

Under the site and structural conditions of large organic farms in North-East Germany, for economic and enviromental reasons, it is especially important to develop and apply special nutrient management strategies concerning the prediction and optimisation of nitrogen inputs so as to prevent deficits and losses. As long-term management strategies the calculation of humus and nitrogen balances are suitable for an overview of the nutrient fluxes of the total farm. To optimise the nitrogen uptake and to minimise nitrate leaching, the effects of specific cultivation methods, selected fields and crops, and short-term management strategies are developed for site- and situation-adapted decision making.

Positive abiotic and biotic environmental effects of organic farming systems, similar to those known from small family farms, can be described or predicted for large farms in North-East Germany. These effects includes significant reduction of nitrate losses through leaching and nitrate contents in the soil water; prevention of pesticide leaching; conservation and increase of the humus content; reduction of CO_2-emission caused by intensive agriculture; conservation and increase of biodiversity of flora and fauna.

To achieve an expansion of organic agriculture, particularly in CEE countries, the development of specific markets with higher prices (especially for milk products) and/or specific economic rewards of the positive environmental effects are necessary.

7. References

1. AGENDA 2000 Bulletin der Europäischen Union, Beilagen 1997
2. AGÖL (1996) Rahmenrichtlinien zum Ökologischen Landbau. Ed. Stiftung Ökologischer Landbau, *SÖL-Sonderausgabe* **17**, 48 S., 14. Aufl., Bad Dürkheim.
3. AGÖL und BUND (1997) Wasserschutz durch Ökologischen Landbau. Leitfaden für die Wasserwirtschaft, ISBN 3-00-001770-4
4. Asmus, F. (1992) Einfluß organischer Dünger auf Ertrag, Humusgehalt des Bodens und Humusreproduktion. *Berichte über Landwirtschaft,* SH **206**: Bodennutzung und Bodenfruchtbarkeit, Bd. 4. Humushaushalt, 127-139
5. Bachinger, J., R. Roth und A. Wurbs (1996) Strategien zur N-Verlustminimierung beim Körnerleguminosenanbau auf Sandstandorten Nordostdeutschlands in Anbausystemen des organischen und integrierten Landbaus, *Mitt. Dt. Bodenkundl. Ges.* **79**, 239-242
6. Berg, M. , G. Haas und U. Köpke (1997) Wasserschutzgebiete: Vergleich des Nitrataustrages bei Organischem, Integriertem und Konventionellem Ackerbau. 4. Wiss.-Tagung Ökol. Landbau, Bonn, *Schriftenreihe Institut für Organischen Landbau*, Verlag Dr. Köster, Berlin, 28-34.
7. Berger, G., K. Richter, K. Schmaler und A. Wurbs (1994) The influence of frost hardiness and nitrogen fertilisation of catch crops on N-conservation and nitrate leaching on sandy soils. *ISTRO proceedings of 13th intern. conference*, 1, 317-322.
8. Brandhuber, R. und U. Hege (1991) Nitratbelastung des Sickerwassers unter Acker- und Grünland viehhaltender Betriebe - Ergebnisse von Tiefenuntersuchungen. *VDLUFA-Schriftenreihe* **33**, 203-208.
9. Diez, T. und M. Krauss (1992) Berechnung von Humusbilanzen. In: *Schule und Beratung,* Bayr. Staatsministerium für Ernährung, Landwirtschaft und Forsten, S. III, 9-11.
10. EC-regulation 2092/91(1996) EG-Verordnung 'Ökologischer Landbau', Ed. Stiftung Ökologie & Landbau, *SÖL-Sonderausgabe* **45**, 3. Auflage, Bad Dürkheim
11. Frieben, B. (1990) Bedeutung des Organischen Landbaus für den Erhalt von Ackerwildkräutern. *Z. Natur und Landschaft,* 68, **7/8**, 379-382.
12. Frieben, B. und U. Köpke (1994) Bedeutung des Organischen Landbaus für den Arten- und Biotopschutz in der Agrarlandschaft. 8. wiss. Fachtagung 'Integrative Extensiverungs- und Naturschutzstrategien'. Lehr- und Forschungsschwerpunkt 'Umweltverträgliche und standortgerechte Landwirtschaft' der Univ. Bonn (Ed.), *Forschungsberichte* Heft **15**, 77-88.

13. Haas, G., U. Geier, D.G. Schulz und U. Köpke (1995) Vergleich Konventioneller und Organischer Landbau - Teil 1: Klimarelevante Kohlendioxid-Emission durch den Verbrauch fossiler Energie. In: *Berichte über Landwirtschaft* **73**, 401-415.
14. Hamm, U. & M. Haccius (1997) Bedeutung der in der AGÖL zusammengeschlossenen Betriebe in der BR Deutschland (Stand 01.01.1997). *Z. Ökologie & Landbau* **3**, 78.
15. IFOAM (1992) International Federation of Organic Agriculture Movements, Ed. Stiftung Ökologischer Landbau, *SÖL-Sonderausgabe* **16**, 46 S., 9. Aufl., Bad Dürkheim.
16. Ingrisch, S., K. Wasner und E. Glück (1989) Vergleichende Untersuchung der Ackerfauna auf alternativ und konventionell bewirtschafteten Flächen. Alternativer und konventioneller Landbau. *Schriftenreihe der LÖLF,* Landwirtschaftsverlag Münster-Hiltrup, Bd. **11**, 113-271.
17. Kaffka, S. and H.H. Koepf (1989) A Case Study on the Nutrient Regime in Sustainable Farming. *J. Biological Agriculture and Horticulture,* Vol. **6** , 89-106.
18. Körschens, M. (1980) Die Abhängigkeit der organischen Bodensubstanz von Standortfaktoren und acker- und pflanzenbaulichen Maßnahmen, ihre Beziehung zu Bodeneigenschaften und Ertrag sowie die Ableitung von ersten Bodenfruchtbarkeitskennziffern für den Gehalt des Bodens an organischer Substanz. Diss. Bad Lauchstädt
19. Lampkin, N. (1998) Organic farming in Europe, *Ökologie & Landbau* **105**, 2, Bad Dürkheim
20. Lang, G. (1995) Berechung von Humusbilanzen. In: *Schule und Beratung,* Bayr. Staatsministerium für Ernährung, Landwirtschaft und Forsten, 5.
21. Leithold, G. (1996) The special qualities of humus and nitrogen budget in organic farming. In: *New Research in Organic Agriculture*, 11th International Scientific IFOAM Conference, Copenhagen, Proceedings Vol. **2**.
22. Leithold, G., K.-J. Hülsbergen, D. Michel und H. Schönmeier (1996) Humusbilanzierung - Methoden und Anwendung als Agrar-Umweltindikator. In: *Initiativen zum Umweltschutz*, Osnabrück, Deutsche Bundesstiftung Umwelt, 3.
23. Leithold, G. und K.-J. Hülsbergen (1997) Grundlagen und Methoden zur Humusbilanzierung im ökologischen Landbau. 4. Wiss.-Tagung Ökol. Landbau, Bonn, *Schriftenreihe Institut für Organischen Landbau*, Verlag Dr. Köster, Berlin, 56-62.
24. Mäder, P., L. Pfiffner, T. Alföldi, A. Wiemken und U. Niggli (1995) Bodenbiologie. In: Biologischer Landbau: Beitrag des DOK-Versuches. *Schriftenreihe der FAC Liebefeld* **21.** Ed. Eidgenössische Forschungsanstalt für Agrikulturchemie und Umwelthygiene, Liebefeld-Bern, 65-79.
25. MUNR (1997) Landschaftsprogramm Brandenburg - Entwurf - Ministerium für Umwelt, Naturschutz und Raumordnung des Landes Brandenburg, Abteilung Naturschutz, Potsdam, 38 S.
26. Nolte, C. and W. Werner (1994) Investigations on the nutrient cycle and its components of a biodynamically-managed farm. *J. Biological Agriculture and Horticulture*, Vol. **10** , 235-254.
27. Roth, R. und Steinbrenner (1992) Vorfrucht- und Fruchtfolgeeffekte auf Sandstandorten in Müncheberg. Tagungsbericht des IV. Fruchtfolgesymposiums am 21. Mai 1992 in Halle a. d. Saale, 70-79.
28. Smilde, K.W. (1989) Nutrient supply and soil fertility. In: *Development of farming systems*, (eds.) J.C. Zadoks, Pudoc, Wageningen.
29. Stein-Bachinger, K. und R. Vögel (1996) Naturschutz und Landwirtschaft in den Großschutzgebieten Brandenburgs. *Z. Lebendige Erde* **6**, 369-375.
30. Stein-Bachinger, K. und W. Werner (1997) Effect of manure on crop yield and quality in an organic agricultural system. *J. Biological Agriculture and Horticulture,* Vol. **14** , 221-235.
31. Stein-Bachinger, K. und J. Bachinger (1997) Nährstoffbilanzen als Grundlage von Optimierungsstrategien für ökologisch wirtschaftende Großbetriebe Nordost-Deutschlands. 4. Wiss.-Tagung Ökol. Landbau, Bonn, *Schriftenreihe Institut für Organischen Landbau*, Verlag Dr. Köster, Berlin , 109-114.
32. Stolze, M. (1997) Analyse von Entwicklungsperspektiven ökologisch wirtschaftender Großbetriebe in Ostdeutschland. In: *Landbauforschung Völkenrode*, **175**, Braunschweig, 107-117.

33. Stolze, M. (1997) The Economic analysis of Large-Scale Organic Farms in East Germany. In: *Proceeding of the 2nd ENOF Workshop 'Steps in Conversion and Development of Organic Farms,* 97-103.
34. Vereijken, P. (1990) Integrierte Nährstoffversorgung im Ackerbau. In: *Schweizerische Landwirtschaftliche Forschung* **29**, 4, 359-365.
35. Weinschenk, G. (1997) Agrarpolitik und ökologischer Landbau. *Agrarwirtschaft* **7**, 251-256

MINIMUM TILLAGE TECHNOLOGY AS AN ALTERNATIVE TO TRADITIONAL SYSTEMS: ENVIRONMENTAL IMPLICATIONS AND THEIR RELEVANCE TO POLAND AND OTHER CEES

B. C. BALL,
Environmental Division,
SAC, West Mains Road, Edinburgh, EH9 3JG, UK

1. Introduction

In North-Western Europe, in the 1960's and 1970's cereal farming increased at the expense of livestock farming. The labour force decreased and, together with the costs of machinery operations, became more expensive. Thus simpler alternatives to conventional plough tillage, termed minimum or conservation tillage, became attractive as offering savings in the cost of establishing crops [1]. A further incentive was the availability of herbicides for weed control in non-ploughing systems. In the UK, the main motivation for the use of minimum tillage is to allow the maximum area of crop, particularly the high-yielding autumn varieties, to be sown in the limited period when the soil conditions are suitable [1]. However, actual use of minimum tillage in the UK, as in most of North-Western Europe is minor because of problems of reliability, straw disposal, weed control, soil drainage status and soil compaction [1]. In Scandinavia, minimum tillage is used mainly for reducing erosion risk by maintaining an undisturbed soil surface during winter and early spring [2]. In drier regions as in USA and the Canadian prairies, conservation tillage is widely used mainly to reverse or slow down soil degradation processes associated with traditional plough tillage and to conserve soil moisture and reduce input costs rather than to give any short-term gain in yields [3]. In Eastern Europe the approach to minimum tillage has differed from that in Western Europe because of the different socio-economic conditions [4]. Minimum tillage in Eastern Europe has still not reached its expected usage because of the perceived risk to crop productivity, lack of research and because suitable soils are confined to the more fertile and well-drained soils in favourable climatic conditions and those which have been agriculturally improved [4].

Most research findings show that tillage fundamentally changes the soil ecosystem through its effect on soil physical, chemical and biological properties; tillage thus has an important influence on soil qualities [5]. These qualities determine the retention of soil particles, soil water and dissolved or adsorbed soil chemicals. These in turn affect crop productivity and diffuse pollution of water and air thereby influencing water quality and atmospheric quality [6]. Tillage influences the loss of soil to

M.J. Wilson and B. Maliszewska-Kordybach (eds.),
Soil Quality, Sustainable Agriculture and Environmental Security in Central and Eastern Europe, **139–151.**

freshwater as sediment [5]. Soil particles clog up lakes and rivers and adsorbed chemicals such as phosphorus, nitrate or herbicide residues can decrease the quality of drinking water [2]. Emissions of carbon dioxide and trace gases from soil increase global warming.

This paper describes minimum and conventional tillage systems and how these influence soil qualities which in turn influence the broader environment through soil erosion and losses of chemicals as particulates, gases or leachates. Finally, some tillage strategies are recommended for sustaining soil quality and crop productivity, thereby achieving environmental benefits.

2. Defining conventional and minimum tillage

The traditional tillage system, here termed 'conventional tillage' (CT) is used to establish most crops in Europe. CT is mouldboard ploughing to 200-300 mm depth (though it is often deeper in Italy) to incorporate weed and crop residues. After ploughing, secondary cultivation, which often requires a powered cultivator, produces a seedbed. After sowing, the seedbed is rolled to yield a level surface. This system optimises soil conditions for crop growth reliably, requires relatively simple agronomic management and permits consistent crop yields. The disadvantages of the system is that it is energy intensive and can encourage erosion and soil structural damage because fine-structured, smooth seedbeds are produced and the crop residues are buried, leaving the soil surface bare. The regular soil disturbance also mineralises the organic material and kills earthworms.

Alternatives to conventional ploughing are often termed minimum or conservation tillage because the soil is less disturbed than by CT and the surface is partly covered by residues, making it less susceptible to erosion and associated soil loss than CT. Here two minimum tillage systems are considered, *viz*. no-tillage (NT) and reduced tillage (RT).

In no-tillage (or direct drilling) soil is disturbed by the drill coulters sufficient only to sow the seed. The residues of the previous crop need to be removed so that the seed can be sown into the undisturbed stubble. The main advantage of this lack of disturbance is that the root system and surface residues of straw and chaff remain at the surface to protect the soil from raindrop impact and to reduce lateral soil movement associated with water runoff. Weeds and volunteers of the previous crop have to be killed by herbicides.

In reduced tillage (RT) soil is tilled with a tined or rotary cultivator to less than conventional ploughing depth. Reduced tillage may involve two or more passes of different cultivators and usually produces a coarser seedbed than conventional ploughing. The seedbeds often have a rough surface containing straw residues.

Farmers use minimum tillage mainly to reduce the time and costs required to establish crops and to increase the trafficability of the soil [7]. Machinery costs may be greater under minimum tillage than under CT if specialist machinery such as a direct drill is required. The main crops grown with minimum tillage are cereals, oil-seed rape, sugar beet and grass. Other crops like potatoes require conventional tillage. Minimum tillage is not widely used in Europe because of problems of slugs, weeds (particularly grass weeds), residues of the former crop and soil compaction [1]. This leads to a requirement for a high standard of management. Currently, the West-European country with the greatest use of minimum tillage is Spain, because of cost savings and improved soil moisture conservation [8].

3. Soil qualities and environmental implications

The two main aspects of the physical, chemical and biological qualities in NT and, to a lesser extent, in RT, which are important in determining soil qualities are the presence of residues at the soil surface and the stability of the undisturbed or coarse soil structure. The importance of tillage in influencing soil qualities and the overall environment has been shown by a wide range of research in Western Europe. Much of this research has been summarised in the reports and database of a European Union Concerted Action on 'Experience with the Applicability of No-Tillage Crop Production in the West-European Countries' led by Dr F. Tebrügge, Justus-Liebig University, Gießen, Germany. This concerted action makes use of data from experiments which include Spain, Portugal, Italy, Germany, Denmark and Scotland [9]. Sites in each country were few, sometimes only one, so the results quoted here are not necessarily representative of the country. In the subsequent Tables there is one site corresponding to each soil type.

3.1. SOIL PHYSICAL PROPERTIES

Compaction is often quoted as a problem with minimum tillage systems, particularly in wet or fine-textured soils [1,13]. Compaction is usually expressed as soil bulk density [10]. At all locations, minimum tillage (NT and RT) increased soil bulk density (Table 1) with the increase being greatest under NT and in the finer-textured soils.

TABLE 1. Soil bulk density (Mg/m^3) at 0-10 cm depth on different soil types. Data from [9].

Location	% sand (average of sites)	CT	RT	NT
Germany (Phaeozem and Fluvisol)	14	1.26	1.35	1.55
Italy (Fluvisol and Regosol)	20	1.4	1.52	1.6
Italy (Cambisol)	51	1.5	1.38	1.65
Portugal (two Luvisols)	72	1.58	-	1.64

These increases in bulk density under minimum tillage are generally accompanied by substantial increases in soil strength. Soil strength is a measure of the ability of the soil to bear applied loads, such as from wheels or tillage implements, without disruption of soil structure. Soil strength also influences the ability of plant roots to explore the soil. Cone resistance (Fig. 1) is a useful empirical indicator of soil strength and its variation with depth. The increased soil strength as commonly found under minimum tillage increases the ability of soils to bear loads but can decrease root exploration.

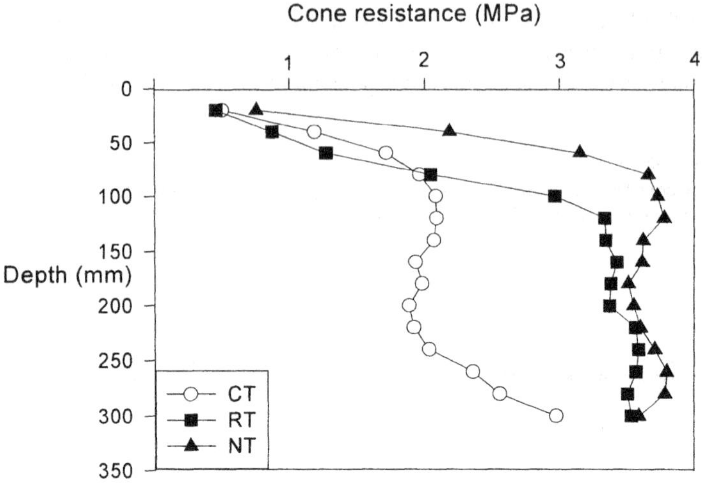

Figure 1. Cone resistance under winter barley in spring, Cambisol/Gleysol, Scotland. Data from [11].

The increased compaction under minimum tillage also decreases the median diameter of the soil porosity [12] and may increase the content of smaller pores which store water extractable by crop roots [9]. This can lead to greater soil water conservation and better plant water supply in periods when water is limiting. However, in wet conditions, these effects can make the soil more prone to waterlogging which may damage the crop or lead to denitrification and losses of nitrogen gases. The more compact soil under NT usually results in the soil water content under NT being typically 1-6 % greater than under CT.

In dry countries, such as southern Spain, NT further increases water conservation in clay soils. This results from the increased surface cover provided by residues and by the delayed development of surface cracks in NT and RT which decrease evaporative losses of water in comparison to CT (Table 2). Later in the season, these cracks remain open and uncovered and help to ensure good water recharge in NT.

TABLE 2. Proportion of soil surface covered by crop residues and cracks (complete cover =1), average of June and July 1995, montmorillonitic clay soil, Spain. Data from [9].

	CT	RT	NT
Residue cover	0.049	0.254	0.605
Surface cracks	0.286	0.24	0.056

Minimum tillage can produce conditions which reduce soil erosivity. On silty clays in Italy, treatments with increased surface roughness reduced runoff volume (Table 3). High surface roughness was best produced by sowing directly into conventionally ploughed soil with a no-till drill i.e. with the secondary cultivation step omitted.

TABLE 3. Conservation tillage effects on soil cover, roughness, run-off and soil losses on silty clay soil in Italy

	Conventional ploughing, smooth seedbed	Conventional ploughing, no-till drill	Reduced tillage (combination cultivator and drill)	No-tillage
Soil surface covered by residues (%)	5	0	65	100
Soil roughness index	7	28	8	8
Total run-off (mm)	43.2	15.7	31.5	20.7
Total soil losses (Mg/ha)	11.2	3.3	2.9	1.1
Sediment concentration (g/l)	25.8	21.0	9.2	5.5

Increasing the area of soil surface covered by residues by using a reduced or no tillage system reduced not only runoff but also the concentration of sediment in the runoff. The marked reductions in total soil losses and sediment concentration under NT were associated with the increased stability of the soil structure and resistance of the soil to erosion. This increased structural stability is not only of the bulk soil but is also found at the scale of individual aggregates, particularly those close to the soil surface where the increased stability also increases resistance to soil compaction [14].

3.2. SOIL CHEMICAL PROPERTIES

Organic matter concentrates near the surface of the topsoil under NT and RT in experiments in most countries (Table 4). This is associated with the presence of residues at the surface and the lack of tillage to stimulate mineralisation of soil carbon. Soil organic matter contents are greater in experiments in the more northern countries, presumably because of the slower rate of mineralisation in the cooler climates. Differences in organic matter between tillage treatments were observed soon after the treatments were applied in southern Europe. In cooler, northern climates accumulation of organic matter may take several years.

TABLE 4. Organic matter and total or soluble phosphorus at one site in each of four European countries. Data from [9].

	Depth (cm)	Organic matter (g/100g)		Total or soluble P (mg P/kg soil)	
		CT	NT	CT	NT
Spain	0-5	1.4	2.0	11	34
	5-10	1.4	1.6	6	9
Portugal	10	1.9	2.5	18*	70*
	20	1.7	2.2	26*	30*
Denmark	0-2	3.1	3.6	43	51
	10-20	2.9	3.1	40	39
Scotland	0-5	3.1	3.9	22	45
	10-25	3.2	3.5	16	15

* indicates soluble phosphorus

However, this delay did not influence the size of the difference in organic matter content between soils of the tillage treatments. Experimentation in Scotland also revealed that carbohydrates concentrated in soil near the surface under NT [14]. The hot water fraction, considered to be produced by microbial activity, was the most enriched under NT. This fraction correlated significantly with aggregate stability [14]. The soil near the surface under NT thus contained more organic matter which was of greater quality in preserving soil structure than under CT. The organic carbon in the topsoil (0-300 mm depth) averaged over 3 Italian and 2 German sites was for CT, RT and NT 85, 91 and 92 t/ha respectively [15]. Thus the overall fixation of carbon in the soil was rather small and unlikely to be sustained.

Phosphorus (Table 4) also tends to concentrate near the soil surface under NT. This increased nutrient retention is a result of the lack of incorporation by tillage and the reduced leaching due to the lower permeability of the soil. Concentration of phosphorus near the soil surface was also observed in experiments in the Corn Belt of the USA [13]. Thus any runoff or particulate soil loss from NT is likely to contain more phosphorus than a corresponding loss from CT. This may reduce the effectiveness of NT in reducing total phosphorus losses through erosion [13]. Another disadvantage of such high nutrient concentrations near the surface is acidification of the surface soil, as indicated by the pH which is typically 0.2 units lower than under CT.

Soil nitrogen is lost either through leaching or as gases. In wet soil conditions, the compact soil structure under minimum tillage increases the likelihood of waterlogging so that gaseous losses are likely to account for a substantial proportion of losses of soil and fertiliser nitrogen. Losses are principally as molecular N_2 and as nitrous oxide (N_2O). N_2O is a radiatively active gas that is 270 times as effective as carbon dioxide in causing global warming [16]. Soils are important sources of emission. In an imperfectly drained soil in Scotland, N_2O emissions under NT (Table 5) were considerably greater than under CT, particularly after heavy rainfall. Emissions were associated with losses of fertiliser.

TABLE 5. Nitrous oxide and carbon dioxide fluxes under spring barley in Scotland from 8 April (Week 1) to 16 June (Week 11) 1997.

Week after sowing and application of fertiliser	Rain (mm)	N_2O (g N/ha/d)		CO_2 (kg C/ha/d)	
		NT	CT	NT	CT
1	0.1	94.8	7.2	6.2	9.3
2	6.4	55.7	4.0	15.8	5.6
3	9.8	154.1	5.1	13.5	12.8
4	47.2	234.1	5.4	16.2	15.7
5	26.2	568.3	154.1	10.8	12.0
6	27.0	679.0	67.0	12.8	13.9
7	23.0	72.4	178.3	1.3	19.8
8	0.1	2.8	25.3	1.1	33.2
9	8.1	17.8	8.9	2.7	16.0
10	72.1	36.7	37.3	3.8	27.6
11	0.6	18.6	23.6	11.3	42.6

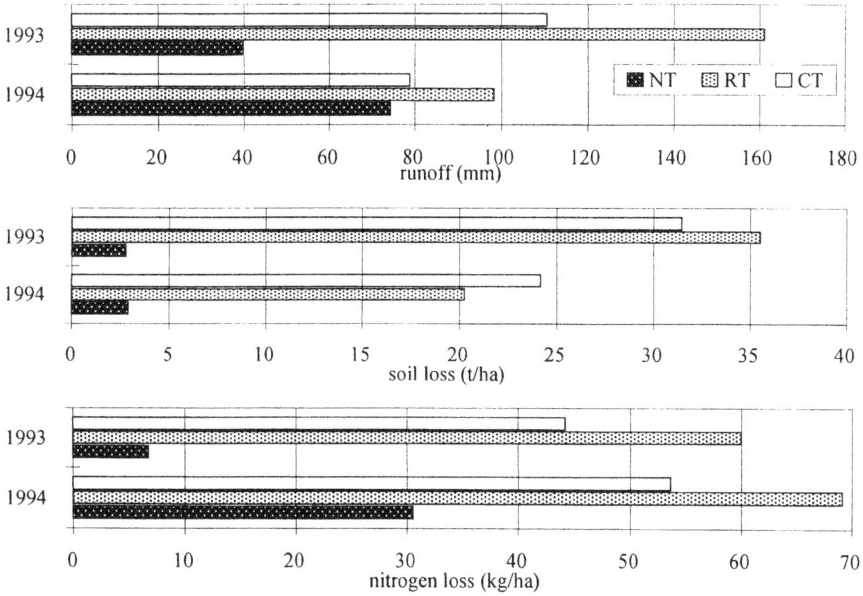

Figure 2. Runoff, soil loss and nitrogen loss, Eutric Regosol, Italy. Data from [15]

Although minimum tillage should reduce the rate of mineralisation of nitrogen and thus the amount at risk from leaching, there is no clear reported evidence of this. However, NT can reduce the amount of soil N lost with runoff and sediment. For example, N losses from an Italian soil (Fig. 2) were less under NT than under CT. This effect was attributed to the greater soil cover by residues and structural stability under NT than under CT [15]. RT gave higher runoff and soil and nitrogen losses than CT, in spite of greater soil cover under RT than under CT. This effect was attributed to the type of cultivator used, here a disc harrow, which incorporated too many residues and caused appreciable compaction of soil at 100 mm depth which reduced water infiltration and increased run-off [15].

3.3. SOIL BIOLOGICAL PROPERTIES

Minimum tillage decreases soil disturbance to habitats and provides cover and food sources in the form of crop residues at the soil surface. Properties which are influenced include microbial activity, enzyme activity and earthworm and arthropod biomass and number [9]. Here microbial activity is assessed as respiration rate, the rate of evolution of carbon dioxide from soil under controlled conditions. In Germany, minimum tillage enhanced respiration rates in soils at 0-120 mm depth, but not at 120-250 mm depth [9]. This effect was attributed to the presence of residues at 0-120 mm depth. However, in an experiment in Portugal, respiration rates (Table 6) were greater under NT than under CT at all soil depths to 300 mm.

TABLE 6. Microbial activity as respiration rate (mg CO_2/50g soil) in a Vertic Cambisol, Portugal. Data from [17]

Depth (cm)	CT	NT
10	3.4	15.8
20	14.4	19.4
30	18.5	29.0

As for microbial activity, use of minimum tillage enhanced enzyme activities, particularly in the 0-100 mm layer [9]. In an experiment in Germany, dehydrogenase activity and respiration rates were greater in the 0-50 mm soil layer than in the 50-100 mm layer and were accompanied by an increase in the rate of mineralisation of N [15]. In Poland, the highest enzyme activity was also found in soil under NT and the enhanced biological activity was considered to be an inhibitor of soil degradation processes [18]

Bulk CO_2 fluxes at the soil surface were measured directly in the field in Scotland concurrently with the N_2O measurements already described. Weekly averages of CO_2 flux (Table 5), mainly attributed to respiration, were not consistently effected by tillage. The decrease in CO_2 flux under NT after week 6 was attributed to soil waterlogging. The overall CO_2 balance under minimum tillage is also influenced by the reduction in fuel usage and the increase in near-surface organic carbon levels which combine to produce significantly less CO_2 emissions than CT [19]. However, the effect due to accumulation of organic matter, as already stated, is minor and will decline with time as the level reaches a new equilibrium.

Carbon is also fixed by uptake of atmospheric methane (CH_4). CH_4 is a significant greenhouse gas [16]. Aerobic agricultural soils usually oxidise rather than emit CH_4 and are important sinks of atmospheric CH_4 [20]. The CH_4 oxidation rate can be reduced by tillage due to the disturbance of the methanotrophic microbes, but is also influenced by gaseous diffusivity as this affects the rate of supply of atmospheric methane [21].

TABLE 7. Methane oxidation rate in a tillage experiment in Scotland, December 1996. The arable was converted from grassland using the tillage treatments on April 1996 or September 1996. Data from [22].

Tillage treatment	Land use	Methane oxidation rate ($mg\ CH_4/m/\ d$)
	Woodland	1.8
	Grass	0.8
Ploughed to 200 mm	Arable for 3 months	0.8
Ploughed to 300 mm	"	0.6
No-tilled	"	0.6
Ploughed to 200 mm	Arable for 8 months	0.2
Ploughed to 300 mm	"	0.5
No-tilled	"	0.8

Initial measurements of CH_4 oxidation rate in a tillage experiment in Scotland where grassland was converted to arable cropping are shown in Table 7. Eight months after the conversion, the CH_4 oxidation rate was reduced by CT to about one-quarter of that under NT, which remained similar to that under the undisturbed grassland. CH_4 oxidation rate is a useful indicator of change in soil quality in response to land use and tillage change because these have long-term effects on methane-oxidising microbes and soil structure [21].

Earthworms are important in establishing continuous macropores which aid water infiltration and in creating stable aggregates rich in microbial activity. In Europe, several experiments have indicated that the ratio of total earthworm numbers per unit area in tillage treatments CT:RT:NT is 1:3:8 [15]. NT commonly increases the number of deep burrowing earthworms (lumbricids) and the proportion of large juvenile and adult worms in comparison to CT [9]. Thus, lumbricid earthworm biomass in a typical experiment in Germany (Fig. 3) under NT and, to a lesser extent, RT, was considerably greater than under CT.

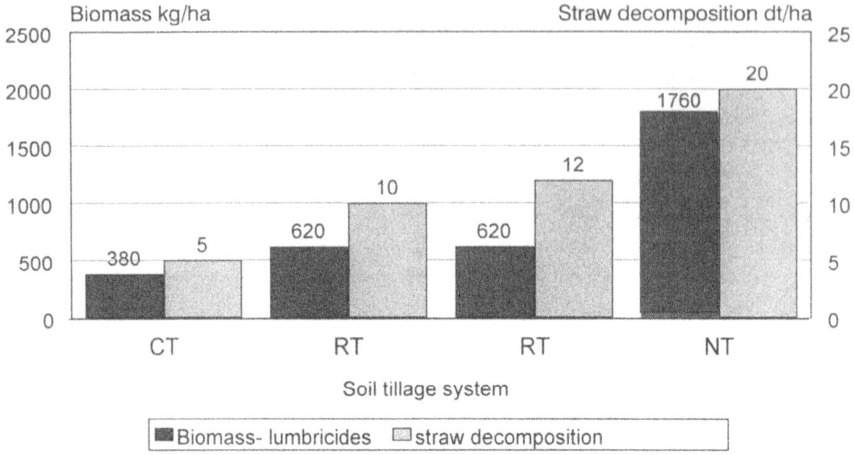

Figure 3. Lumbricid earthworm biomass and straw decomposition associated with these worms, Luvic Phaeozem, Germany. Data from [15].

In this experiment the straw decomposition attributable to earthworms was closely related to the earthworm biomass (Fig. 3). In the same experiment the number of earthworm burrows in the topsoil was related to treatment differences in infiltration rates and in the population of earthworms [15]. Thus earthworm activity under NT is important for incorporating and decomposing straw and maintaining drainage in the topsoil. Other fauna influenced by tillage are carabid beetles and slugs. In Scotland, beetle populations under NT were almost double those under CT. Beetles are important in low input arable systems because they may help to control cereal aphids, thereby giving some natural control of virus disease borne by aphids [1]. In cool, moist conditions, slugs may damage crop seedlings, though the crop usually recovers [1].

A common feature of most of these changes in soil properties, particularly those related to improvements in soil structure, is that they are relatively slow and appear only after several years' continuous use of minimum tillage.

4. Recommendations for minimum tillage in Poland and other CEES

The benefits of minimum tillage to soil quality in Western Europe depended mainly on enhanced biological activity and changes in soil structure associated with more compact soil. The enhancement in biological activity appeared most quickly in warm areas and the changes in soil structure were of most benefit in areas prone to drought. Thus minimum tillage is most suitable for general use in drier and warmer regions. A similar conclusion was also reached in Germany for the suitability of soils for sequential conservation tillage [23]. Land areas with an annual rainfall of < 700 mm were considered suitable, but in these areas, soils subject to periods of wetness were unsuitable [23]. In soils subject to wetness, the principal environmental advantage of minimum tillage was the reduction in the likelihood of soil and particulate phosphorus loss by water erosion. Minimum tillage can also reduce the risk of wind erosion on lighter soils in drier areas. These advantages need to be balanced against possible disadvantages to crop growth caused by compaction and waterlogging and possible gaseous losses of nitrogen. However any long-term benefits of minimum tillage to the soil and to the environment need to take into account the greater need for chemical weed control, particularly pre-sowing, and the greater risks of reduced crop productivity than under CT.

Recent experience with minimum tillage (RT to 80-100 mm depth) in Poland has indicated that even with good control of weeds, compaction persists as a problem which, along with inadequate incorporation of residues of the previous crop, caused reduction in crop yields [24]. In Hungary, both compaction and weeds were problems under minimum tillage, particularly NT, where the weed problem was aggravated by low levels of fertiliser application [25]. Soils in Poland are considered to be of medium quality; more than half of the soils are coarse-textured and many contain hard pans [26]. Field traffic resulting from agricultural mechanization has brought a substantial compaction problem to Poland [26]. This limits the applicability of minimum tillage. Indeed, it could be argued that deep tillage might be more applicable in response to the need to improve many soils [4, 24]. Environmental benefits of minimum tillage seem to be less in East-European than in West-European countries and are confined to the reduction of erosion risk [4]. Widespread usage of minimum tillage would not appear to be advisable at present.

Adoption of 'rational tillage' where tillage is chosen according to soil constraints and crop requirements but includes the need to reduce erosion losses in vulnerable areas might be the best approach. Experience with tillage in Bulgaria has shown that a mixture of ploughing, deep loosening and minimum tillage techniques specific to crop rotation and to soil type best preserves crop productivity and soil physical properties [27]. In areas where minimum tillage is considered advantageous, such as for erosion control, a variety of novel tillage techniques, such as ridge tillage, rotational tillage [28] and the possibilities of farming-by-soil in which tillage is varied according to soil type are available [29] as are novel direct drill coulters and drill types to cope with trash [30]. The wider application of techniques to limit field traffic (such as tramlines) or to reduce the compactive effort of vehicles such as using light machinery or low-ground pressure tyres and wheels might also be considered [31]. Novel herbicides and herbicide systems are continually being developed. For erosion control, specific tillage and cropping practices [32,33] need to be integrated with minimum tillage. However little knowledge is available on how to apply these techniques successfully at the farm level. The use and timing of appropriate crop rotations, tillage techniques and compaction control within minimum tillage systems requires further assessment in R & D programmes and subsequent inclusion in systems of advising farmers. The lack of availability of appropriate advice to farmers partly explains the low uptake of minimum tillage technology.

5. Conclusions

1. Minimum tillage provided most benefits to soil quality in warm, dry areas, where it is likely to be most suitable for general use.
2. Improvements in soil quality observed after long periods of minimum tillage in Western Europe do not appear to be common in Poland.
3. The widespread adoption of minimum tillage in Poland is unlikely to conserve and improve soil quality, because of problems with compaction, weeds and crop residues.
4. The main environmental benefit of minimum tillage in Poland and other CEES is the reduction of erosion risk in localised vulnerable areas, though supplementary measures to control compaction and runoff would also be required.
5. More R & D on various aspects of crop and tillage rotation and novel machinery and tillage techniques and the supply of appropriate advice to farmers are required in the CEES.

6. References

1. Christian, D.G. and Ball, B.C. (1994) Reduced cultivation and direct drilling for cereals in Great Britain, in M.R. Carter (ed.), *Conservation Tillage in Temperate Agroecosystems*, CRC Press, Florida, pp. 117-140.
2. Riley, H., Børreson, T., Ekeberg, E. and Rydberg, T. (1994) Trends in reduced tillage research and practice in Scandinavia, in M.R. Carter (ed.), *Conservation Tillage in Temperate Agroecosystems*, CRC Press, Florida, pp. 23-45.
3. Larney, F.J., Lindwall, C.W., Izaurralde, R.C. and Moulin A.P. (1994) Tillage systems for soil and water conservation on the Canadian prairie, in M.R. Carter (ed.), *Conservation Tillage in Temperate Agroecosystems*, CRC Press, Florida, pp. 305-328.
4. Butorac, A. (1994) Conservation tillage in eastern Europe, in M.R. Carter (ed.),*Conservation Tillage in Temperate Agroecosystems*, CRC Press, Florida, pp. 357-374.
5. Karlen, D.L., Wollenhaupt, N.C., Erbach, D.C., Berry, E.C., Swan, J.B., Eash, N.S. and Jordahl, J.L. (1994) Long-term tillage effects on soil quality, *Soil Tillage Research* **32**, 313-327.
6. Ball, B.C. and Dickson, J.W. (1998) The potential for application of conservation tillage in UK, in A.M. Petchey, B.D'Arcy and A.Frost (eds.), *Diffuse Pollution and Agriculture II*, Proceedings of a Conference held by SEPA and SAC, April 1997, Edinburgh, pp. 61-71.
7. Tebrügge, F. and Wagner, A. (1995) Soil structure and trafficability after long-term application of no-tillage, in F. Tebrügge and A, Böhrnsen (eds.), *Experience with the applicability of no-tillage crop production in the West European countries*, **II** Wissenschaftlicher Fachverlag, Gießen, pp. 49-57.
8. Costa, J. (1997) Farmer acceptance and public support for no-tillage - conservation farming in Spain, in F. Tebrügge and A, Böhrnsen (eds.), *Experience with the applicability of no-tillage crop production in the West European countries*, **III** Wissenschaftlicher Fachverlag, Gießen, pp. 127-135.
9. Ball, B.C., Tebrügge, F., Sartori, L., Giráldez, J.V. and Gondález, P. (1998) Influence of no-tillage on physical, chemical and biological soil properties, in F. Tebrügge and A, Böhrnsen (eds.), *Experience with the applicability of no-tillage crop production in the West European countries*, Final Report Fachverlag Kohler, Gießen, pp. 7-27.
10. Carter, M.R. (1990) Relative measures of soil bulk density to characterize compaction in tillage studies of fine sandy loams. *Canadian J. Soil Science* **70**, 425-434.
11. Ball, B.C. (1995) Soil responses to tillage and their environmental implications in Scotland, in F.Tebrügge and A, Böhrnsen (eds.), *Experience with the applicability of no-tillage crop production in the West European countries*, **II** Wissenschaftlicher Fachverlag, Gießen, pp. 7-23.
12. O'Sullivan, M.F. and Ball, B.C. (1993) The shape of the water release characteristic as affected by tillage, compaction and soil type, *Soil Tillage Research* **25**, 339-349.

13. Lal, R., Logan, T.J., Eckert, D.J., Dick, W.A. and Shipitalo, M.J. (1994) Conservation tillage in the corn belt of the United States, in M.R. Carter (ed.), *Conservation Tillage in Temperate Agroecosystems*, CRC Press, Florida, pp. 73-114.
14. Ball, B.C., Cheshire, M.V., Robertson, E.A.G. and Hunter, E.A. (1996) Carbohydrate composition in relation to structural stability, compactibility and plasticity of two soils in a long-term experiment, *Soil Tillage Research* **39**, 143-160.
15. Tebrügge, F., Borin, M., Mazzoncini, M. and Basch, G. (1996) Effects of tillage system on physical, chemical and biological soil characteristics, in M. Borin, L. Sartori, C. Giupponi, M. Mazzoncini, R. -A. Düring and G. Basch (eds.), *Effects of tillage system on herbicide dissipation: an experimental approach at field scale*, Unipress, Padova, pp. 41-74.
16. IPCC (1996) *Climate Change 1995:The Science of ClimateChange:The Second Assessment Report of the Inter-Governmental Panel on Climate Change*, Cambridge University Press, UK
17. Carvalho, M. and Basch, G. (1995) Long term effects of two different soil tillage treatments on a vertisol in Alentejo region of Portugal, in F. Tebrügge and A, Böhrnsen (eds.), *Experience with the applicability of no-tillage crop production in the West European countries*, **II** Wissenschaftlicher Fachverlag, Gießen, pp. 17-23.
18. Domzal, H., Malicki, L., Slowinska-Jurkiewicz, A. and Bielinska, E.J. (1997) The effect of differentiated methods of arable cultivation on enzymatic activity in the arable layer of soils, *Bibliotheca Fragmenta Agronomica* TOM **2A**, 189-192.
19. Borin, M., Menini, C. and Sartori, L. (1997) Effects of tillage systems on energy and carbon balance in north-eastern Italy, *Soil Tillage Research.* **40**, 209-226.
20. Dobbie, K.E., Smith, K.A., Priemé, A., Christensen, S., Dégorska, A. and Orlanski, P. (1996) Effect of land use on the rate of methane uptake by surface soils in northern Europe, *Atmos. Environ.* **30**, 1005-1011.
21. Ball, B.C., Dobbie, K.E., Parker, J.P. and Smith, K.A. (1997) The influence of gas transport and porosity on methane oxidation in soils, *J. Geophys. Res.* **102**, 23,301-23,308.
22. Ball, B.C., Scott, A. and Parker, J.P. (1999) Field N_2O, CO_2 and CH_4 fluxes in relation to tillage, compaction and soil quality, *Soil Tillage Research* (in press).
23. Ehlers, W. and Clauphein, W. (1994) Approaches toward conservation tillage in Germany, in M.R. Carter (ed.), *Conservation Tillage in Temperate Agroecosystems*, CRC Press, Florida, pp. 141-165.
24. Fotyma, M., Lipiec, J. and Naglik, E. (1997) The impact of soil tillage methods on crops yield and soil density, *Bibliotheca Fragmenta Agronomica* TOM **2A**, 227-230.
25. Birkás, M., Nyárai, F. and Szalai, T. (1997) Experiments with direct drilling in Hungary - impacts on soil condition, weed infestation and crop yields, *Bibliotheca Fragmenta Agronomica* TOM **2A**, 79-86.
26. Domzal, H., Glinski, J. and Lipiec, J. (1991) Soil compaction research in Poland, *Soil Tillage Research* **19**, 99-109.
27. Dimitrov, I., Stoynev, K., Borissov, G. and Mitova, T. (1997) Ecological, agrotechnical and economical aspects of soil tillage systems in Bulgaria, *Bibliotheca Fragmenta Agronomica* TOM **2A**, 185-188.
28. Sommer, C. and Zach, M. (1992) Managing traffic-induced soil compaction using conservation tillage, *Soil Tillage Research* **24**, 319-336.
29. Allmaras, R.R., Copeland, S.M., Power, J.F. and Tanaka, D.L. (1994) Conservation tillage systems in the northernmost Central United States, in M.R. Carter (ed.), *Conservation Tillage in Temperate Agroecosystems*, CRC Press, Florida, pp. 255-284.
30. Baker, C.J., Saxton, K.E. and Ritchie, W.R. (1996) *No-tillage seeding: Science and practice*, CAB International, Oxford.
31. Larson, W.E., Eynard, A., Hadas, A. and Lipiec, J. (1994) Control and avoidance of soil compaction in practice, in B.D. Soane and C. van Ouwerkerk (eds.), *Soil Compaction in Crop Production*, Elsevier, Amsterdam, pp. 597-625.
32. Klima, K. (1997) Impact of varying soil cultivation techniques on water erosion volume in particular hydrologic periods, *Bibliotheca Fragmenta Agronomica* TOM **2A**, 185-188.
33. Pabat, J.A. (1997) Water runoff and soil erosion control in crop rotations, *Bibliotheca Fragmenta Agronomica* TOM **2B**, 499-502.

SOIL PHYSICAL QUALITY AND THE EFFECTS OF MANAGEMENT

A.R.DEXTER and E.A.CZYŻ
Institute of Soil Science and Plant Cultivation (IUNG)
ul.Czartoryskich 8, 24-100 Puławy, Poland

1. Introduction

The use to which a soil is to be put determines which physical properties are desirable or undesirable. In this paper we examine aspects of soil physical quality which are important for plant growth and for certain aspects of environmental protection. Some soil physical properties and their consequences are discussed using examples from Poland.

The key physical properties of soil are the soil strength, stability, hydraulic properties and aeration status. All of these are governed by soil structure. However, although these properties are interesting in their own right, it is the effects that they have on soil function which is of primary importance. The proper functioning of the soil as a biological reactor and as an environmental buffer is strongly dependent on its physical properties.

In this paper, the physical properties are considered in turn in relation to soil function, but firstly it is necessary to describe the way that soil is organised in terms of its structure.

2. Introduction to soil structure

Soil structure may be defined as „the geometric arrangement of particles and the pore spaces between them where the term „particles" includes not only the primary mineral particles or grains, but also compound particles". There is another important aspect of structure which is discussed in section 4, and that is the stability of the structure. The key feature of the geometrical definition, above, is it's reference to compound particles. Many of the physical properties which are important for soil function depend strongly on the presence of these compound particles.

Micro-aggregates are perhaps the most important structural feature in most soils. These are compound particles of about 20 - 250 μm diameter. They are usually

M.J. Wilson and B. Maliszewska-Kordybach (eds.),
Soil Quality, Sustainable Agriculture and Environmental Security in Central and Eastern Europe, 153–165.

stabilised by organic matter which acts like glue to hold the primary particles together. Micro-aggregates also act to „physically-protect" some of the soil organic matter against microbial degradation. This happens because the organic matter becomes coated by clay particles and then is not accessible by the micro-organisms which could decompose it [1]. This mechanism accounts for the great age of organic matter in soils which have not been subjected to frequent or intensive tillage [2]. The main importance of micro-aggregates is the pore spaces between them. These pores, which often take the form of micro-cracks, are of the size range which is important for the storage of water for plant use and for the transport of water to plant roots in response to transpirational demand.

Aggregates may be defined as compound particles with sizes in the range 250 μm. - 25 mm. In many soils, the aggregates are stabilized by plant roots and mycelia [1]. The pore spaces between aggregates are important for rapid infiltration of water , for aeration and for providing paths of low mechanical impedance for root growth.

Aggregates larger than about 25mm diameter are called clods. Clods are often formed during tillage of compacted soil or other soil of low friability. Clods are usually undesirable, but may have a role in reducing erosion by wind of tilled soils.

For many soils it is useful to consider the structure to be hierarchical with clods being composed of incipiant aggregates which are composed of micro-aggregates, which are composed of clay domains which are composed of clay mineral particles. Larger primary particles, such as silt or sand grains become incorporated into the hierarchy at the appropriate size scale. The recursive nature of the hierarchy has led to the application of the concept of „self-similarity" to soils. That is, for some purposes they may be considered to look the same when examined on any size scale. Fractal mathematics has been developed to describe this type of situation, and has been applied to the study of soils in an attempt to handle the effects of soil structure over the wide range of size scales which occur in real soils [3]. However, there is no inherent reason why soils should exhibit self-similarity, because different processes are operating on the different size scales in the structural hierarchy. Nevertheless, much can be learned from these attempts to take account of soil structure [3].

3. Soil strength

There are several different measures of soil strength depending on the mode of soil deformation, and these are all important for different aspects of soil use. However, any soil deformation can always be resolved into components of the two basic modes: pure shear and pure compression. In pure shear, the sample changes shape but not volume, whereas with pure compression, the sample changes volume but not shape. Therefore, uniaxial compression (compression in the vertical direction only), which is very common and important in environmental considerations, has components of both pure shear and pure compression.

Soil strength is strongly influenced by water content which must always be measured accurately and taken into consideration. It is usually found that the logarithm of some measure of soil strength varies linearly with gravimetric water content. For example, soil strength may double when the water content decreases by 2.5 $kg(100kg)^{-1}$. The plasticity index, PI, provides a useful measure of how rapidly the strength of freshly-moulded soil varies with water content. This is defined by

$$PI = LL - PL\,, \tag{1}$$

where LL is the liquid limit (or upper Atterberg limit) and PL is the plastic limit (or lower Atterberg limit). These limits, which are determined by simple laboratory tests, are the water contents at which the behaviour of mechanically-disturbed soil changes from liquid to plastic and from plastic to brittle, respectively. It has been found that the shear strength of soil is about 2 kPa at the LL, and about 200 kPa at the PL. Therefore, a small value of PI indicates that strength increases very rapidly with decreasing soil water content, whereas a large value of PI indicates that strength increases only slowly with decreasing water content.

The above information can be combined to produce a simple equation for estimating the shear strength, τ, of disturbed soil at any gravimetric water content, w, in terms of the Atterberg limits:

$$\tau = \exp\left[\, 5.298 - 4.605\frac{(w - PL)}{(LL - PL)}\,\right], \ \text{kPa} \tag{2}$$

Penetrometer resistance provides a useful indication of the mechanical impedance which must be overcome by elongating plant roots. It has been found that penetrometer pressures greater than about 2.5 MPa can reduce root growth to the extent that crop yield is reduced [4]. If the friction and adhesion between the penetrometer probe and the soil are taken into account, this method can provide estimates of the pressures experienced by root tips [5].

Compressive and shear strengths are important for the ability of the soil to support the weight of vehicular and other traffic, and to provide the necessary tractive forces for machinery operations. The steadily increasing weight of agricultural machinery has led to the problem that the compactive stresses from such machines can produce soils which are too strong to be readily penetrated by plant roots as described above.

Soil tillage requires that the soil will crumble into aggregates to provide, for example, a fine bed for the uniform germination of seeds, emergence of seedlings and the early growth of plants. This crumbling propensity of soils is expressed by the friability which can be measured in simple laboratory tests [6]. Soils which have been degraded by loss of organic matter, compaction or increasing sodicity have a low friability which means that they are difficult to cultivate and tend to produce large clods. In contrast, soils which have a high organic matter content, which are not compacted, and which are calcarious and not sodic tend to have high values of friability and are easily cultivated to form fine seed beds. Measurements of friability have been used to identify soils which are suitable for direct-drilling of seeds [7].

4. Soil structural stability

By stability we mean the ability of a soil to maintain its structure in the presence of water. There are two main causes of lack of stability. Firstly, there is clay dispersion in which, for example, the clay particles in sodic soils repel and disperse spontaneously in the presence of pure water [8]. Secondly, there is slaking in which the soil breaks down into micro-aggregates. This can occur either on rapid wetting of the soil during which the soil may fragment as a result of the pressure of entrapped air or as a result of differential swelling, or on prolonged wetting when cementing substances between micro-aggregates may dissolve.

Lack of stability has several important consequences for plant growth, and for water infiltration, and run-off with associated erosion and flood risk. Crusts often form on the surface of unstable soils and these can mechanically impede crop emergence [9,10]. Soils which contain dispersed clay when wet can become cemented by the same clay when they dry, and can have high strength and low friability when dry. Such soils can usually be tilled only over a narrow range of water contents and tend to form large clods. A stable soil is one in which the clay particles remain flocculated in the presence of water in the form of stable compound particles.

5. Hydraulic properties

Soils for plant production need to be able to absorb water rapidly, to store large quantities of water, and to be able to release the water in response to plant demand. An example of a water retention curve is shown in Fig.1.

Storage of water for plant use requires pores which drain at suctions between -100 hPa (which corresponds approximately to „field capacity") and -1.5 MPa (which corresponds approximately to the „wilting point" of many plant species). These two suctions correspond to the drainage of pores of 30 μm to 0.2 μm diameter.

Rapid infiltration requires that there is a network of large diameter pores which exhibit continuity both through the soil volume and to the soil surface. Transport of water at suctions greater than about -200 hPa is rather slow and becomes much slower as the soil dries further. Therefore, in most situations, plant roots must be able to elongate into zones of soil containing previously „untapped" water. That is, access of plants to soil water usually depends more on root growth than on water movement.

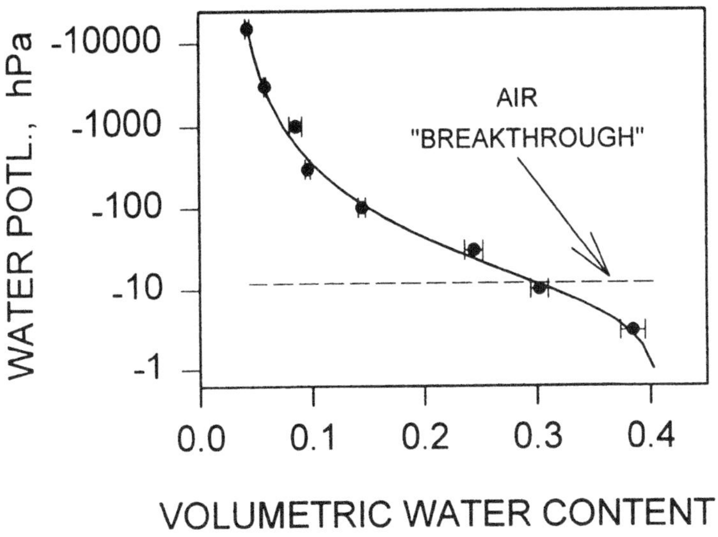

Figure 1. An example of a water retention curve for a Polish sandy soil. The start of air -breakthrough at the point of inflection is shown.

The water retention characteristics and the unsaturated hydraulic conductivity of soils have beeen described by many different equations which have been proposed by many researchers over the years. Currently, the most popular are the equations of van Genuchten [11]. In these, the water retention function is given by

$$\frac{\theta - \theta_r}{\theta_s - \theta_r} = [1 + (\alpha\Psi)^n]^{-m} , \tag{3}$$

and the hydraulic conductivity function is given in terms the same parameters by

$$\frac{K(\Psi)}{K_s} = \frac{[(1 + (\alpha\Psi)^n)^m - (\alpha\Psi)^{n-1}]^2}{(1 + (\alpha\Psi)^n)^{m(l+2)}} , \tag{4}$$

where θ is the water content, θr and θ_s are the residual and saturation water contents, ψ is the modulus of the water potential (in hPa), and α, n and m are adjustable parameters. The parameter l usually takes a value of 0.5. $K(\psi)$ and K_s are the hydraulic conductivities at potential ψ and at saturation, respectively.

The enormous effects of sodicity and salinity on the hydraulic conductivity of soil have been discussed by Quirk [8]. In order to maintain permeability it is necessary to keep the soil clay flocculated and this requires that the electrolyte concentration be above a certain threshold level. This is well-illustrated by the work of Reeve and Bower [12], who reclaimed a sodic soil in California. In order to keep the soil permeable, they mixed progressively-decreasing proportions of sea water with river water and in this way were able to achieve reclamation in only 12 days.

Compaction too can cause large reductions in the hydraulic conductivity of soil [13]. Compaction tends to occur mainly at the expense of the largest pores present in the soil. Since hydraulic conductivity decreases rapidly with decreasing pore diameter, the removal of the largest pores has a large effect on the hydraulic conductivity. The effects of mechanical stress on saturated and unsaturated hydraulic conductivity have been studied in terms of the resulting changes in the parameters of the van Genuchten equations [14]. Compaction, therefore, can reduce infiltration of water and can increase run-off with the consequences of increased soil erosion and risk of flooding.

6. Aeration

Plant roots and most soil micro-organisms consume oxygen for their metabolic activity and respire carbon dioxide. With many crops and soils, the amounts of oxygen per unit area of soil required by the crop roots and the micro-organisms are similar. If the oxygen supply at the soil surface is cut-off, then the soil oxygen will become depleted after a few days, and plant growth will cease.

Good aeration status is therefore required for plant growth and for the activities of most important microbes which are involved in mineralization of organic matter and in the decomposition of organic pollutants.

A saturated soil, of course, contains no air-filled pores. As the potential of the pore water decreases, the largest pores drain and become air-filled first, as described by the Young-Laplace equation. Initially, only large pores which connect to the soil or sample surface drain. When a certain water potential is reached, „air breakthrough" occurs and a continuous network of air-filled pores permeates the soil. Air breakthrough occurs at the point of inflection (i.e. the point where the curvature is zero) of the water characteristic curve which may be described by the van Genuchten equation, above. The point of inflection is shown in Fig.1 and occurs at a volumetric water content of

$$\theta = (\theta_s - \theta_r)\left(1+\frac{n-1}{1+nm}\right)^{-m} + \theta_r \qquad (5)$$

or, at a water potential of

$$\Psi = \frac{1}{\alpha} \left(\frac{n-1}{1+nm}\right)^{\frac{1}{n}} \tag{6}$$

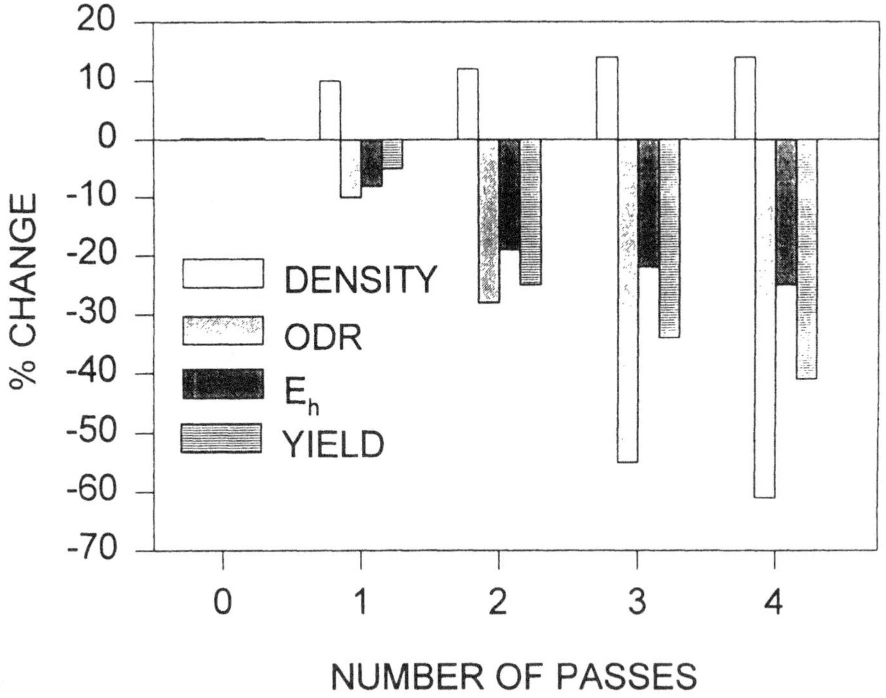

Figure 2. Effects of compaction of a Polish clay soil, as measured by the number of passes of tractor wheels, on the soil bulk density, oxygen diffusion rate, redox potential and on the yield of spring barley.

Adequate aeration is usually achieved when the soil has at least 10-15% air-filled porosity at field capacity [15,16]. That is, this proportion of the soil volume must comprise continuous pores of diameter greater than 30 μm. However, soils differ and the requirements of different organisms differ. The aeration status of soils is usually better described in terms of oxygen diffusion rates and redox potentials [16,17]. Compaction of soil by tractors and other agricultural vehicles can reduce aeration as shown in Fig.2. In non-swelling soils, the water-filled and air-filled pore spaces are complementary, but in swelling/shrinking soils the situation is more complicated.

7. Soil management factors

7.1 GENERAL CONSIDERATIONS

In order to have soil with a high physical quality, as described above, it is necessary for the soil to have two main structural attributes:

Firstly, the soil must be stable. That is,the clay particles must have a tendency to flocculate and not to disperse. Flocculation can be achieved by keeping the soils calcarious instead of sodic and by maintaining a sufficiently high electrolyte concentration in the pore water. Also, a high content of organic matter is very important in maintaining the stability of micro-aggregates and other compound particles.

Secondly, the soil must have a high total porosity which must be distributed so that there is a wide distribution of pore sizes. This gives the soil the required ability to absorb, store and release water as required. Such a soil will also be readily penetrable by plant roots and will have good aeration status. In clay soils, an important part of the pore network consists of micro-cracks which are important not only for water infiltration and storage, and for aeration but which also make the soil friable and easy to cultivate over a wide range of water contents.

The above combination of characteristics usually cannot be achieved with any single change in soil management practice, but usually a whole new system of management needs to be introduced:

7.2 PHYSICAL - CHEMICAL REQUIREMENTS

Stability at the level of clay particles depends on the physical chemistry of the surfaces of the clay particles, and on the electroytic composition of the pore water between them.

A suitable level of divalent cations can be maintained by the addition of calcium ions to the soil in the form of quick lime (CaO), lime ($Ca(OH)_2$), gypsum ($CaSO_4$) or crushed limestone or chalk ($CaCO_3$). Both CaO and $Ca(OH)_2$ increase the pH markedly, and so are useful only when the soil is initially strongly acidic. $CaCO_3$ dissolves only very slowly, and tends to be removed from the soil profile by leaching as quickly as it dissolves. This leaves $CaSO_4$ as the most obvious choice of calcium compound for soil physical amelioration. Application rates need to be quite high (for example, 10 t ha^{-1}), and depending on the amount of leaching by „excess" rainfall, applications may have to be repeated perhaps once every 4 years. However, $CaSO_4$ is cheap and plentiful in many areas as it is a by-product from the process of de-acidification of exhaust gases from coal-burning power stations.

7.3 ORGANIC MATTER REQUIREMENTS

Stability at the level of micro-aggregates and aggregates requires organic matter and the biological activity with which it is usually associated. For a soil in equilibrium, the annual input of organic matter is equal to the annual output. The organic matter content may be increased to a new, higher equilibrium content by increasing the inputs, decreasing the outputs, or both. Inputs may be increased by retaining crop residues instead of burning them or removing them, and by having plant cover and hence photosythesis for as much of the year as possible. It is instructive consider the following simple equation:

$$F = T + R + E, \tag{7}$$

where, F is the amount of carbon fixed by photosynthesis, and T, R and E are the amounts of carbon being partitioned into plant tops, roots and root exudates into soil, respectively. Amounts of exudation can be quite large, and for the purpose of this exercise, we shall assume that T=0.45F, R=0.15F, and E=0.40F. For a cereal crop, we shall assume that the proportions of T partitioned into grain, G, and straw + leaves, S, are the same (i.e. the harvest index, HI = 0.5) and therefore that

$$G = S = 0.225F. \tag{8}$$

In U.K. conditions, when a cereal crop is harvested and the straw is removed, only about 78 percent is removed and the remaining 22 percent of the S remains as stubble , ST [18]. Therefore, ST = 0.05F, and the amount removed is 0.175F. Of course, the grain,G, is usually removed.

Using this information , we can make a simple comparison of the amounts of carbon going into the soil with and without removal of crop residues:

with residues removed: $\quad \text{carbon} = R + E + ST = 0.60F$, and (9)

with residues retained: $\quad \text{carbon} = R + E + S = 0.78F.$ (10)

Therefore, the retention of crop residues and their subsequent incorporation into the soil can potentially increase carbon inputs by around 30 percent. The above analysis is a considerable simplification as a number of additional factors influence the partitioning of the photosythesized carbon. Nevertheless, it gives us some idea of what can potentially be achieved by crop residue retention. In spite of this, an 8-year trial comparing the effects of straw incorporation and straw burning on five different soils showed no significant effects on organic matter content [19,20]. Part of the explanation, at least, may be the effects of the tillage used for incorporation of the straw on the output of carbon from the soil. This is discussed below.

Systems such as „no till", in which crop residues are not incorporated, but are retained on the soil surface avoid any adverse effects of tillage. Part of the residues may then subsequently become incorporated by earthworms or other soil meso- and macro-fauna [4].

7.4 TILLAGE FACTORS

Tillage can have beneficial or adverse effects on soil physical quality: the key factors here being the water content at which tillage is done and the mechanical energy input to the soil. For example, there is an optimum water content for tillage at which the finest seedbed is produced. In many soils, this optimum water content is at about 0.9 PL, where PL is the plastic limit of the soil as discussed in section 3. At this water content, soil is most friable [6].

When soil is tilled at water contents greater than the optimum, the soil does not crumble, but is compressed and sheared into large clods which have a plastic consistency. In the process of doing this, clay is mobilized within the soil matrix. This reduces the stability of the soil in water [21,22] and causes the clods to become very hard and non-friable when they dry.

High intensity tillage, also removes the physical protection from a proportion of the soil organic matter. It is then accessible to microbes, and decomposition processes can proceed. This is evidenced by the increased respiration of CO_2 from tilled soils [23,24]. This increased respiration following tillage can be a significant proportion of the annual input of organic matter from conventional cropping, and probably is a major contributory factor in the decline of organic matter content in arable soils. It has also been estimated that the increased respiration of soil following tillage contributes significantly to global emissions of CO_2, and hence to the greenhouse effect.

Low intensity tillage causes less disturbance to the soil, and results in lower losses of organic carbon.

In these days of herbicides, one of the principal reasons for tillage is to attempt to undo the damage caused by wheel traffic. Perhaps it is better to concentrate our efforts on the cause of the problem rather than on cures.

7.5 WHEEL TRAFFIC AND COMPACTION

The passage of heavily-loaded wheels can have profound effects on soil physical properties and on the functioning of the soil as shown in Fig.2. This „wheeling" is the operation which we perform the most frequently on soil. Measurements have shown that on average, the ratio of wheel track area per annum to field area is 3.5-5 in Sweden [25], up to 3 on experimental plots in The Netherlands [26] and 1.65 in the semi-arid climate of South Australia [27].

Håkansson [25] defined the term „traffic intensity" in an attempt to integrate the effects of the different field traffic over the year. The traffic intensity is obtained by multiplying the mass of each vehicle (t) by the distance that it travels (km) per unit area of field (ha) and by summing this for all the passes of all the vehicles over the year. For typical Swedish cereal farms, he found intensity values in the range 110-230 t km ha^{-1}, whereas in South Australia, Holloway and Dexter [27] obtained a mean value from two cereal farms of 63 t km ha^{-1}.

Soil compaction may be defined as the reduction of the volume of a given mass of soil in response to applied stresses. These stresses may be externally-applied, such as by wheel traffic, or may be internally-applied as „effective stresses" by the soil water menisci and the water potential as soil dries [28,29].

The levels of stress which occur in soil below different wheels with different tyre inflation pressures and with different loads in different soils at different water contents have been studied in detail [30]. There is not space to discuss all the results from these experiments. However, it is possible to simplify some of the conclusions as follows:

(i) compaction in the topsoil depends mainly on the inflation pressure of the tyres,
(ii) compaction in the sub-soil depends mainly on the total wheel load,
(iii) wheel slip increases the amount of shearing within the soil which increases the amount of compaction.

The first two conclusions, above, are consistent with the equations of Newmark [31], and Söhne [32]. The third is consistent with Critical State Soil Mechanics [33].

An important concept in relation to soil compaction in the field is the „pre-compaction stress"[30]. Since soil compaction is an essentially irreversible process, the soil „remembers" the maximum level of stress, or pre-compaction stress, to which it has been previously exposed. Further compaction will not occur until the pre-compaction stress is exceeded. However, because soil strength is very sensitive to water content, as shown in Eqn (2), the pre-compaction stress at one water content is not the same as the pre-compaction stress at another water content. Therefore, in order to standardize the measurements, they should always be made at a given water matric potential.

8. Optimum soil physical quality

From the above discussion, it is possible to summarize the physical properties which are consistent with the concept of a high level of soil physical quality. Unfortunately, no single property can be used as an index of soil physical quality and a combination of properties usually needs to be taken into consideration.

In spite of this, the required combination of physical properties can often be achieved by control of the physical chemistry of the clay surfaces (for example, by addition of $CaSO_4$), by maintaining a high level of organic matter (by retaining crop residues, by keeping plant cover for a greater proportion of the year, and by using low-energy tillage techniques), by not tilling the soil when it is wetter than the Plastic Limit (PL), and by reducing compaction (by reducing the factors which contribute to „traffic intensity", or by adopting controlled- or zero-traffic techniques) [34,35].

None of these practices is mutually incompatible with any of the others, and so there is, in principle, no technical barrier to the adoption of these techniques. Even the long-term economics are probably moderately attractive in view of the associated lower levels of machinery and energy usage, and the greater crop yields which are often obtained.

Perhaps the greatest barriers to the adoption of these management techniques are that:

(a) many farmers use crop residues as animal feed which reduces the returns of organic matter to the soil,

(b) conservatism on the part of the farmers, and

(c) the high initial capital investment required to introduce a new system.

Yet, we must realise that there are many soils in Central and Eastern Europe which are physically degraded in various ways and that improvements in their physical quality are required in order to restore or maintain their proper functioning not only for agricultural production but also as key components of the environment.

9. References

1. Oades, J.M. (1984) Soil organic matter and structural stability: mechanisms and implications for management, *Plant and Soil* **76**, 319-337.
2. Jenkinson, D.S. and Rayner, J.H. (1977) The turnover of soil organic matter in some of the Rothamsted classical experiments, *Soil Sci.* **123**, 298-305.
3. Rieu, M. and Sposito, G. (1991) Fractal fragmentation, soil porosity, and soil water properties, *Soil Sci. Soc.Amer.J.* **55**, 1231-1244.
4. Whalley, W.R., Dumitru, E. and Dexter, A.R. (1995) Biological effects of soil compaction, *Soil and Tillage Research* **35**, 53-68.
5. Bengough, A.G., Mullins, C.E. and Wilson, G. (1997) Estimating soil frictional resistance to metal probes and its relevance to the penetration of soil by roots, *European J. Soil Sci.***48** (4), 603-612.
6. Watts, C.W. and Dexter, A.R. (1998).Soil friability: theory, measurement and the effects of management and organic matter content, *European J.Soil Science* **49** (1), 73-84.
7. Macks, S.P., Murphy, B.W., Cresswell, H.P. and Koen, T.D. (1996) Soil friability in relation to management history and suitability for direct drilling, *Australian J. Soil Research* **34**, 343-360.
8. Quirk, J.P. (1986) Soil permeability in relation to sodicity and salinity, *Phil.Trans. Roy.Soc. Lond.* A**316**, 297-317.
9. Le Bissonais, Y. (1990) Experimental study and modelling of soil surface crusting processes, *Catena* (supplement) **17**, 13-28.
10. Sumner, M.E. and Stewart, B.A. (1992) *Soil Crusting: Chemical and Physical Processes*, CRC Press, Boca Raton.
11. van Genuchten, M.Th. (1980) A closed-form equation for predicting the hydraulic conductivity of unsaturated soils, *Soil Sci. Soc.Amer. J.* **44**, 892-898.

12. Reeve, R.C. and Bower, C.A. (1960) Use of high-salt waters as a flocculant and source of divalent cations for reclaiming sodic soils, *Soil Science* **90**, 139-144.
13. Horton, R., Ankeny, M.D. and Allmaras, R.R. (1994) Effects of compaction on soil hydraulic properties, in B.D.Soane and C.van Ouwerkerk (eds.), *Soil Compaction in Crop Production*, Elsevier,Amsterdam, pp. 141-165.
14. Horn, R., Baumgartel, T., Gräsle, W. and Richards, B.G. (1995) Stress induced changes of hydraulic properties in soils, in E.E.Alonso and P.Delage (eds.), *Unsaturated Soils/Sols Non Saturés*, Balkema Publ., Rotterdam, pp. 123-127.
15. Dexter, A.R. (1988) Advances in characterization of soil structure, *Soil and Tillage Research* **11**, 199-238.
16. Gliński, J. and Stępniewski, W. (1985) *Soil Aeration and its Role for Plants*, CRC Press, Boca Raton.
17. Czyż, E. and Tomaszewska, J. (1993) Changes of aeration conditions and the yield of suger beet on sandy soil of different density, *Polish J. Soil Sci.* **XXVI**, 1-9.
18. Addiscott, T.M. and Dexter, A.R. (1994) Losses of chemicals from soils related to tillage and crop residue management, *Soil and TillageResearch* **30**, 125-168.
19. Johnson, P.A. and Chambers, B.J. (1996) Effects of husbandry on soil organic matter, *Soil Use and Management* **12**, 102-103.
20. Nicholson, F.A., Chambers, B.J., Mills, A.R. and Strachan, P.J. (1997) Effects of repeated straw incorporation on crop fertiliser nitrogen requirements, soil mineral nitrogen and nitrate leaching losses, *Soil Use and Management* **13**, 136-142.
21. Watts, C.W., Dexter, A.R., Dumitru, E. and Arvidsson, J. (1996) An assessment of the vulnerability of soil structure to destabilisation during tillage. Part I - A laboratory test, *Soil and Tillage Research* **37**, 161-174.
22. Watts, C.W., Dexter, A.R. and Longstaff, D.J. (1996) An assessment of the vulnerability of soil structure to destabilisation during tillage. Part II - Field trials, *Soil and Tillage Research* **37**, 175-190.
23. Reicosky, D.C., Dugas, W.A. and Torbert, H.A. (1997) Tillage-induced soil carbon dioxide loss from different cropping systems, *Soil and Tillage Research* **41**, 105-118.
24. Reicosky, D.C. and Lindstrom, M.J. (1995) Impact of fall tillage on short-term carbon dioxide flux, in R. Lal, J. Kimble, E. Levine and B.A.Stewart (eds.), *Soils and Global Change*, CRC Lewis Publishers, Boca Raton, pp. 177-187.
25. Håkansson, I. (1985) Soil compaction research in Sweden. Proc. Conf. Soil Compaction, Columbus, Ohio.
26. Lumkes, L.M. (1984) Traffic intensity, in *Experiences with three tillage systems on a marine loam soil. II: 1976-1979.* Agric. Res. Reports No.925, Pudoc, Wageningen.
27. Holloway, R.E. and Dexter, A.R. (1990) Traffic intensity on arable land on the Eyre Peninsula of South Australia, *J. Terramechanics* **27**(3), 247-259.
28. Mullins, C.E. and Panayiotopoulos, K.P. (1984) The strength of unsaturated mixtures of sand and kaolin and the concept of effective stress, *J.Soil Science* **35**, 459-468.
29. Snyder, V.A. and Miller, R.D. (1985) Tensile strength of unsaturated soils, *Soil Sci.Soc.Amer.J.* **49**, 58-65.
30. Horn, R. (1988) Compressibility of arable land, *Catena* (supplement) **11**, 53-71.
31. Newmark, N.M. (1942) *Influence Charts for the Computation of Stress in Elastic Foundations*, Engng. Exp. Station Bull. Series 338, Univ. Illinois Bull **40**(12), Urbana.
32. Soehne, W. (1958) Fundamentals of pressure distribution and soil compaction under tractor tyres, *Agric.Engng.***39**, 276-281, 290.
33. Kirby, J.M. (1989) Measurements of the yield surfaces and critical state of some unsaturated agricultural soils, *J.Soil Sci.* **40**, 167-182.
34. Chamen, W.C.T. and Longstaff, D.J. (1995) Traffic and tillage effects on soil conditions and crop growth on a swelling clay soil, *Soil Use and Management* **11**, 168-176.
35. Chamen, W.C.T., Watts, C.W., Leede, P.R. and Longstaff, D.J. (1992) Assessment of a wide span vehicle (gantry) and soil and cereal crop responses to its use in a zero traffic regime, *Soil and Tillage Research* **24**, 359-380.

HERBICIDE MANAGEMENT TO MAINTAIN ENVIRONMENTAL QUALITY

Lessons to be Learned from North American Herbicide Management Practices

SHARON A. CLAY
Professor, Plant Science Dept.
South Dakota State University, Brookings, SD 57007 USA

1. Introduction

The introduction of organic-based herbicides in the late 1940's revolutionized weed management with low rate chemical applications that selectively control weeds within crops. After 50 years of using herbicides as the major weed management technique in the Western World, we can reflect on benefits and costs associated with herbicide applications. Benefits are easy to list. Herbicide applications reduce labour requirements, mechanical cultivation (thereby reducing soil erosion and compaction potential), and increase crop productivity, harvest efficiency, and grazing capacity of pastures. However, in spite of numerous positive benefits associated with herbicides, there also are problems.

Two major problems with herbicides that are frequently cited are: 1) environmental concerns (ground and surface water quality, and offsite movement to sensitive areas); and 2) the development of herbicide resistant weeds. This overview will address these two concerns and give some practical management solutions that have been implemented to alleviate these problems.

2. Environmental Quality

The quest for greater crop yields to feed expanding populations and drive export economies has led to several environmental concerns including deforestation, soil erosion (due to the increase in marginal land use for crop production), loss of biodiversity, and herbicide contamination of offsite or nontarget areas [1]. In the United States, the use of soluble pesticides in the Mississippi River Basin is estimated at over 100,000 metric tons annually [2]. The ten most used herbicides by weight (based on 1992 survey results) in the Mississippi River Basin region are presented in Table 1 [2].

M.J. Wilson and B. Maliszewska-Kordybach (eds.),
Soil Quality, Sustainable Agriculture and Environmental Security in Central and Eastern Europe**, 167–176.**

Table 1. Top ten herbicides used in the Mississippi River Basin of the United States in 1992. (adapted from data from [2]).

Herbicide	Total amount used (1,000's kg)
Atrazine	32,000
Metolachlor[a]	26,000
Alachlor	21,000
2,4-D	21,000
Cyanazine	14,000
Trifluralin	11,000
Pendimethalin[b]	8,000
Glyphosate	5,400
EPTC	5,000
Butylate[c]	3,600

[a]Metolachlor (2-chloro-N-(2-ethyl-6-methylphenyl)-N-(2-methoxy-1-methylethyl)acetamide)
[b]Pendamethalin (N-(1-ethylpropyl)3,4-dimethyl-2,6-dinitrobenzeamine)
[c]Butylate (S-ethyl diisobutyethiocarbamate)

Eight out these 10 herbicides on this list are applied directly to soil [the two exceptions are 2,4-D (2,4-dichlorophenoxyacetic acid) and glyphosate (N-(phosphono-methyl) glycine)]. Soil-applied herbicides give excellent control of weeds germinating at the time of application, provide residual chemical for later germinating weeds thus eliminating one or several late flushes of weeds, and reduce the need for early cultivation. However, it is estimated that less than 10% of a soil-applied herbicide is used for weed control, leaving 90% for other environmental mischief. The rates of applied herbicide must be high enough to counteract physical and environmental processes such as sorption to soil particles, uptake by crop, microbial degradation, volatilization, photo-decomposition, leaching losses, and soil dilution [3]. In addition, several soil-applied herbicides [including atrazine (2-chloro-4-ethylamino-6-isopropylamino-s-triazine) and trifluralin (alpha, alpha, alpha-trifluoro-2,6-dinitro-N,N-dipropyl-p-toluidine)] are considered persistent or long-lived, i.e., herbicide residues in the soil from one season often injure sensitive crops during the following season. The longer the persistence in the environment, the greater the potential for problems. These facts indicate why soil-applied herbicides are cited as having a high potential to cause environmental quality problems. The complex interactions of the pesticide in the environment and their rates of reaction are beyond the scope of this paper. However, "Pesticides in the soil environment: Processes, impacts, and modeling" [4] is an excellent resource.

In watershed areas, several key factors contribute to the behavior of the herbicide. These factors include the total amount of hectarage treated, amount of product applied in the area, management input including formulation, frequency of application, application methodology, water input and outflow, and watershed system properties (soils, topography, drainage) [3]. A United States Geological Survey (USGS) study examined pesticide contamination of streams and tributaries of the Mississippi River throughout the year. Atrazine was detected in 100% of the collected samples [5]. The detection limit for atrazine in this study was 0.002 ug/L or 2 parts per trillion (2 X 10^{-12}g/L), of which about 50% of the detections were near the detection level. The high use level of atrazine in the Mississippi basin and the low detection levels contributed to atrazine being found in all the samples collected for that study. Atrazine has also been detected at these low concentrations in rain [6], groundwater [7, 8], and in samples of windblown sediments deposited on snow taken at least 6 months after application [9]. Herbicide metabolites also have been detected in groundwater [10], surface water [11], and other offsite areas.

Offsite movement of even of very low concentrations of anthropomorphic chemicals has raised societal concern and has led to herbicide label changes and regulation. These changes have included reducing herbicide use rates, regulating timing of application, banning some herbicides [i.e., US ban on cyanazine (2-(4-chloro-6-ethylamino-s-triazin-2-ylamino-2-methylpropionitrile) by the year 2000], and creating areas where some herbicides cannot be applied (groundwater protection zones).

The loss of herbicides due to regulatory restrictions can lead to further problems including reduced crop production and monetary loss, increased production costs, and greater environmental problems if more marginal land is used to make up for reduced production. For example, Bridges [12] estimated the monetary loss in crop production due to weeds in the U.S. and Canada would be over $19 billion if herbicides were not available, compared to the estimated $4.1 billion loss with the use of herbicides. These losses not only reflect the cost of increased management (cultivation, hand-labor, etc.) but also a true loss in crop production. Therefore, to maintain the use of herbicides in production agriculture, agronomists must work in concert with herbicide users and manufacturers to find methods to reduce or eliminate herbicide movement to non-target areas.

3. Management Practices to Reduce Herbicide Environmental Impacts

As soil-herbicide interactions are understood and the major controlling factors are defined, scientists can begin to manipulate agricultural management systems to limit or eliminate environmental risks from the pesticide. A set of Best Management Practices (BMP) have been proposed to reduce or eliminate herbicide contamination of surface water (through runoff) and ground water (through leaching). These BMPs are being adopted voluntarily by concerned growers. In some cases, regulations have been implemented that incorporate these practices.

1. Never add chemicals or clean tanks near the water supply source. Contamination of water supplies in agricultural areas has been linked to back siphoning and chemical spills near wells. Direct injection sprayers currently are marketed that add chemical inline during spraying to eliminate much of the handling and all of the "extra" (overage) solution.

2. Soil-applied herbicides can be incorporated shallowly (from 1 to 5 cm) without reducing efficacy. Incorporation takes the herbicide off the surface soil where it

can be moved by water (runoff) or wind erosion [13]. Surface reductions in herbicide with shallow incorporation can lower the offsite movement by about 70% [14].

3. Apply herbicides when there is a low probability for runoff and leaching events to occur. In South Dakota, rainfall amounts of single storms that occur in late May and early June can exceed 15 cm. Application prior to this will allow the chemical to become bound to soil constituents and for some metabolism to occur thus reducing the amount present of off-site movement. The objective is to keep the herbicide in the seed germination zone (top 10-15 cm) and certainly in the root zone of the crops and weeds. Herbicides are usually sorbed more in surface soil than in the subsurface [15], resulting in less movement with leaching events. Once out of the root zone, downward movement increases and degradation slows. In the root zone, degradation is much faster than in subsurface areas due to higher microbial populations [16] and microbial activity [17, 18] compared to lower depths. Applications to areas above shallow groundwater aquifers should be avoided also. Muller-Wegener et al.[19] reported that more herbicide leached into unconfined aquifers located 3 m below the surface compared to aquifers located at 3.6 m below the soil surface.

4. Apply herbicides in bands to only the crop row. If the area treated is only 50% of the entire field, this will limit the amount of herbicide present for offsite movement. Leaching losses were reduced by 43% when herbicides were band- vs. broadcast-applied [20] and runoff losses can be reduced by about 70% [21]. Other application techniques include reducing broadcast rates from one-half to one-quarter of the recommended rates thus reducing total loading. Researchers throughout the U.S. have reported that weed control is usually satisfactory at these lower rates [22]. Banding over rows and using cultivation in inter-row areas also will reduce loading by 50% or more (depending on the band vs row width). This technique, along with applying the herbicide to ridges in ridge-tillage systems, has been shown to reduce herbicide movement in soil [23].

5. Apply post-emergence herbicides. Post-emergence applications allow growers to choose herbicides based on known weed infestations and limits soil-herbicide interaction. Problems may occur due to offsite movement with drift to sensitive crops, runoff with rainfall, or leaching into soil if rainfall occurs soon after application.

6. Plant filter strips at the edges of fields. This will slow water movement, allow soil particles to settle out of the runoff, and create an area high in organic matter to sorb herbicides from the water. Herbicide concentration in runoff has been reduced up to 64% as a result in runoff water flowing through filter strips [24].

7. Allow an adequate buffer between application areas and streams. Label restrictions that require unsprayed areas of 20 m from field to stream with slightly soluble herbicides are often common.

8. Use the lowest rate that will give adequate control and/or apply split herbicide applications. Sprayer calibration should not be overlooked. Surveys in several states have reported that over 50% of the sprayers tested were not within 20% of the desired (or intended) application rate. Over application occurred in a number of sprayers due to worn nozzles and resulted in greater application costs due in part to more herbicide usage. Plugged and or mismatched nozzles also were common resulting in uneven spray patterns and patchy weed control.

9. Do not apply soil-applied herbicides late in the season (post-harvest applications). The herbicides applied late in the season are applied to cool soil that has slowed microbial degradation rates. This allows almost the entire amount of applied herbicide to overwinter at or near the soil surface. There is a high potential for herbicides to move via wind or water erosion to surface waters or leach out of the root zone with snowmelt or early spring rains [25].

4. Herbicide Sorption in Soil

In order to understand why some of these practices work, principles of herbicide movement in and from soils must be understood. Herbicide sorption or retention to soil is one of the most important mechanisms that influence herbicide fate in the environment [26]. Herbicide sorption will influence bioavailability, leaching potential, and the rate of transformation to metabolites.

Herbicide sorption to soil is controlled by physical and chemical properties of both the herbicide and soil. Cationic herbicides, those with permanent positive charge such as paraquat (1:1-dimethyl-4,4'-bipyridinium dichloride), are tightly sorbed and (usually) irreversibly bound to soil. They exhibit no soil activity and are not a concern for surface and groundwater contamination, although often implicated in drift injury. Anionic herbicides [ie salts of acids including 2,4-D and picloram (4-amino-3,5,6-trichloro-picolinic acid)] are not tightly bound to soil due to their negative charge and there have been infrequent reports of their presence in surface and groundwater samples. Two reasons that curtail their presence in offsite areas are that 1) most of the anionic herbicides are applied as foliar application, thus limiting their contact with the soil; and 2) due to their limited sorption, they are more available in the soil matrix for microbial degradation (although this is not the case with picloram). Neutral [alachlor (2-chloro-2',6'-diethyl-N-(methoxymethyl)acetanilide)] and slightly basic herbicides (triazines) are sorbed to soil to a greater extent than anionic herbicides. These herbicides tend to be more tightly sorbed to than the anionic type herbicides but less sorbed than the cationic herbicides.

Soil texture (sand, silt, and clay content), soil pH, soil organic matter (amount and type) influence the neutral and slightly basic herbicide groups. There are sorption differences between sandy soil types and soil horizons (Table 2) but all have a high potential for herbicides to leach through to subsurface areas. The use of long-lived persistent herbicides such as atrazine should be carefully managed on these soils. Rainfall and timing of irrigation are extremely important to limit leaching from surface soil. Surface runoff is also affected by time of application and rainfall.

Soil management with tillage and soil fertilization affects soil chemical and physical properties that in turn influence herbicide sorption, leaching, and degradation patterns. Systems with well-developed macropores, in general, have been shown to have greater leaching losses than where the soil has been tilled and few continuous pores remain [29]. Another example of increased leaching loss is atrazine-ammonia fertilizer interaction [30, 31]. The application of ammonia fertilizer increases soil pH to about 9 for extended periods of time. Atrazine sorption to soil, which decreases as pH increases, is less sorbed

Table 2. Location and soil characteristics of four midwestern soils. Kd is the partition coefficient for atrazine determined by batch adsorption techniques with mol g-1. The higher the Kd number the more atrazine is sorbed to soil.

Site	Soil type	Horizon	%sand	%clay	Kd
Minnesota[a]	Zimmerman fine sand	A	93.4	1.1	1.09
		B	95.8	0.9	1.09
		C	98.2	1.2	0.37
Wisconsin[b]	Sparta sand	A	95.7	0.9	0.59
		B	97.5	0.9	0.26
		C	99.7	0.2	0.02
North Dakota	Hecla sandy loam	A	69.2	13.2	4.66
South Dakota[c]	Brandt silt loam	A	13.8	26.8	5.80
		B	15.3	20.7	0.90
		C	85.8	5.3	0.10

[a]Rochette and Koskinen, 1996 [27]
[b]Seybold et al., 1994 [28]
[c]Clay et al., 1997 [16]

to soil and movement with water to deeper depths has been reported. In addition, the slot created by the ammonia injection equipment, if not closed, acts as a direct conduit to the ammonia band, increasing the possibility of atrazine interacting with the ammonia treated soil [23].

Organic matter (amount and distribution) is influenced by tillage. More organic matter at the soil surface in no-till or reduced tillage systems has been shown to decrease herbicide bioavailability by increasing sorption or microbial herbicide degradation. Manure applications appear to have a similar effect on herbicide fate. While this may be useful to lessen the herbicides' impact on the environment by reducing leaching and run-off losses, weed control is often reduced, requiring more herbicide for controlling the weed or shifting to a post-emergence herbicide strategy where soil properties do not impact herbicide efficacy.

Knowledge of movement, degradation, and sorption potential of a herbicide can help agricultural scientists design management practices that will result in protection of offsite areas from herbicide contamination. However, these techniques must be accepted and implemented by producers to do any good. Therefore, demonstration and outreach activities should be planned to help the grower become familiar with and use suitable weed-herbicide management strategies for their areas.

5. Herbicide Resistant Weeds

Specific weeds have become resistant to older chemicals such as atrazine, 2,4-D and others and to new chemistries such as immidiaziliones [imazethapyr (2-[4,5-dihydro-4-methyl-4-(1-methyllethyl-5-oxo-1H-imidazol-2-yl] -5-ethyl-3- pyridinecarboxylic acid)] and sulfonylureas [chlorsulfuron (2-chloro-N-((4-methyoxy-6-methyl-1,3,5-triazin-2-yl) aminocarbonyl)benzenesulfonamide)] (ALS-inhibitors) and grass herbicides such as diclofop (methyl, 2-[4-(2,4-dichlorphenoxy)phenoxy]propanoate) and difenzoquat (1,2-dimethyl-3,5-diphenyl-1H-pyrazolium)[32]. It was thought that the "new" generation of herbicides [those used at very low rates (g/ha) and having a single target enzyme] would provide post-emergence herbicides that would be very selective and innocuous in the environment. In some cases the promised control has been realized but resistant kochia (*Kochia scoparia*), ryegrass (*Lolium* sp.), common cocklebur (*Xanthium pennsylvanicum*), velvetleaf (*Abutilon theophrasti*), common lambsquarters (*Chenopodium album*) and other weeds have turned the panacea into a nightmare. Not only that, but carryover of even the low herbicide rates may be a problem on sensitive crops. Immidiazilone resistant maize (*Zea mays*) will allow the use of these chemicals on both the maize and soybean (*Glycine max*) crop.

The introduction of glyphosate-resistant soybean and (in the not too distant future) maize may have the same heightened effect on the development of resistant weeds or shifts to tolerant species. Shifts to tolerant weed species already have been reported in the midwestern U.S. Common and tall waterhemp (*Amaranthus rudis* and *A. tuberculatus*, respectively) have a high tolerance to glyphosate and peak emergence is later than most common agronomic weeds. The lack of residual activity allows these species to thrive and, with the potential to produce over one million seeds per plant, become a long-term agronomic problem [33].

6. Best Management Practices to Reduce Development of Herbicide Resistant Weed Populations

Best management practices have been suggested to reduce the possibility of herbicide resistant weeds developing, as well as treatment of areas where herbicide resistant weeds are already present. Long-term weed management strategies should be implemented prior to the outbreak of herbicide-resistant weed populations. It becomes essential to change farm management practices once herbicide-resistant weeds become problem and limit crop production. Changes should include the use of nonspecific control techniques such as crop rotation (perhaps into forage production, green manuring, or fallow), and cultivation to break the herbicide cycle [34]. Other practices include:

1. Apply different herbicide chemistries with different modes-of-action frequently. The yearly use of herbicides that target the same target site in the plant can result in resistant plant development very quickly. For example, immidiazilone and sulfonyl urea herbicides have the same mode of action in the plant (ALS inhibitors). Only one enzyme is affected in a plant and one of six different mutations at the site can render the site

resistant. It has been documented that ALS-resistance can occur with only two consecutive applications of ALS-herbicides [35].
2. Use tank mixes that contain chemicals with multiple modes of action. This tactic helps control weeds that are resistant to one or more modes of action [35].
3. Have a good long-term rotation to break the weed cycle. Growing and harvesting alfalfa (*Medicago sativa*) for two years has been shown to reduce foxtail (*Setaria* sp.) density, competition [36], and seed production [37] in the following maize crop.
4. Include tillage and mechanical weed control to lessen the dependence on herbicides.
5. If herbicide resistant plants are suspected, get these identified as soon as possible. Things to look for include having problems with weeds that used to be (easily) controlled with the herbicide. Resistant weed problems in the field can develop in as little as 2 cycles of herbicide use (especially ALS resistant kochia). Atrazine-resistant broadleaf weeds have taken longer to develop but can still be a significant problem. Once identified, control measures should be implemented quickly. Hand removal, cultivation, or applying a different chemical treatment should be done. Plants should not be allowed to seed. If the plants have shed seed, avoid moving the seeds with harvest equipment, cultivators, etc. If the crop from the field is fed to animals, manure from these fields should only be applied back to the same field to avoid further spread to uncontaminated areas. Correctly composting the manure has been shown to reduce seed viability but temperatures must reach a minimum of 80C for 8 to 12 days for numerous weed seeds to be killed by heat treatment [38].

7. Weed Management Systems

Broadcast herbicide applications are most effective when weeds are uniformly distributed over fields at population levels at or above economic thresholds. However, uniform population distributions are the exception rather than the norm [39,40, 41]. Therefore, applying herbicides only where they are needed is, in theory, a better allocation of resources. At the present time, the spatial and temporal patterns of weed population dynamics are not well understood, but that should not deter a grower from using knowledge about weed distributions in his fields.

Yearly scouting to determine problem species, problem areas, and most effective treatment(s) should be conducted. Maps generated from these data will give a historic record that will be used in present and future management decisions, point to problems, and allow successes to be observed. Scouting and mapping techniques are useful in determining the correct herbicide(s) for a field and will, most likely, limit the amount of herbicide applied. Weed seed bank data and weed seedling densities also have been used as decision aids for application with consequent reduction in herbicide inputs and environmental contamination. Rather than using broadcast herbicide applications, site specific application techniques, by dead-reckoning, or using global positioning systems (GPS) and geographic information system analysis (GIS), for mapping and relocating areas of infestation, are strategies that should be considered. GPS/GIS techniques are being linked to other procedures to improve weed management.

Integrated systems are essential for long-term effective weed management programs. Sanitation (i.e., cleaning equipment and materials to reduce seed and propagule movement from area to area), preventing establishment of new invading species, crop rotations, and cultural techniques [using best management practices (proper fertility, disease management, insect control) to grow a vigorous, highly competitive crop] will help limit weed problems. Herbicides should be used sparingly, with human and environmental safety at the forefront. Using integrated management practices will reduce unnecessary herbicide applications and improve environmental quality.

8. References

1 Altieri, M.A. 1995. Agroecology: The Science of Sustainable Agriculture. 2nd Edition. Westview Press. Boulder, CO. 433 pg.

2 Gianessi, L.P. and J.E. Anderson. 1995. Pesticide use in U.S. crop production, National data report 1992. National center for food and agricultural policy: Washington, D.C.

3 Himel, C.M., H. Loats, and G.W. Bailey. 1990. Pesticide sources to the soil and principles of spray physics. In Cheng, H.H. ed. Pesticides in the soil environment: Processes, impacts, and modeling. Soil Sci. Soc. Amer. Book Series: 2. Madison, Wi. pg. 7-50.

4 Cheng, H.H. ed. Pesticides in the soil environment: Processes, impacts, and modeling. Soil Sci. Soc. Amer. Book Series: 2. Madison, Wi.

5 Meyer, M.T., E.M. Thurman, and D.A. Goolsby. 1996. Cyanazine, atrazine, and their metabolites as geochemical indicators of contaminant transport in the Mississippi River. In Meyer, M.T. and E.M. Thurman (eds.) Herbicide metabolites in surface water and groundwater. ACS Symposium Series 630. American Chemical Society. Washington, D.C. 320 pp.

6 Richards, R.P., J.W. Kramer, D.B. Baker, and K.A. Krieger. 1987. Pesticides in rainwater in the northeastern United States. Nature. 327:129-131.

7 Richards, R.D., D.B. Baker, N.L. Creamer, J.W. Kramer, D.E. Ewing, B.J. Merryfield, LK. Wallrabenstein. 1996. Well water quality, well vulnerability, and agricultural contamination in the Midwestern United States. J. Environ. Qual. 25:389-402.

8 Koplin, D.W., E.M. Thruman, and D.A. Goolsby. 1996. Occurrence of selected pesticides and their metabolites in near surface aquifers of the Midwestern United States. Envrion Sci. Technol. 30:335-340.

9 DeSutter, T.M., S.A. Clay, and D.E. Clay. 1998. Atrazine, alachlor, and total inorganic nitrogen concentrations of winter wind-eroded sediment samples. J. Environ. Sci. Health. B33:683-691.

10. Barrett, M.R. 1996. The environmental impact of pesticide degradates in groundwater. In Meyer, M.T. and E.M. Thurman (eds.) 1996. Herbicide metabolites in surface water and groundwater. ACS Symposium Series 630. American Chemical Society. Washington, D.C. pg 200-225.

11. Barcelo, D., S. Chiron, A. Fernandez-Alba, A. Valverde, and M.F. Alpendurada. 1996. Monitoring pesticides and metabolites in surface water and groundwater in Spain. In Meyer, M.T. and E.M. Thurman (eds.) 1996. Herbicide metabolites in surface water and groundwater. ACS Symposium Series 630. American Chemical Society. Washington, D.C. pg. 237-253.

12. Bridges, D.C. (ed.) 1992. In Crop losses due to weeds in Canada and the Uniter States. WSSA. Champaign, IL. 403 pp.2. Anonymous. 1993. Reducing herbicide use in field crops. Iowa State University Extension Service. IPM-36.

13. DeSutter, T.M. 1998. Agrichemical concentrations on wind and nonwind erodible sediment sizes as influenced by time, sediment size, and sorption-desorption characteristics. South Dakota State Univ. MS thesis. 93 pg.

14. Regehr, D.L., D.L. Devlin, P.L. Barnes, and S.L. Watson. 1996. Reducing atrazine runoff from crop fields. Kansas State CES publication MF 2208. Manhattan, KS.

15. Koskinen, W.C. and S.A. Clay. 1997. Factors affecting atrazine fate in north central U.S. soils. Rev. Environ. Contam. Toxicol. 151:117-165.

16. Clay, S.A., T.B. Moorman, D.E. Clay, and K.A. Scholes. 1997. Sorption and degradation of alachlor in soil and aquifer material. J. Environ. Qual. 26:1348-1353.

17. Pothuluri, J.V., T.B. Moorman, D.E. Obenhuber, and R. D. Wauchope. 1990. Aerobic and anaerobic degradation of alachlor in samples from a surface-to-groundwater profile. J. Environ. Qual. 19:525-530.
18. Yen, P.Y., W.C. Koskinen, and E.E. Schweizer. 1994. Dissipation of alachlor in four soils as influenced by degradation and sorption processes. Weed Sci. 42:223-240.
19. Muller-Wegener, U., R. Schmidt, C. Ehrig, B. Ahlsdorf, and G..Milde. 1993. Groundwater contamination by pesticides: field experiments in shallow and deeper groundwater. In Mansour, M. (ed.) Fate and prediction of envrionmental chemicals in soils, plants, and aquatic systems. Lewis Publishers. Boca Raton, FL pg 101-110.
20. Clay, S.A., D.E. Clay, W.C. Koskinen, and R.K. Berg, Jr. 1998. Application method: Impacts on atrazine and alachlor movement, weed control and corn yield in three tillage systems. Soil and Tillage Res. 48:215-224.
21. Gaynor, J.D. and I.J. van Wesenbeeck. 1995. Effects of band widths on atrazine, metribuzin, and metolachlor runoff. Weed Technol. 9:107-112.
22. Anonymous. 1993. Reducing herbicide use in field crops. Iowa State Univ. Exten. Serv. IPM-36.
23. Clay, S.A., K.A. Scholes, and D.E. Clay. 1994. Fertilizer shank placement impact on atrazine movement in a ridge tillage system. Weed Sci. 42:86-91.
24. Hall, J.K., N.L. Hartwig, and L.K. Hoffman. 1983. Application mode and alternative cropping effects on atrazine losses from a hillside. J. Environ. Qual. 18:439-445.
25. Clay, S.A., W.C. Koskinen, and J.M. Baker. 1995. Alachlor and metoachlor movement during winter and early spring at three midwestern sites. J. Envion. Sci. Health B(30):637-650.
26. Koskinen, W.C. and S.S. Harper. 1990. The retention process: mechanisms. In Cheng, H.H. (ed) Pesticides in the Soil Environment: Processes, Impacts, and Modeling. Soil Sci. Soc. Amer. Book Series 2. Madison, WI. pp 51-77.
27. Rochette, E.A. and W.C. Koskinen. 1996. Supercritical carbon dioxide for determining atrazine sorption by field-moist soils. Soil Sci. Soc. Am J. 60:453-460.
28. Seybold, C.A., K. McSweeney, and B. Lowery. 1994. Atrazine adorption in sandy soil of Wisconsin. J. Environ. Qual. 23:1291-1297.
29. Isensee, A.R., and A.M. Sadeghi. 1994. Effects of tillage and rainfall on atrazine residue levels in soil. Weed Sci. 42:462-467.
30. Liu, Z., S.A. Clay, D.E. Clay, and S.S. Harper. 1995a. Ammonia fertilizer influences atrazine adsorption-desorption characteristics. J. Agric. Food Chem. 43:815-819.
31. Liu, Z., S.A. Clay, D.E. Clay, and S.S. Harper. 1995b. Ammonia impacts on atrazine leaching through undisturbed columns. J. Environ. Qual. 24:1140-1173.
32. Powles, S.B. and J.A.M. Holtum (eds). 1994. Herbicide resistance in plants: Biology and biochemistry. 353 pg. CRC Press, Boca Raton, FL.
33. Hartzler, R. 1998. Are roundup-ready weeds in your future? Proceedings 18th Annual Crop Pest Management Short Course. Univ. Minnesota pg 12-19.
34. Matthews, J.M. 1994. Management of herbicide resistant weed populations. In Powles, S.B. and J.A.M. Holtum (eds). Herbicide resistance in plants: Biology and biochemistry. CRC Press, Boca Raton, FL. Pg. 317-335.
35. Wolf, R. 1998. Management strategies for controlling ALS-resistant kochia in wheat and soybean. South Dakota State Univ. MS thesis. 64 pg.
36. Clay, S.A. and I. Aguilar. 1998. The influence of two-years of alfalfa on weeds in the following corn crop. Agronomy J. 90:813-818.
37. Kegode, G.O., F. Forcella, and S.A. Clay. Weed seed production influenced by crop rotation, tillage, and management inputs. Weed Sci. (In press)
38. Turnipseed, E.B. and S.A. Clay. 1995. Weed seed analysis from grain/seed screenings from South Dakota. Proc. S.D. Acad. Sci. 74:107-112.
39. Cardina, J., D. Sparrow, and E.L. McCoy. 1995. Analysis of spatial distribution of common lambsquarters (Chenopodium album) in no-till soybean (Glycine max). Weed Sci. 43:258-268.
40. Johnson, G.A., D. Mortensen, L.J. Young, and A. Martin. 1995. The stability of weed seedling popultaion models and parameters in Eastern Nebraska corn (Zea mays) and soybean (Glycine max) fields. Weed Sci. 43:604-611.
41. Lems, J. 1998. Weed spatial variability and management on a field-wide scale. South Dakota State Univ. MS thesis. 104 pg.

CONTROLLING ANIMAL WASTE MANAGEMENT IN DENMARK AS RELATED TO SOIL AND WATER QUALITY

Carl Aage Pedersen
Chief Adviser
The National Department of Plant Production
Udkaersvej 15, DK-8200 Aarhus N, Denmark

1. Introduction

This paper contains a short background on the rules for use of organic manure and fertiliser in Denmark, as well as some facts on soil type, precipitation, nitrogen leaching, land use and livestock.

The paper outlines the most important rules and regulations as they were in 1995, but the Danish Parliament is constantly discussing how to tighten these rules. Thus, there are proposals on how to elevate the required utilisation of total nitrogen in organic manure by approx. 10 percent units, and it is further proposed to impose taxes on nitrogen fertilisers. These new proposals are not dealt with in this article.

2. Some facts concerning Denmark and Danish agriculture

Table 1 shows some key figures on Danish agriculture. Agricultural land covers 2.7 mill. ha of the total 4.3 mill. ha Danish land area.

Danish agriculture is characterized by its animal production and, for example, the number of pigs raised for slaughter is more than 20 million per year while the number of milking cows is around 800,000.

TABLE 1. Facts about Danish Agriculture (1)

	1960	1996
Number of farms	196,000	66,000
Full time farms		33,000
Employed in agriculture	16%	4%
Average size of farms (ha)	16	42
Farms with cattle and pigs		9%
Farms with cattle only		24%
Farms with pigs only		17%
Farms without cattle and pigs*		50%
Number of cows per cow-holding farm		44
Number of pigs per pig-holding farm		512

*42% of these have less than 12 ha each

M.J. Wilson and B. Maliszewska-Kordybach (eds.),
Soil Quality, Sustainable Agriculture and Environmental Security in Central and Eastern Europe, 177–187.

Figure 1 shows the average precipitation for Denmark and that the western part of the country is much wetter than the eastern part. This has an impact on the calculated amount of nitrogen leached from the root zone (Figure 2), although in addition to the higher rainfall, the more sandy soils and higher density of livestock in the western part of the country are also factors.

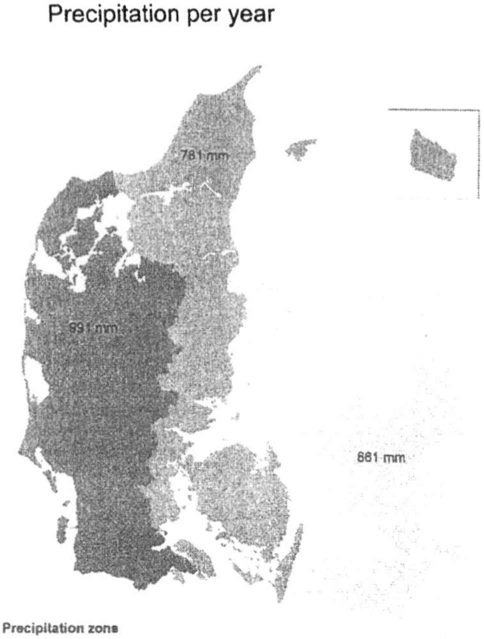

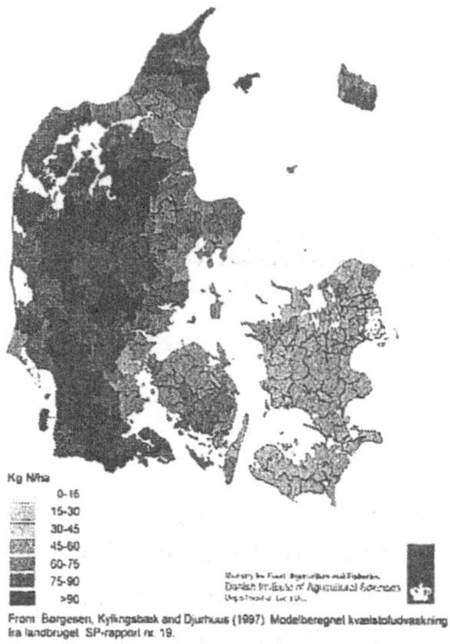

Figure 1. Average precipirataion in Denmark. Figure 2. N leaching from arable land

Due to the increasing content of nitrogen in some ground water reservoirs, the Danish Parliament decided in 1984 to introduce a harmonization rule stating that a sufficient area must be available for the spreading of animal manure. Figure 3 shows that the nitrogen content of surface water generally increases with increasing proportion of agricultural land in river catchments. As more than 60 % of Danish land is used for agricultural purposes, this is one of the reasons why Danish streams also have a relatively high nitrogen content. In addition to the high percentages of agricultural land, the number of livestock units per hectare is also relatively high. Figure 4 shows a comparison with other western European countries. What is interesting is that around 1990 there was a strong correlation between the calculated surplus of nitrogen per hectare and the amount of nitrogen in animal manure.

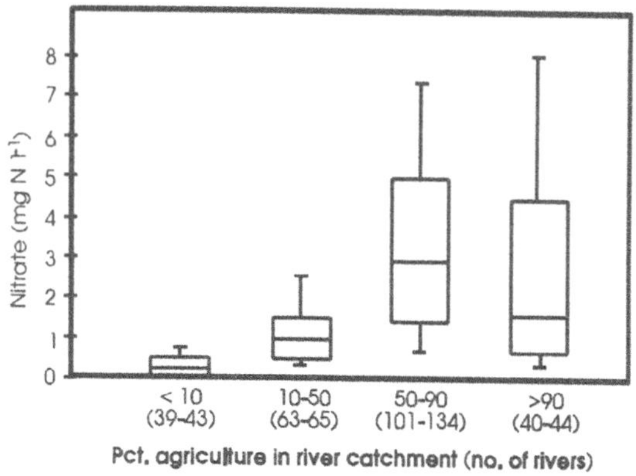

Figure 3. Nitrate content in rivers depending on % agricultural land in catchment. (3)

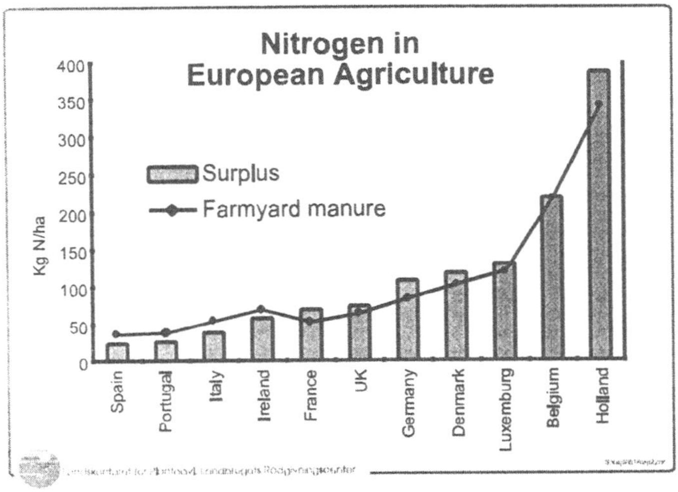

Figure 4. Nitrogen surplus in European agriculture(4)

3. Harmonization rule

In 1984, the Danish Parliament introduced a harmonization rule which forced the farmers to spread animal manure on sufficient land, and stipulated how to match the size of the herd with the land available. A sufficient area must be available for the spreading of animal manure. The spreading of manure from more than the following numbers of livestock units per ha (average per farm) is not permissible

Cattle farms (more than 2/3 cattle)	2.3
Pig farms (more than 2/3 pigs)	1.6
Other livestock farms	2.0
Farms with no livestock	1.7

The cultivated area of a farm that needs fertilising is included in the statement of area versus livestock production. Areas with no need for fertiliser and areas where animal manure cannot be spread under normal conditions are excluded.
If a herd is too big for the owned and leased land, written agreements must be available ensuring that surplus animal manure can be disposed of and utilized on other farms or disposed of to common facilities or biogas plants. The agreement must be valid for at least 1 year at a time with effect from 1 August. One-year agreements can be made for a maximum of 35 ha. For the share of an area which is not owned or covered by one-year agreements, written agreements must be made which at any time show that the livestock producer is capable of disposing to the necessary area at least 3 years ahead. The rule about 3-year agreements must be met by 1 September 1997. The municipality must have a copy of the agreement and forms are available from the Danish Agricultural Advisory Centre.

If the amount of manure produced on a farm exceeds the amount that can be spread on the land of the farm, the supervising authority must have access to an up-dated account showing manure supplies, name of receiver, date of delivery, amount delivered, and type of manure. Suppliers and/or receivers of animal manure with more than 3 agreements are to send copies of their accounts of manure supplies to the municipal council by 1 August.If a holding that purchases livestock manure is located in another municipality, a copy of the balance is to be sent to this municipality as well.

4. Oxygen deficiency in 1986

In 1986, the Kattegat Sea was strongly affected by oxygen deficiency resulting in serious fish death. The Parliament at once decided to reduce the nitrogen leaching from agricultural land by 50% within 3 years. It was decided too, that the emission of phosphorus should be reduced dramatically, namely by 80%, but this in fact did not affect agriculture as the main sources of phosphorus were from sewage from cities and from some industries. Despite the fact that agronomically it is impossible to reduce nitrogen leaching to this extent within 3 years so that the effect becomes measurable in streams, the politicians, nevertheless, have several times since then tightened the rules as they could not see such an effect.

5. Current Rules

Under Danish harmonization rules in the year 2002, it is expected that the rule due to the EU nitrate directive will be tightened so that only 1.7 animal unit per hectare will be allowed, the only exception being cattle farms with enough grassland to allow probably 2.3 animal units per hectare.

5.1 Spreading of Animal Manure

Animal manure and silage effluent applied to the soil must only be used for fertilizing purposes. Animal manure must not be spread in forests, except for areas with greenery cuttings and afforestation where the planting is less than 10 years old (limited rates of application). Animal manure must not be spread in such a way or on areas where there is a risk of run-off to lakes, streams or drains following thaws or heavy showers. Slurry must not be spread on Saturdays, Sundays or public holidays on areas sited less than 200 m from urban zone, summer house areas or rural areas zoned for housing purposes.

From harvest to 1 February it is not allowed to spread liquid manure, except for the period from harvest to 1 October in overwintering grassland crops or areas due to be sown to winter rape the following winter. In cultures of greenery cuttings and on afforestation areas liquid manure must not be spread from 1 September to 1 February. Liquid manure and silage effluent spread on areas with no vegetation are to be incorporated as soon as possible and within 12 hours. In the period from harvest to 20 October solid manure may be spread only on areas covered by crops the following winter. Solid manure spread on areas with no vegetation is to be ploughed in immediately after spreading. In the period from harvest to 1 November silage effluent may be spread only on vegetated areas or areas due to be sown to winter crops.

5.2 Storage Capacity for Farmland Manure

Holdings with commercial livestock with facilities for storing animal manure must observe the rules concerning spreading of animal manure and the rules concerning utilization of nitrogen from animal manure. At least 6 months' capacity is required for storage facilities. Normally, sufficient capacity corresponds to at least 9 months' supplies - for cattle farms normally at least 7 months when the cattle are pasturing during the summer.If any changes are made after 1 January 1995 which have an effect on the need for storage capacity, then a new statement must be sent to the municipal council. A statement on the sufficient storage capacity of a farm in terms of animal manure must be made on the basis of manure quantities, crop rotation plans, rules for manure spreading and requirements concerning the utilization of the nitrogen in manure. The Danish Plant Directorate has sent out guidelines stipulating how to calculate a sufficient storage capacity. Tanks for liquid manure with no tight floating layer must be covered by another type of tight cover. Dung pits must be laid out so that run-off from the pit is lead to and collected in a tank and so that surface water from surrounding areas and from roof surfaces cannot run off into the dung pit.

Required storage facilities on other farms or in common biogas plants etc. necessitate written agreements with a duration of a minimum of 5 years. The Danish Agricultural Advisory Centre has worked out a set of contracts which can be used to notify contracts. The quantity of animal manure on a farm is (primarily) dependent on the size of the livestock production. When calculating the storage capacity and the utilization of the animal manure in the fertilizer plans, the amount of wash water and water spillage etc. from livestock houses, surface water from consolidated areas and silage effluent running-off to manure tanks must be considered. The guide on commercial livestock keeping, animal manure, silage etc. prepared by the Environmental Group shows the amount of manure produced by the most common livestock types and the guiding values for the amount of water spillage etc. in manure stores.

In order to ensure that the required strength, quality, construction and durability is met for new manure stores, a form agreement (contract) has been prepared under the Danish agricultural building recommendations, no. 103.0 4-23. This form is recommended where manure stores are to be established. A number of manufacturers of manure stores have subjected themselves to a voluntary control scheme, including manufacturing and erection of manure stores. This scheme ensures that the supplies of the manufacturers meet the requirements in terms of quality and security.

The user of a store facility for liquid manure, silage seepage or waste water is responsible for having the store inspected every tenth year for strength and tightness. The inspection is carried out by a person selected by the municipal council. Store facilities established before 1988 were due to be inspected as from 1997.

5.3 Compulsory Crop Rotation and Fertilizer Plans - Fertilizer Balance

Farms exceeding 10 ha and farms with less than 10 ha producing more manure than can be spread on the farm's own land according to the rules, are required to have compulsory fertilizer and crop rotation plans. Leased land is included in the farm's area. The farmer is responsible for the preparation of the plans and he/she may choose to have a crop adviser prepare them.

Every year the users of farm holdings are to prepare crop rotation and fertilizer plans for the planning period, comprising (a)a crop plan, including green cover areas, and (b) a fertilizer plan, stating the needs of the crops for nutrients and how the needs are met using animal manure and/or commercial fertilisers. The planning period covers the period from 1 August - 31 July and also comprises the following period from 1 August - 31 July in those cases where the fertilisers are applied to crops. This includes cases where fertilisers are to be applied before harvest but not later than 30 September and where crops are to be harvested before 31 December. The fertiliser applied in the extended planning period is not included in the crop rotation and fertiliser plans for the current planning period.

When estimating the needs for nutrients the standards fixed by the Danish Plant Directorate are to be applied. The standards are optimal amounts calculated by The Danish Agricultural Advisory Centre/Dep. of Plant Production based on numerous field trials.The farmer has to adjust, for instance, according to yield level, annual nitrogen prognosis or a special statement from a crop adviser. The plan has to be prepared in approved forms. By 1 September every year, crop rotation and fertiliser plans covering the period from 1 August to 31 March the following year concerning the farm's area with green cover crops, set-aside and the planned applications of animal manure, must be available. Crop rotation and fertiliser plans for the present planning period must be available by 31 March, as well as a fertiliser balance for the past planning period. The fertiliser balance must state information on the need for nitrogen and the consumption of nitrogen from animal manure, other organic manures and commercial fertilisers. Furthermore the fertilizer balance has to include the utilization of nitrogen in animal manure. The total nitrogen application must not exceed the amount needed in order that the following minimum utilization percentage of total-N can be met for nitrogen from animal manure:

Pig slurry:

Until 31 July 1997:	Min. 45%
From 1 August 1997:	Min. 50%

Cattle slurry:

Until 31 July 1997:	Min. 40%
From 1 August 1997:	Min. 45%

Manure from deep-litter systems

From 1 August 1993:	Min. 15%

Other types of animal manure:

From 1 August 1995:	Min. 40%.

In addition to that, the calculated crop need has to be reduced by 10 % of the total nitrogen applied in animal manure in the previous year. On request by the Plant

Directorate, a key figure sheet covering the last planning period is to be filled in and sent to the Plant Directorate before a date fixed by the Directorate. 30,000 holdings have to send in key figure sheets annually. The key figure sheet must contain a statement of the farm's total need for and consumption of nitrogen and also the utilization percentage of the animal manure spread.

5.4 Green Cover Area

Sixty five percent of the land of farm holdings covered by the rules on compulsory crop rotation and fertilizer plans plus fertilizer balances must be sown to green cover crops. Neighbours within a distance of 5 km can meet this requirement on a joint basis. Green cover means that farmland is sown to one of the following crops:

Winter cereals	winter wheat, winter rye, winter barley, triticale.
Maize	forage maize.
Grassland crops	grasses and clovers included in grass mixtures, lucerne, permanent grassland, Italian ryegrass for feeding,
Root crops	beet, industrial potatoes
Seed crops	seed grass, clover seeds and other legume seeds, winter rape, sunflower, beet seeds, caraway,garden seeds and medicinal plants.
Vegetables, fruit and berries	late outdoor vegetables and fruit and berry cultures
Christmas trees and greenery cuttings	
Aftermath	All cultivated crops growing in the autumn used as aftermath due to wilt or to be ploughed in. The aftermath must be established according to normal farming principles so that the aftermath can develop in the best possible way.
Other crops	Crops comparable to the above-mentioned.

Other points are that areas sown to grassland crops, seed crops and aftermaths are not considered green cover areas if ploughed before 20 October. Twenty out of the 65 percent units of green cover may be substituted by incorporation of grain and rape straw. Straw is to be incorporated on 1.6 ha of land to substitute 1 ha of genuine green cover area. Thus, a 100 ha-farm can substitute at a maximum 20 ha of green cover area (20 percent) by straw incorporation. Straw has to be incorporated on 32 ha of land to substitute for the 20 ha.

6. Involvement of the farmers and the organisations

In Denmark, the farmers themselves run the advisory system with all disciplines including more than 400 local plant production advisers and a staff of about 50 at a national level. The advisory service and therefore also the farmers' organisations agreed that animal manure was not utilised efficiently. Therefore, in 1989 these organisations introduced an action plan for better utilisation of the nutrients in animal manure. The action plan included 200 field trials and 10 demonstration farms.

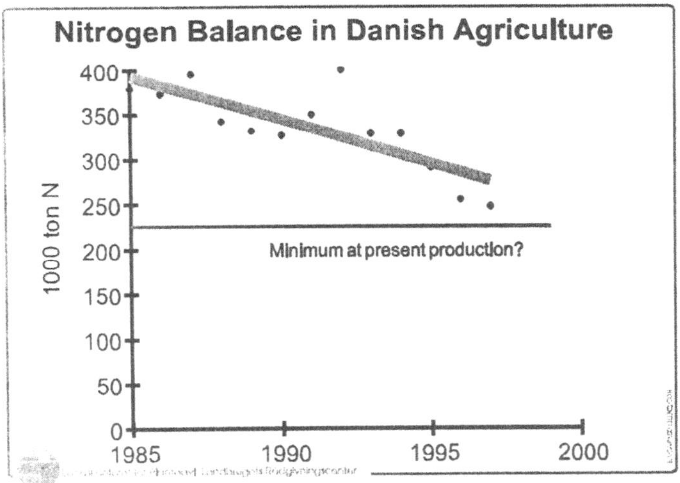

Figure 5. Crop need for N and application of inorganic N in chemical fertilisers.

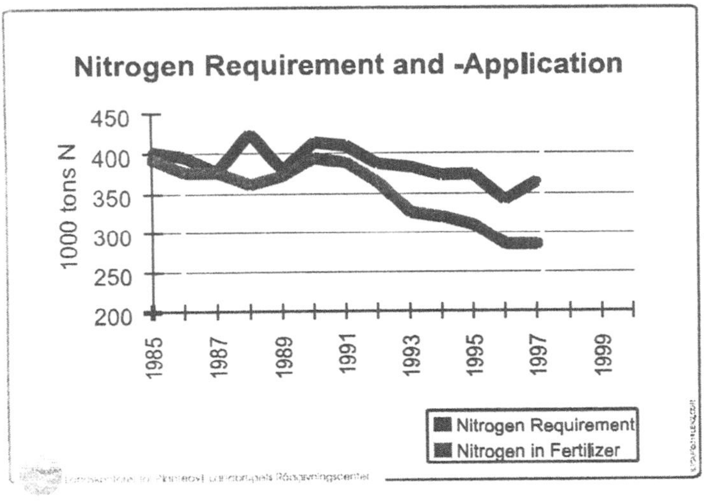

Figure 6. N surplus on field trial in Danish agriculture. (7)

7. The effect so far

The overall effect of the rules and the action plan has been remarkable. Fig. 5 shows the calculated total need of nitrogen and the application of inorganic fertilizers (6). The gap between the curves indicates the utilization of nitrogen in animal manure, which also had an effect on the overall nitrogen surplus. This increased utilization of animal manure is one of the main reasons why Danish farmers - despite a slightly increased production - have reduced the overall nitrogen surplus from Danish agriculture by approximately 25 % (Fig. 6) As far as is known, a similar reduction has not been made in any other nation with a significant production of pork, meat, and milk. Nitrogen leaching has yet to be reduced by the 50% that Parliament decided, but in some years it will be reduced by more than 30%. It is then up to the Parliament to decide whether or not this is acceptable as further reduction can only be achieved at the expense of production.

8. Phosphorus

Due to the early introduction of the Harmonization rule there are in Denmark no phosphorus-saturated soils or phosphorus "hot spots" like there are in the Netherlands, Belgium, France, and Italy.

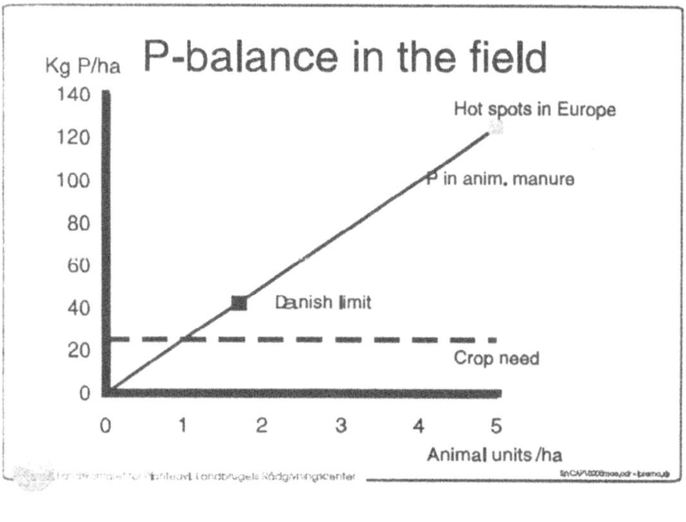

Figure 7. P-balance in the field

Nevertheless, it is necessary to achieve a higher P-utilization in the animals so that the content of P in animal manure can be reduced. Researchers are working hard on this problem and it looks as if they will be successful. The alternative is that it will be necessary to demand that more land is made available for the spreading of manure.

9. Copper

The Harmonization rules have had - as a side effect - the reduction of problems caused by a too high copper content in agricultural soils. Pig producers have lowered the amount of copper added to the diet, but in the long run we have to be aware that the accumulation of heavy metals in agricultural soils is not acceptable.

10. Conclusions

As in other countries with a significant livestock production, the situation in Denmark 10 years ago was that nitrogen losses were higher than acceptable from an environmental point of view. Thanks to the early regulation of spreading of animal manure and a combination of motivation and legislation these losses have been reduced significantly from year to year. In the long run, the utilization of phosphorus in livestock-feeding has to be improved and it is also necessary to reduce the amount of copper and zinc in the feed. But as far as nitrogen is concerned, it seems that Danish agriculture can continue to produce large amounts of animal products for the world market in a sustainable way.

11. References

1. Anon. (1997), Landøkonomisk Oversigt. De danske Landboforeninger, København.
2. Børgesen, C.D., Kyllingsbæk and Djurhuus, J. (1997). Modelberegnet kvælstofudvaskning fra landbruget. Sprapport nr. 19. Danmarks JorbrugsForskning, Foulum.
3. DMU (1994), European rivers and lakes. National Environmental Research Institute, Silkeborg
4. Knudsen, L. (1997), Personal communication. The Danish Agricultural Advisory Centre, Department of Plant Production, Aarhus
5. Anon (1995), Environment '95, Rules applying to Danish agriculture. The Danish Agricultural Advisory Centre, the Environmental Group, Aarhus
6. Pedersen, C.A. (1997). Oversigt over Landsforsøgene 1997. The Danish Agricultural Advisory Centre, Department of Plant Production, Aarhus.
7. Grant, R., Blicher-Mathiasen, G., Andersen, H. E., Laubel, A. R., Jensen, P. G. and Rasmussen, P (1997). Landovervågningsoplande. Faglig Rapport fra DMU, nr. 210. National Environmental Research Institute, Silkeborg.

THE USE OF SOIL DATA TO PREDICT ENVIRONMENTAL SENSITIVITY TO POLLUTION

E. PATERSON, W. TOWERS and S.J. LANGAN

Macaulay Land Use Research Institute
Craigiebuckler, Aberdeen AB15 8QJ
Scotland, UK

1. Introduction

Research over the past twenty years has demonstrated clearly that the environment is not infinitely resilient to anthropogenic activities, such as land management and use, fugitive emissions from industry and waste utilisation on land. Consequently, in national and international strategies aimed at sustainable development, where the maintenance of environmental capital is a key feature, environmental protection plays an important role. Thus, air, water and soil quality may be viewed as indicators of sustainable development. Of the three environmental media, environmental standards have been developed for both air and water but soil has received far less attention [1], despite the fact that it interacts strongly with the other media and frequently performs an important role in buffering living systems from environmental pollution. A number of factors have contributed to this situation including the multifunctional role played by soil in the environment [2], the complex interactions of chemical, physical and biological processes that contribute to soil quality [3], and, finally, the spatial and temporal scales over which these processes operate. However, it is the authors' view that soil quality can provide a valuable conceptual framework within which many of these aspects can be integrated.

Within Europe, a number of countries are making significant contributions to the development of environmental standards for soils and the recent formation of the European Soil Bureau and the Soil Topic Centre of the European Environmental Agency should assist in their further development and wider application. In the UK, the Department of the Environment, Transport and Regions has taken a lead role through the development of the UK Strategy for Sustainable Development [4] and the publication of a series of Indicators for Sustainable Development [5]. For soils, two sets of indicators have been proposed, one based on soil quality which includes organic matter content, soil pH, and the content of major nutrients and another set based on the content of potentially toxic elements, as measured by *aqua regia* extraction. These indicators form an important starting point but there are some difficulties with such compositionally based indicators. Firstly, composition *per se* may not have any direct relationship to

M.J. Wilson and B. Maliszewska-Kordybach (eds.),
Soil Quality, Sustainable Agriculture and Environmental Security in Central and Eastern Europe, 189–206.

functionality or future responses to stresses; both of these aspects are more strongly related to soil processes and their maintenance. The second is the extent to which quality indicators should be aimed at specific uses of soil or whether they should be generic [6]. The third is to develop an appropriate scoring method that takes account of baseline and threshold values to enable overall quality to be evaluated [7]. Clearly, in order to undertake a rigorous, comprehensive assessment of soil quality considerable data would be required, although efforts are being made to define minimum data sets for such purposes [8].

In order to tackle the issues related to soil quality and environmental protection, progress on a number of fronts is required. One of the principal requirements is the development of a conceptual framework in which soil quality in relation to environmental stresses may be defined. In the longer term, this must be based on an understanding of key processes taking place in soil and the identification, collation and retrieval of soil attributes that can best indicate conditions required to sustain those processes. However, in view of **current** requirements to provide useful information in applied and strategic areas relating to soil quality using our **existing** knowledge-base it is important to examine methods by which indicators may be derived that relate to processes rather than simply composition. In this paper two indicators of soil quality will be addressed using compositional and process-based approaches and will be illustrated using data on Scottish soils. Attention will be focussed on soil quality indicators associated with the assessment of soil acidification and the behaviour of potentially toxic elements in soils but the case studies raise generic issues on the effective use of soil data that should be of value elsewhere in the world and allow an objective assessment of the relevance and value of soil data in predicting the impact of anthropogenic pollution on the environment.

2. Data sources

Within the UK, separate soil survey activities were based in England and Wales, Scotland and Northern Ireland. This has influenced both the nature of the information available, in terms of soil maps, classification systems [9] and databases and also the organisations currently responsible for holding these data. In England and Wales, the former Soil Survey of England and Wales has now evolved into the Soil Survey and Land Research Centre at Cranfield University, who hold both soil maps and a soils database for England and Wales. The Soil Survey of Scotland, a department of the Macaulay Institute for Soil Research, has now ceased its survey activities and has been incorporated into the Macaulay Land Use Research Institute. The Macaulay Institute now holds soil datasets for the whole of Scotland. In Northern Ireland, the soil survey is now complete [10] and has included the publication of a geochemical atlas for the Province [11].

In order to provide information for the case studies, data from the activities of the Scottish Soil Survey will be used. A considerable amount of information on the soils of Scotland has been collected over the past fifty years and much has been published on paper media in the form of maps, memoirs and reports [12]. In addition, the data on which these publications are based forms the core of the Scottish Soils Database held at the Macaulay Land Use Research Institute [13]. In common with many soil databases throughout the world, much of this information is not immediately understandable to the non-specialist. Therefore, one of the main purposes of this paper is to show how a detailed knowledge of the process of data collection, organisation and classification can contribute to the effective and appropriate utilisation of soil data.

A European perspective is available from a recent conference, organised by the Joint Research Centre at Ispra and by the Institut National Recherche Agronomique. The proceedings from this meeting have summarised data availability for soils in the countries within the European Union and, indeed, some European countries who are not yet members [14]. In addition, a number of pan-European initiatives are described including the European Soils Bureau, the activities of the Soils and Geographic Information Systems (GIS) Support Group, the development of a pedotransfer rule base for environmental interpretations of the EU Soil Database and the establishment of a Soil Profile Analytical Database for the European Union. These initiatives will initially lead to the development of a European Union perspective but, with enlargement of the Union currently under way, will ultimately incorporate data from a much wider range of European countries, including many considered at this workshop.

2.1 SCOTTISH SOIL DATABASE

Briefly, the Scottish Soils Database is held in a relational database management system with four principal tables giving information on

- the sampling site and the soil profile;
- morphological features of the mineral soil horizons;
- morphological features of the organic soil horizons;
- analytical data for a limited number of physical and chemical properties.

There is, in total, information on over 13,000 soil profiles obtained from two major sources - the systematic mapping of the Soil Survey of Scotland, which ended in 1987, and the National Soil Inventory project carried out in the late 1970s and early 1980s [13]. In the former case, free survey techniques were used and the soils were mapped using the classification system first described by Glentworth [15] whereas in the latter, a systematic grid at 5km intervals was used and at each intersect the profile was described and classified [16]. At every alternate point, i.e. on a 10km grid, samples were taken from the centre of the major soil horizons and analysed for a range of attributes. In addition to the major tables, there are others which contain supplementary information, either derived from the main tables [17, 18] or from new measurements on archived samples. An example of the latter is a geochemical dataset derived from the National Soils Inventory sample set and analysed using similar techniques to those used by McGrath and Loveland [19] in their work on the topsoils of England and Wales.

2.2 SOIL MAPS

In order to use published soil maps it is important to recognise that there are different soil classification systems in use throughout the world, although most categories have analogues in other classifications to allow correlation between them [20]. However, the philosophy underpinning all soil classifications is that they are based on the nature and sequence of soil horizons within the soil profile; the level of precision and definition differs between them but the properties used should 'to the greatest extent possible should be measurable and observable in the field'. Thus, for a particular map series it is important to understand the classification scheme implicit in it.

In Scotland, the basic mapping unit is the soil series based on the major soil groups or sub-groups which are aggregated together into larger units, the Soil Association, which groups together soils developed on the same parent material and which, characteristically, occur together in the landscape. Thus, at the series level, dominant pedogenic processes, such as the turnover of soil organic matter and leaching of Al, Si and Fe are represented, whereas, at the association level, lithogenic influences are dominant which influence both soil mineralogy and geochemistry.

In England and Wales, the classification system has a hierarchical structure, with four categories; major soil group, soil group, soil subgroup and soil series [21, 22] to give more rigorously defined classes at the lower levels. For example, at the major soil group level, divisions are based on the presence or absence of major diagnostic horizons that have an important agronomic, hydrological, ecological or engineering significance whereas at the group or subgroup level, divisions are based on more specific features which reflect differences of soil behaviour within major soil groups.

A World Reference Base for Soil Resources [23] has recently been published, which describes the soil classification used by the Food and Agriculture Organisation of the United Nations. It is not intended to replace national soil classification systems, but to provide a tool for better correlation between them. There are two levels within the classification comprising thirty reference soil groups which are subsequently refined into soil subunits using a range of diagnostic horizons, properties and materials.

Irrespective of the classification system used it is critical that the user of soil data understands the system employed so that legitimate comparisons may be made between soils and that the relationship between analytical data for point-based observations and spatial representations on maps are rigorous.

3. Compositional and process-based indicators

The key indicators that have been selected in the UK Strategy for Sustainable Development [5] to assess the current status of the soil resource and to monitor changes in response to anthropogenic stresses are soil quality, which includes soil pH and the contents of major nutrients, and the contents of soil organic matter and of potentially toxic elements (PTEs). Of these, the Scottish Soil Inventory contains analytical data which covers soil pH and organic matter content for all the major soil horizons at the inventory points while the geochemical data refers to the surface horizons only. However, it is important to note that these are point based datasets and that their representation on a map should be used only in order to give a national perspective rather than be interpreted on a detailed regional basis. In particular, they are different in nature to conventional soil maps and should not be combined with such datasets without prior statistical examination.

As indicated earlier, however, indicators based on composition can present some difficulties in interpretation and often require the incorporation of expert judgement in order to be of value [7]. An alternative approach is to develop process-based indicators which provide an indication of the capacity of soil to perform a particular function. In the assessment of soil response to acid inputs the approach used in the first phase was one which combined both of these approaches. The key process in long-term response to acid inputs was identified as mineral weathering, expert judgement was then used to classify soil parent materials in terms of their susceptibility to weathering and to provide a linkage to soil map units. Thus, soil sensitivity to acidification could be mapped. In relation to soil response to heavy metal inputs, instead of simply mapping the PTE content of soils specific adsorption can be identified as the key process which controls availability. By classifying the binding capacity in terms of the soil attributes that contribute to it the capacity of the soil to bind PTEs can be mapped. Thus, with existing knowledge it is possible to develop process-based indicators of soil quality that can contribute to environmental protection and, thus, to sustainable development.

3.1 SOIL ACIDITY

Soil pH is a key soil property that influences many important processes that occur in soils. These include soil development processes, such as mineral weathering and carbon turnover, soil responses to contaminants as well as influencing biological processes that significantly affect plant growth.

3.1.1 Direct measurement

The measured pH values found across Scotland show a broad correlation with carbon content and clearly delineate the highly organic soils of the Highlands and the Southern Uplands from the mineral soils of the lowland areas (Figure 1). Few soils in Scotland are alkaline with only three sites in the National Soil Inventory where the pH >7. All of these are located on the coast and probably arise from calcareous materials in the coastal sands on which these soils have developed. In the cultivated soils of the drier north-east and east of the country, soil pH's lie the range 5-7 with the majority between 5 and 6. In the organic soils of the north and west the values of pH are almost all below 5 with many falling below 4. For the mineral soils in Scotland the median pH is 4.80 whereas for the organic soils a median value of 2.67 is found. The median value for the mineral soils is more than one pH unit lower than those recorded for England and Wales [19].

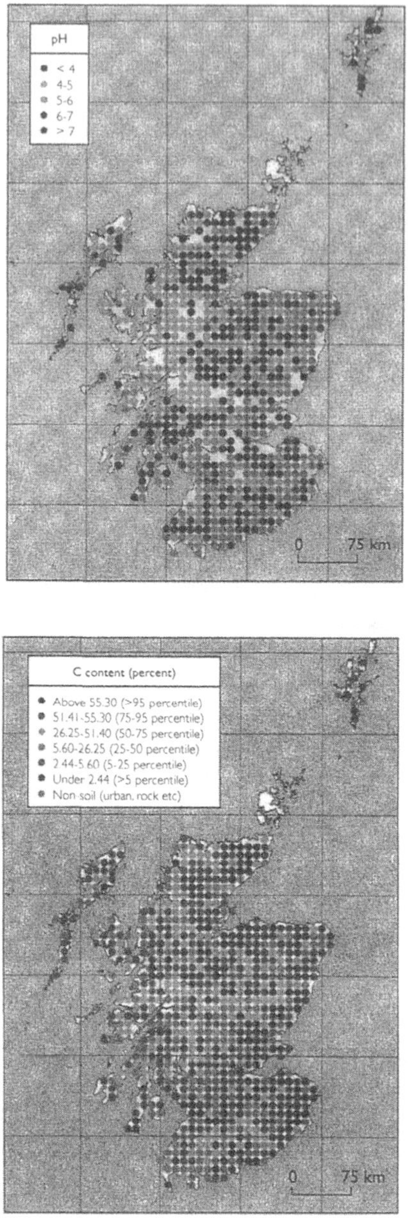

FIGURE 1. pH values (upper map) and carbon contents (lower map) for topsoils in the National Soil Inventory of Scotland plotted on a percentile scale. (see also coloured plate at end of book)

In order to use pH as a soil quality indicator, soil function must be considered. In soils that are used for the growth of arable crops, such as cereals, pH values between 6 and 7 are considered the optimum and, in Scotland, land management for arable cropping almost invariably includes liming. However, for some soils, particularly those managed to achieve natural heritage and conservation objectives, much more extreme values of pH may represent the optimum, for example, to sustain acidic or calcareous grasslands or blanket bog communities. Thus, in evaluating the relationship between soil pH and soil quality for arable agriculture and natural heritage management the standard scoring functions, 'optimal value' and 'undesirable range', respectively, would be appropriate [7]. However, even using such an approach does not give any indication of the sensitivity of the soil resource to acid inputs or the resilience of the system in its response. In order to make predictions about likely change an approach based on the dominant soil processes involved must be used.

3.1.2 Soil acidification and the critical loads concept

Acidification of soils is a natural process which occurs when nutrients such as calcium and magnesium are depleted at a rate faster than they can be replenished by mineral weathering, deposition and decomposition processes. Such a natural process occurs very slowly over periods of thousands of years. However, with the addition of acid deposition inputs, the depletion of nutrients and concurrent increase in acidity may enhance the process of acidification with adverse ecological effects occurring over much shorter (decadal) time periods because of the liberation of toxic metals, such as aluminium.

The first policies aimed at reducing the impact of acid deposition adopted a blanket reduction of all emission sources [24]. However, such an approach does not recognise the spatial variability in the sensitivity of ecosystems to acidification and, perhaps more importantly, does not allow the assessment of the efficacy of the abatement taken. With this background, in 1990, the United Nations-Economic Commission on Europe (UN-ECE) set out to renew the sulphur protocol. In order to provide a scientific basis for negotiation of the second protocol, it was agreed that advances in the understanding of the processes of acidification were sufficiently quantified to permit a targeted approach for controlling acidifying emissions.

This approach, utilised within the framework of critical loads, encapsulates a dose-response relationship to acid inputs. Thus, ecosystem sensitivity to acidification is built on the principle that if the soil is protected from acidification caused by acid deposition, then those ecosystems developed on it (trees, shrubs, moorland vegetation and surface drainage waters) will likewise not be damaged. A critical load can be generally defined as 'the highest deposition of acidifying compounds that will not cause chemical changes leading to long term harmful effects on ecosystem structure and function' [25]. The relationship described is illustrated in Figure 2. Initially, with increasing deposition, there is no damage or effect on the ecosystem. However, if deposition inputs are greater than the ecosystem tolerance, expressed as a critical load, there will be a negative effect or damage to the ecosystem.

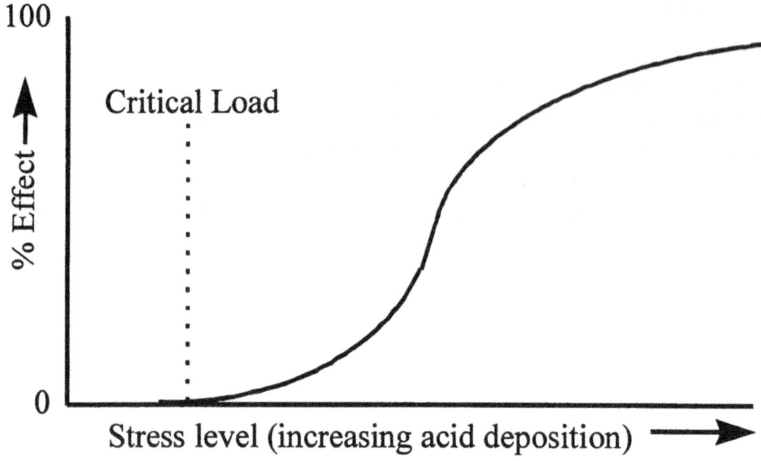

Figure 2 Generalised relationship between atmospheric acid inputs and ecosystem damage as described by the critical load concept

Under the auspices of the UN-ECE, all member states have been requested to produce national critical load maps. To fulfil this obligation, the Macaulay Land Use Research Institute (MLURI) was commissioned to research and produce a soil critical load map for Scotland. The method considered most appropriate was to set the critical load equal to the rate at which base cations were replaced at a rate which equalled their loss through leaching (and vegetation uptake) from inputs of atmospherically derived acidity. For mineral soils it is suggested the principle source of base cations is that of the base cations release from minerals by chemical weathering. Soils which have a high content of readily weatherable minerals, will have a high critical load, which is unlikely to be exceeded at a regional level by current levels of acid deposition across Europe. Those soils, which are developed from parent materials dominated by slowly weatherable minerals, are likely to have very low weathering rates and low critical loads. Therefore the calculation and assessment of the release of base cations can be seen to be fundamental to reliable estimates of critical loads.

In order to apply this concept it has been necessary to consider the availability and appropriateness of data sets which would permit the delineation of soils and ecosystems into different critical load sensitivities. Given that the exercise being undertaken was to advise international policy and variations to be detected are at a regional level meant that the data used for critical loads had to be available at a national scale and be able to carry information which related to variations in base cation production. For these reasons the 1:250,000 national soil survey was considered most appropriate. At this scale of soil survey the individual mapping units describe soils which are compound descriptors of soils and associated chemical information which reflect parent material, major soil sub group and landform attributes as well as, to a lesser extent, land cover. In the legend for the 1:250,000 map there are over 580 such units [16]. Using this map data together with the associated chemical and mineralogical information held on the national soil data base each of the 580 units was allocated to one of five sensitivity classes. The spatial distribution of these five classes across the country was then mapped according to the dominance of the soil at a 1 km^2 resolution. As each of the classes has a numerical critical load on the basis of base cation production it is possible to map those areas in which the critical load (based on weathering rate) is either less than or greater than the pollutant input. For those areas in which deposition of pollutants is greater than the critical load, it is suggested that damage to the ecosystem will be occurring in the areas. A more detailed description of the development of the critical loads and some of the underlying problems with the approach is given in Langan and Wilson [26] and Hodson and Langan [27].

Similar maps have also been constructed for England and Wales [28] and other countries of Europe [29]. These maps show large areas of Northern and central Europe to have extensive areas with low critical loads and are particularly sensitive to acidification. By combining these data with existing and future scenarios of deposition, following emission abatement, it is possible to show regions of the country where the critical load is exceeded and, thus, where ecosystem damage is likely to be occurring Within a policy context the maps and underpinning science have been used to guide the negotiations on the second UN-ECE protocol on sulphur which has resulted in the call for emission abatements of SO_2 in excess of 70 and 80% in some countries.

However, it should be noted that despite the success of critical loads at the national and European scale some difficulties have been encountered when attempting to interpret the mapped outputs at, for example, the catchment scale. Current work at MLURI is attempting to address this issue and also to strengthen the causal link between chemical presence and biological impact through a more rigorous examination of the evidence for biological impacts in areas designated as highly impacted within the national critical loads programme. Clearly, a number of scientific issues require to be resolved before the approach can be refined for operation at the spatial scale where land management decisions are made.

3.2 POTENTIALLY TOXIC ELEMENTS

Within the context of current, compositionally defined soil quality indicators for potentially toxic elements, geochemical datasets, which form a component of the work of the Forum of Geological Surveys [30], are an important source of information that can have particular environmental significance. The FOREGS Geochemistry Task Force has recently published meta-data which describes the datasets available for soils, sediments and waters.

3.2.1 Compositionally based indicators

The simplest of the compositionally based indicators used for potentially toxic elements is their content as determined by *aqua regia* digestion and spectroscopic analysis. As mentioned previously, a Topsoil Geochemical Atlas has been published for England and

Wales [19], one for Northern Ireland [11] and one for Scotland is at an advanced stage of preparation. In addition, other datasets are also available from British Geological Survey (http://www.bgs.ac.uk) including the G-Base dataset and several regional Geochemical Atlases, based on stream sediment analysis.

For soils, in order to use data on PTE content as a quality indicator expert judgement is required to set thresholds at which the PTE content may represent a hazard to the environment. Although not designed for general application specific values from the range recommended by EC in its guidelines for PTE content of soils to be treated with sewage sludge are widely used to provide such thresholds (Table 1). Within the UK, the values selected are under constant review as new evidence becomes available on the impacts of PTEs on soils as well as on plant uptake and entry into the food chain.

Table 1 EC and UK limits for PTE's in soils receiving sewage sludge

PTE	EC mandatory	EC recommended	UK limit
Zinc	300	150	200
Copper	140	50	80-135
Nickel	75	30	50-75
Cadmium	3	1	3
Lead	300	50	300
Mercury	1.5	1	1

Note:
1. All values in mg/kg dry matter
2. Limit values in EC Directive is for a soil with a pH between 6 and 7
3. UK limit values are relevant for soils with a pH between 5 and 7; higher limits are allowed for calcareous soils above pH7.

In Figure 3 the aqua-regia extractable contents of copper, nickel, lead and zinc are shown for the Scottish national soil inventory sample set. As can be seen, the data show some evidence for spatial structure but attempts to correlate PTE content with other spatial map units, such as Soil association, were unsuccessful. Previous work at the Macaulay Institute has shown clearly the influence of soil parent material on the PTE contents [31, 32, 33], particularly for metals such as nickel [34], but in most cases the strong lithogenic influence was observed for subsoils not for topsoils. In topsoils the confounding influences of anthropogenic deposition, pedological redistribution and dilution by organic matter presumably obscures the anticipated relationships. When the data are replotted using class intervals determined by the UK implementation of the EC limits (Table 1) then it can be readily seen that of the four metals, lead gives most cause for concern (Figure 4).

However, it is important to recognise that although this approach has value within a legislative framework it has less significance in an ecological or biological context where the action of soil processes, such as specific adsorption, can immobilise metals on the solid phase and reduce their bioavailability. It is in this latter context that process-based indicators come into their own.

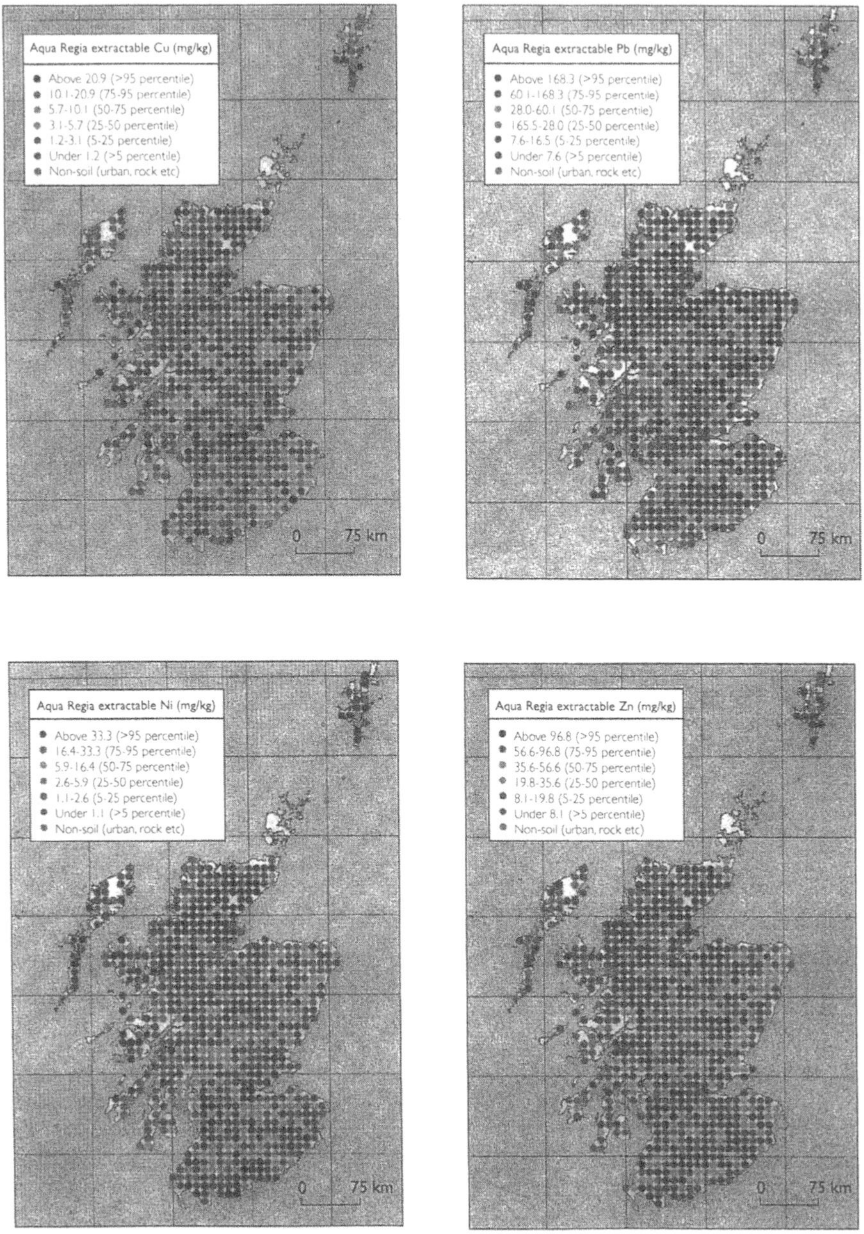

FIGURE 3. Aqua regia extractable contents of copper, lead, nickel and zinc in topsoils in the National Soil Inventory of Scotland plotted on a percentile scale. (see also coloured plate at end of book)

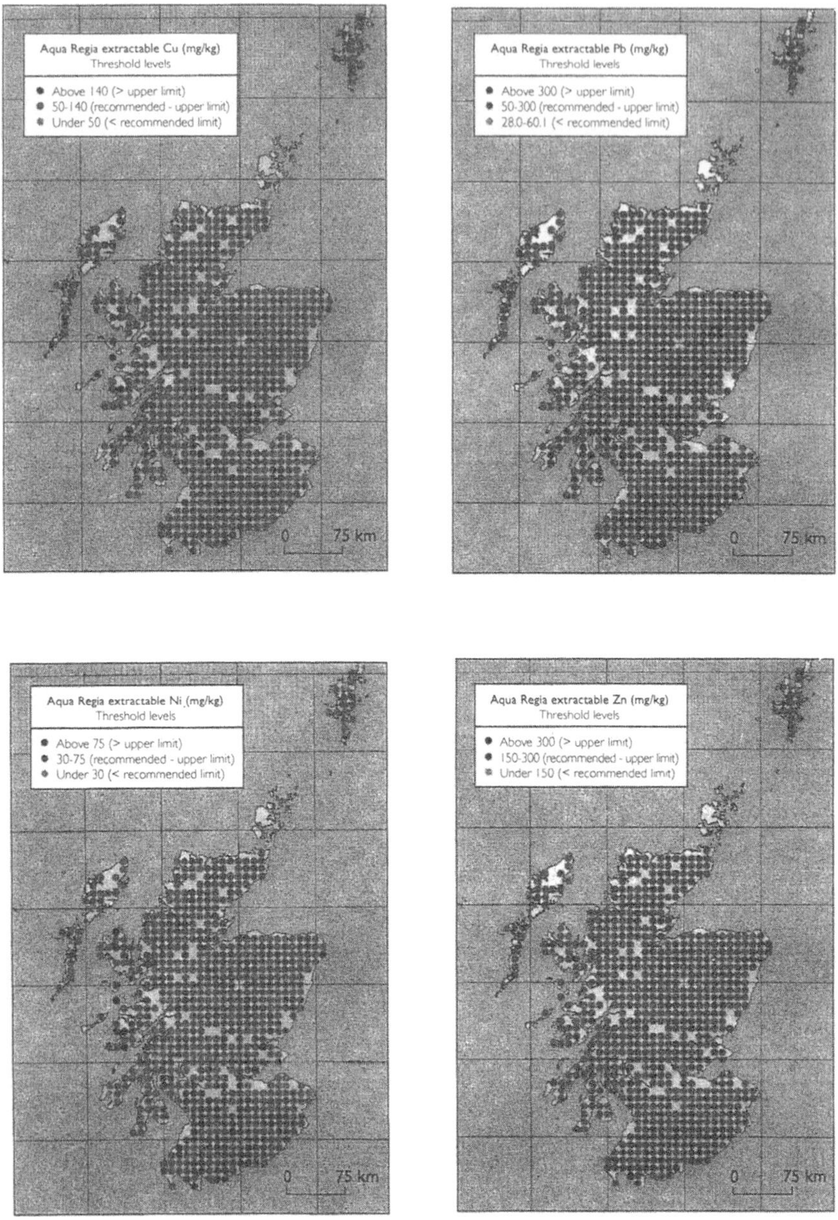

FIGURE 4. Aqua regia extractable contents of copper, lead, nickel and zinc in topsoils in the National Soil Inventory of Scotland plotted in classes based upon EC limits. (see also coloured plate at end of book)

3.2.2 Process based indicators

There are many routes by which heavy metals may enter the soil environment including atmospheric deposition, the application of metal-containing wastes to land and redistribution from mining wastes by aerial and fluvial transport. Although measurement of the total contents is of value in a regulatory context it is less useful in an ecological one where the key factor is the bioavailability of the potentially toxic element. Many factors may contribute to this [35] but one of the key chemical processes is specific adsorption. This process largely controls the partition of metal between the solid and liquid phases and many studies have shown that particular soil components, including the contents of clay, organic matter and iron oxide, and soil conditions, such as pH, play a major role in determining the extent to which adsorption occurs [38].

Blume and Brummer [39] have developed a scoring system - effectively weighting different soil parameters - to predict the metal binding capacity of mineral (< 20% organic matter) topsoils based on relatively easily measured or assessed soil properties. The properties used are soil pH (in $CaCl_2$), soil organic matter content, clay content and sesquioxide concentration using the degree of redness of the soil as a surrogate. The procedure also accounts for the different physicochemical properties of the different metals. An initial rating is made using soil pH and this is subsequently adjusted to reflect the influence of the other soil properties. Examples of the approach when applied to two contrasting mineral soils are shown in Table 2. The final "scores" are then converted into six metal binding capacity classes: none, very weak, weak, medium, strong and very strong.

Table 2 Assessment of Zinc Binding Capacity in two contrasting mineral soils

SOIL A		SOIL B	
pH (CaCl) of topsoil	6.1	pH (CaCl) of topsoil	4.9
Binding strength score	**4.5**	**Binding strength score**	**2.0**
Organic matter content %	9	Organic matter content %	6
Binding strength score	**0.5**	**Binding strength score**	**0**
Topsoil texture	SCL	Topsoil texture	LS
Binding strength score	**0.5**	**Binding strength score**	**0**
Topsoil colour	7.5YR3/3	Topsoil colour	10YR3/2
Binding strength score	**0.5**	**Binding strength score**	**0**
Total score	6.0	Total score	2.0
BINDING STRENGTH	**very strong**	**BINDING STRENGTH**	**weak**

Initially, the method was applied to soils in Scotland likely to receive sewage sludge, an important entry route of metals into the soil environment [38]. However, the method has been extended to include organic soils where, although the key process remains specific adsorption, the soil attributes that contribute to the binding capacity differ. Thus, the methodology has been applied to the soils at the 10km intersects of the Scottish National Soils Inventory to give a national assessment of the sensitivity of Scottish soils to potentially toxic elements [36]. In addition, a number of general observations may be made concerning the interaction of heavy metals and Scottish soils e.g.

- All the metals are much more strongly bound to mineral horizons than to organic ones. For example, only 1% of the mineral soils have a very weak or no binding capacity for cadmium compared to 65% for the organic soils. This finding would be consistent with metal adsorption being controlled by pH as the median pH value for mineral horizons is pH 4.9 and for organic horizons pH 3.3.
- The order of metal binding is $Zn<Cd<Ni<Pb$ in the mineral soils. However, Cd is the most weakly bound if soil pH is the only criterion used. Cadmium and nickel have a stronger binding to organic matter than zinc although this is not catered for in the existing UK sludge regulations; soil pH is the soil criterion.
- The metal binding assessments are very sensitive to a reduction in soil pH particularly for the more mobile metals such as cadmium and zinc. In contrast, the behaviour of lead does not vary within this pH range. This is of particular relevance for those agricultural soils which may receive sewage sludge, where the pH is maintained by periodic liming.
- The structured sampling framework of the National Inventory dataset, and the consistency of the data within it, have provided a valuable insight into the sensitivity of Scottish soils to inputs of potentially toxic elements.
- Finally, the approach allows for a comparison between soil quality indicators based on metal content with those based on binding capacity (Figure 5).

It is clear that, on the basis of composition, lead poses a greater hazard to soil quality in Scotland than zinc. However, when the capacity of soils to immobilise any added metals is considered a different picture emerges. In particular, the strong adsorption and hence immobilisation of lead compared to the relative mobility of zinc shows that zinc represent a greater hazard and, consequently, will have a greater impact on soil quality. Current work is now directed at assessing whether the different binding classes of soils result in different biological impacts as assessed by microbial assays. This work clearly leads towards a critical loads approach for potentially toxic elements. This may ultimately have as great an influence on policies for sustainable management of soils with respect to potentially toxic elements as the critical loads approach for acidifying inputs has had on policies to control gaseous emissions.

4 Conclusions

The concept of soil quality has considerable potential as a unifying framework in which the role of soils in the wider environment may be integrated and one whereby soils may be afforded the same level of protection as that enjoyed by other environmental media. In addition, bearing in mind the time-scale often involved in the recovery of damaged

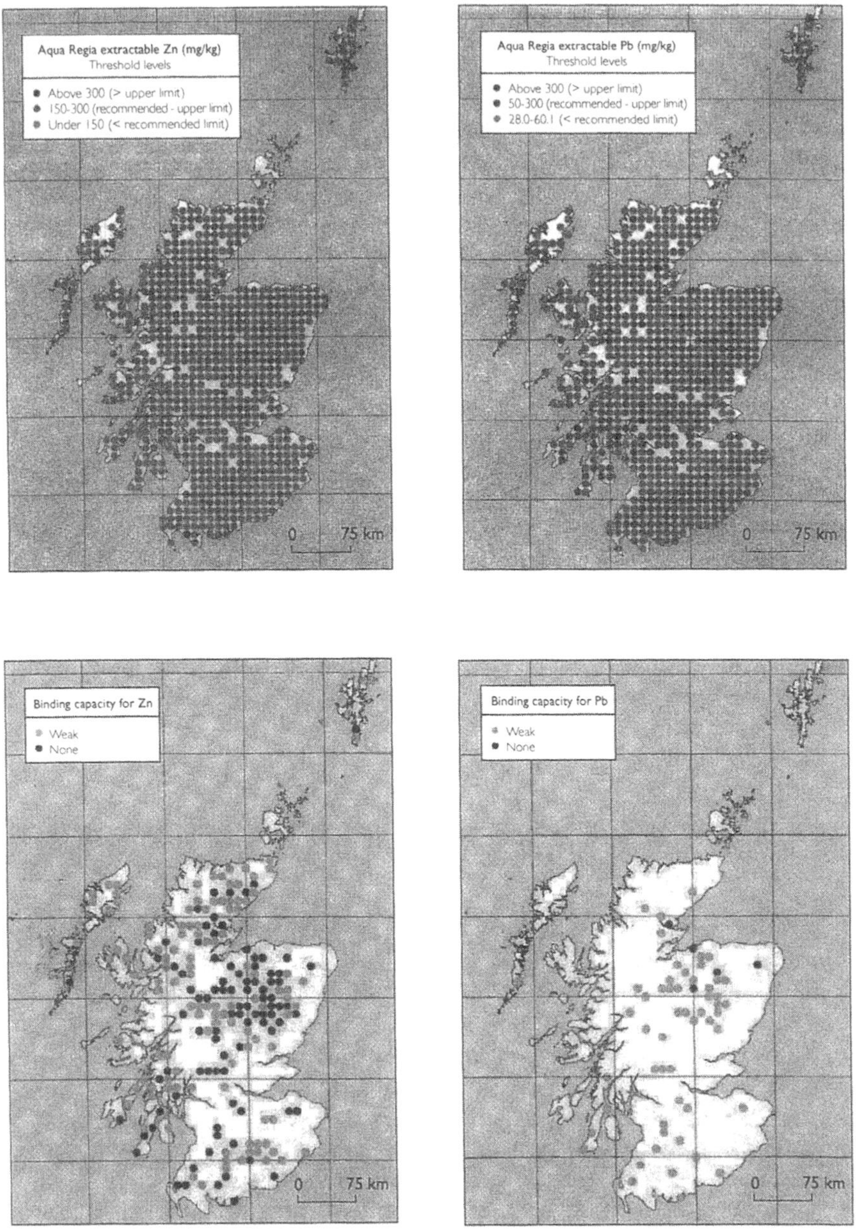

FIGURE 5. Comparison of compositional and process based indicators for lead and zinc in topsoils in the National Soil Inventory of Scotland. (see also coloured plate at end of book)

soil ecosystems soil protection may have much to offer strategies for sustainable development throughout the world. However, on the basis of past experience of land management, where wise stewardship attempts to harness natural processes rather than ignore them or even work against them, it may be more useful in the longer term to develop the concept of process-based indicators rather than to expend efforts in refining compositionally based indicators. That is not to say that composition cannot be used as a first attempt but rather that the acquisition of such data should not hinder our attempts to increase our knowledge base at the process level and devise methods to deploy that knowledge more effectively than we have hitherto.

5 References

1. Royal Commission on Environmental Pollution (Nineteenth Report) (1996). *Sustainable Use of Soils*, HMSO, London.

2. Blum, W.E.H. (1993). Soil Protection Concept of the Council of Europe and Integrated Soil Research, in H.J.P. Eisjackers and T. Hamers (eds.), *Integrated Soil and Sediment Research: A Basis for Proper Protection*, Kluwer Academic Publishers, Dordrect, pp37-47.

3. De Haan, F.A.M. (1998) Soil Quality Evaluation, in: F.A.M. de Haan and M.I. Visser-Reynevald (eds.) *Soil Pollution and Soil Protection*, International Training Centre (PHLO), Wageningen Agricultural University, Wageningen, Netherlands, pp. 1-17.

4. Department of the Environment (1994) *UK Strategy for Sustainable Development*, HMSO, London.

5. Department of the Environment (1996) *Indicators of Sustainable Development for the United Kingdom*, HMSO, London.

6. Eisjackers, H. (1998) *Soil Quality Assessment in an International Perspective: Generic and Land-use Based Quality Standards*, Ambio, 27, 70-77.

7. Karlen, D.L. and Stott, D.E. (1994) A Framework for Evaluating Physical and Chemical Indicators of Soil Quality, in J.W. Doran, D.C. Coleman, D.F. Bezdicek and B.A. Stewart (eds.), *Defining Soil Quality for a Sustainable Environment*, Soil Science Society of America Special Publication Number 35, Madison, Wisconsin, USA, pp53-72.

8. Doran, J.W. and Parkin, T.B. (1996) Quantitative Indicators of Soil Quality: A Minimum Data Set, in J.W. Doran and A.J. Jones (eds.), *Methods for Assessing Soil Quality*, Soil Science Society of America Special Publication Number 49, Madison, Wisconsin, USA, pp25-37.

9. Ragg, J.M. and Clayden, B. (1973) *The Classification of some British Soils according to the Comprehensive System of the United States*, Soil Survey Technical Monograph No3, Rothamsted Experimental Station, Harpenden.

10. Cruickshank, J.G. (1997) *Soil and Environment: Northern Ireland*, Agricultural and Environmental Science Division, Department of Agriculture, Northern Ireland and The Agricultural and Environmental Science Department, Queens University of Belfast, Belfast, Northern Ireland.

11. Jordan, C., Cruickshank, J.C., Higgins, A.J. and Hamill, K.P. (1997) *The Soil Geochemical Atlas of Northern Ireland*, The Department of Agriculture for Northern Ireland, pp107.

12. Gauld, J.H. and Paterson, E. (1996) The information base for Scottish soils, in C. Le Bas and M. Jamagne (eds.) *Soil Databases to Support Sustainable Development*, European Commission Joint Research Centre & INRA, pp 93-97.

13. Langan, S.J., Paterson, E. and Taylor, A.G. The Scottish Soil Resource: Its current status and future priorities for its management, in A.G.Taylor, J. Gordon and M. Usher (eds.), *Soil, Sustainability and the Natural Heritage*, HMSO, Edinburgh, pp69-92.

14. Le Bas, C. and Jamagne, M. (1996) *Soil databases to support sustainable development*, European Commission Joint Research Centre & Institute National Recherche Agronomique.

15. Glentworth, R. (1962) The principal genetic soil groups of Scotland, in *Proceedings International Society of Soil Science Conference, Transactions Commissions IV and V*, pp 2-9. New Zealand.

16. Macaulay Institute for Soil Research (1984) *Organisation and methods of the 1:250 000 Soil Survey of Scotland*, Macaulay Land Use Research Institute, Craigiebuckler, Aberdeen.

17. MacDonald, A.M., Matthews, K.B., Paterson, E. and Aspinall R.J. (1994) The impact of climate change on the soil/moisture regime of Scottish mineral soils, *Environmental Pollution*, **83**, 245 - 250.

18. Matthews, K.B. MacDonald, A.M. Aspinall, R.J. Hudson, G. Law A.N.R. and Paterson E. (1994) Climatic soil moisture deficit - climate and soil data integration in a GIS, *Climatic Change*, **28**, 273 - 287.

19. McGrath, S.P. and Loveland, P.J. (1992) *The Soil Geochemical Atlas of England and Wales*, Blackie Academic, Glasgow.

20. Birse, E.L. (1980) *Plant Communities of Scotland A Preliminary Phytocoenonia*. Soil Survey of Scotland Bulletin No 4. The Macaulay Institute for Soil Research, Craigiebuckler, Aberdeen

21. Avery, B.W. (1980) *Soil Classification for England and Wales (Higher Categories)* Soil Survey Technical Monograph No 14, Rothamsted Experimental Station, Harpenden.

22. Clayden, B. and Hollis, J.M. (1984) *Criteria for differentiating soil series*. Soil Survey Technical Monograph No 17, Rothamsted Experimental Station, Harpenden.

23. Food and Agriculture Organization (FAO) of the United Nations (1998) *World Reference Base for Soil Resources* World Soil Resources Reports 84. FAO Rome 1998.

24. Bull, K.R. (1992) The critical loads/levels approach to gaseous emission control. *Environmental Pollution*, **69**, 105-123.

25. Nilsson, J., and Grennfelt, P. (1988) *Critical loads for sulphur and nitrogen*, Report from a workshop held at Skokloster, Sweden March 1988, Published by Nordic Council of Ministers, Copenhagen.

26. Langan, S.J., and Wilson, M.J. (1994) Critical loads of acid deposition to Scottish soils, *Water, Air and Soil Pollution*, **75**, 177-191.

27. Hodson, M.E. and Langan, S.J. (1999) Considerations of uncertainty in setting critical loads of acidity of soils: the role of weathering rate determination, *Environmental Pollution*, **106**, 73-81.

28. Hornung, M., Cresser, M., Bull, K., Langan, S.J., Loveland, P., Hall, J. and Smith, C. (1995) An empirical map of critical loads of acidity for soils in Great Britain, *Environmental Pollution*, **90**, 301-310.

29. Posch, M., Hettelingh, J.-P., de Smet, P.A.M. and Downing, R.J. (1997) *Calculation and mapping of critical thresholds in Europe*, Status report 1997, Centre for Co-ordinating Effects, RIVM, Netherlands, pp163.

30. Plant, J.A., Klaver, G., Locutura, J., Salminen, R., Vrana, K. and Fordyce, F.M. (1996) *Forum of European Geological Surveys (FOREGS) Geochemistry Task Group 1994-96 Report*, Technical Report WP/95/14, British Geological Survey, Keyworth, Nottingham, pp52.

31. Reaves, G.A. and Berrow, M.L. (1984a) Total copper contents of Scottish soils, *Journal of Soil Science*, **35**, 583-592.

32. Reaves, G.A. and Berrow, M.L. (1984b) Total lead concentrations in Scottish soils, *Geoderma*, **32**, 1-8.

33. Berrow, M.L. and Reaves, G.A. (1986) Total chromium and nickel contents of Scottish soils. *Geoderma*, **37**, 15-27.

34. Wilson, M.J. and Berrow, M.L. (1978) The mineralogy and heavy metal content of some serpentinite soils in North-East Scotland, *Chemie der Erde*, **37**, 181-205.

35. Peijnenburg, W.J.M., Posthuma, L., Eisjackers, H.J.P. and Allen, H.E. (1997) A conceptual framework for implementation of bioavaialability of metals for environmental management purposes, *Ecotoxicology and Environmnetal Safety*, **37**, 163-172

36. Paterson, E., Towers, W., Lumsdon, D.G. and Meeussen, J.C.L. (1997) *Responses of Scottish soils to heavy metal inputs*, Scottish Natural Heritage Review No 61, pp1-55.

37. Blume, H.P. and Brummer, G. (1991) Prediction of heavy metal behaviour in the soil by means of simple field tests. *Ecotoxicology and Environmental Safety*, **22**, 164-174.

38. Towers, W. and Paterson, E. (1997) Sewage sludge application to land - a preliminary assessment of the sensitivity of Scottish soils to heavy metal inputs, *Soil Use and Management*, **13**, 149-155.

PART IV

MANAGEMENT PRACTICES RELATING TO AGRICULTURAL SUSTAINABILITY AND ENVIRONMENTAL SECURITY, EMPHASIZING EXAMPLES FROM CENTRAL AND EASTERN EUROPE

Synopsis and Overview

This section deals with agricultural management practices that have been adopted in the CEECs to counteract the threats posed by soil degradation of various kinds. It is in this context that *Fotyma* discusses the situation in Poland. Soil quality is simply defined as the sum of its productivity and resilience, the latter term being understood to mean the soil's ability to recover following degradation either by displacement processes such as erosion or *in situ* deterioration such as acidification. *Fotyma* points out that soil quality in Poland is rather low, with inherently low fertility arising from pedological processes acting upon nutrient-poor soil parent materials under unfavourable climatic conditions and generally poor management practices. About a quarter of Poland is threatened by water and wind erosion but at present this does not appear to seriously affect soil productivity and effective counter measures are well-understood, even if they are not widely applied because of economic and other reasons. However, the single most important factor limiting soil quality is the high degree of soil acidification, mostly through natural causes but significantly exacerbated by anthropogenic inputs through fertiliser applications and atmospheric inputs. Poor soil physical properties are also a major cause of loss of soil quality. The most effective measures that have been adopted to maintain soil quality are, in the main, the traditional ones, such as rational crop rotation, organic and mineral fertilisation, liming and soil tillage. Tillage practices are based on the mouldboard plough, which seems appropriate to local conditions. Methods such as direct drilling have had a negative influence upon soil productivity, at least in the short term.

At present, municipal sewage sludge is little used in Poland as a soil amendment, particularly to agricultural land, despite the well-known benefits of the material as a fertiliser. The reasons for this are to be found in concern that misuse of sewage sludge could be ultimately detrimental to the quality of agricultural soils as well as the very stringent conditions and criteria regulating its use. However, *Stuczynski et al*, show that even metal-contaminated sludges may be viably and safely used for purposes of land reclamation, such as in the revegetation of smelter wastes. It was found that the sludge provided essential nutrients such as N, P and other elements needed to establish vegetation, as well as improving soil structure and tilth, increasing organic matter content and enhancing moisture retention and biological activity. A land reclamation case study from a metal smelting area in Silesia is described, based upon the use of recycled biosolids. The results have been encouraging enough to stimulate interest in other CEECs which face similar

M.J. Wilson and B. Maliszewska-Kordybach (eds.),
***Soil Quality, Sustainable Agriculture and Environmental Security in Central and Eastern Europe*, 207–210.**

problems of soil reclamation and waste management and could be a first step towards the implementation of large scale programmes for the wider use of sewage sludge in agriculture.

Efforts to analyse the conditions necessary for the development of sustainable agriculture in Hungary are based upon long-term plant nutrition studies involving fertilisation statistics and nutrient balance calculations and in this respect a historical perspective has proved to be useful. *Németh* shows that from the turn of the last century to the late 1950s, N-balances for Hungarian agriculture were strongly negative, the amount applied being much less than that removed in the harvest. With the introduction of fertilisation in the 1960s, balances became positive and increasingly so for the period 1970-1990. From 1991 onwards, however, the nitrogen balance became strongly negative, a trend that has continued to the present day. *Németh* describes the Hungarian fertiliser recommendation system which was developed in order to satisfy the crop's demands but at the same time to diminish or eliminate the contamination of the environment. The system is based upon grouping soils into a limited number of genetic categories and then classifying their nitrogen supplying capacity on the basis of the organic matter content in the plough layer as well as on soil texture and other factors. Evaluation of the data sets from long-term experiments shows that it is possible to improve nitrogen management practices so as to secure stable and sufficient yields whilst protecting the environment at the same time.

In Russia, the loss of optimal soil physical properties has led to extensive soil degradation and erosion and tillage systems have been developed aimed at minimising these processes. *Bondarev* describes how new practices are oriented towards minimum tillage, as well as systems involving sub-surface cultivation. Generally, a widely used system involves cultivation of the surface with a stubble plough and repeated loosening of the deep sub-surface. This effectively decreases wind and water erosion, helps conservation of soil moisture through the accumulation of stubble at the soil surface, and decreases the rate of loss of soil humus through mineralisation. One of the main problems is an increase in weeds but this can be dealt with by the use of bare fallow in rotations in conjunction with the use of herbicides. *Bondarev* recommends that cultivation should take into account the difference between the actual and optimal parameters of the soil physical properties. Such optimal parameters have been determined for properties such as density, structure, porosity, water permeability and water availability for many of the important soil types in Russia, but usually the Russian farmer does not know the actual state of his soil in terms of its physical properties. It is, therefore, important that such data can be provided by a Soil Advisory Service and that methods be developed enabling soil physical properties to be determined rapidly.

As a result of loss of organic matter, soil degradation leading to a decline in natural fertility and ultimately to water and wind erosion is a serious problem affecting the agricultural soils in the Ukraine. According to *Shikula,* this is a consequence of the disruption of the natural self-regulating processes that promote the maintenance of fertility in typical Ukrainian chernozemic soils. The process involves the assimilation of carbon dioxide and carbonic acid from the soil by heterotrophic microorganisms, leading to the formation of new protein-like substances. These substances restore the nutrient status of the soil and also improve its physical properties. Evidence is presented to show that mouldboard ploughing is an inappropriate method for cultivating Ukrainian chernozems and *Shikula* recommends a form of soil tillage that does not involve inversion of the furrow slice. This is more in keeping with natural soil formation processes and maintains the process of humification in crop production systems. It is concluded that a better understanding of the self-regulatory processes in soils is essential for their better management.

Building on Shikula's ideas, *Kapshtyk et al* describe some field experiments in Ukraine aimed at taking advantage of the natural processes for the reproduction of soil fertility described above. The experiments involved non-plough tillage and mulching the soil surface with straw, corn/sunflower stalks and other plant residues, together with the application of organic and mineral fertilisers and comparison with plots that had been conventionally tilled. The experiments were continued over a period of 21 years and it was concluded that real benefits were obtained in terms of higher yields, fewer fertiliser and agrochemical requirements and reduced labour costs. The field experiments suggested that fundamental changes to the soil occurred in distinct stages involving seasonal and long-term cycles of labile organic matter. After 5 years of conservation farming, it was concluded that the natural processes and interconnections that had been destroyed by mouldboard ploughing had been restored and after 9 years the soil had acquired the ability to self-regulate its productivity. Finally, a third change occurred after 15 years when it was considered that the soil was now functioning in a completely natural way. The changes were marked by higher crop yields, when compared to the mouldboard ploughing system and there was less dependence upon weather and fertilising conditions.

This section concludes with three papers from Lithuania describing various management techniques that have been used to combat various types of soil degradation. *Jankauskas and Jankauskiene* describe control measures that have been effectively used in field experiments on soils that are particularly susceptible to erosion. Selection of long-term perennial grasses in conjunction with grass-grain crop rotations were particularly effective in decreasing soil loss on slopes up to 10° and additional improvements were obtained through the use of reduced tillage and the application of mineral and organic fertilisers.

Another threat to the sustainability of Lithuanian soils is through acidification, arising through a combination of natural soil processes and anthropogenic inputs. *Pleseviciene and Jankauskas* show that a large proportion of Lithuania's soils are being affected in this way and that liming, in conjunction with the long term periodic use of cattle manure was the most effective way of dealing with the problem.

Finally, *Staputis* examines the interacting effects of liming and mineral fertilisation in order to determine the soil factors which have the greatest influence on plant growth and to investigate the accumulation and distribution of nutrients in the plant biomass under different conditions. Optimal soil reactions varied widely according to soil type, depending to a large extent upon soil texture and organic matter content.

The main conclusion to be drawn from this set of papers is that the main threat to soil quality in the CEECs is a decline in inherent fertility through acidification and erosion. This decline is reflected in loss of organic matter content and has been exacerbated by the virtual collapse of inputs of fertilisers and lime since 1989, combined with the employment of inappropriate management practices. Traditional soil management techniques should go a long way to rectify this situation, although in some countries it seems that reduced tillage systems have much to recommend them.

IMPACT OF MANAGEMENT PRACTICES ON SOIL QUALITY IN CEE COUNTRIES WITH PARTICULAR REFERENCE TO POLAND

M. FOTYMA
Institute of Soil Science and Plant Cultivation
24-100 Pulawy, Czartoryskich 8

1. Introduction

Soil quality has been defined as its capacity to produce healthy and nutritious crops, resist erosion and reduce the impact of environmental stresses on soil and on plants [12]. In a more practical approach, soil quality is the sum of its productivity and its resilience. The simplest definition of resilience is the soil's ability to recover after disturbance caused by external and internal stresses [6]. The antithesis of soil quality is soil degradation, which in an agricultural context may be thought of as the loss of capacity to produce high yields of nutritious crops. Two main groups of soil degradation processes are recognised. The first one relates to the displacement of soil materials, mainly due to erosion. The second group deals with soil deterioration *in situ* by chemical, physical or biological soil degradation processes. Soil quality can be improved by good management practices , while soil degradation may be hastened by bad ones. The main management practices briefly described in this paper are anti-erosion measures, crop rotation, mineral and organic fertilisation, and soil tillage. So-called canopy management practices, i.e. crop protection, growth regulator application, leaf fertilisation and others are outside the scope of this paper.

M.J. Wilson and B. Maliszewska-Kordybach (eds.),
***Soil Quality, Sustainable Agriculture and Environmental Security in Central and Eastern Europe*, 211–226.**

2. Erosion and anti- erosion measures.

A substantial part of Poland is threatened by water, gully and to a smaller extent by wind erosion (Fig. 1, 2).

Figure 1. Water erosion hazard in Poland [4]

About 30% of the total area including 28% of agricultural land and 2% of forests is potentially prone to water erosion, while very high and high erosion risks occurs on 13% of agricultural land. On these areas special anti-erosion measures should be performed to avoid soil degradation. Gully erosion threatens 18% of the total country area including 14% of agricultural land and 4% forest. Very dense networks of gullies exceeding 0.5 km of gullies per 1km^2 of land cover almost 6% of agricultural land and 1.5% of forests in Poland. The potential water erosion risk is the highest in five regions: Carpathian Mountains, Sudety Mountain, Beskidy Upland, Loessial Plateaus and the Baltic Lake District (Fig. 1). Gully densities in the same regions are the highest (Fig. 2)

Figure 2. Intensity of land break up by gullies in Poland [4]

Water erosion leads to loss of topsoil which is commonly known as surface wash or sheet erosion. This process is usually preceded by soil compaction or crusting, resulting in a decrease of water infiltration rate and hence increasing the runoff processes. The impact of rills and gullies is far less with respect to the total affected area but is a much more spectacular and obvious threat to soil quality compared to the loss of topsoil which may often be almost imperceptible. Control of active gully is very difficult and full land restoration or reclamation is almost impossible. Anti-erosion measures can be divided into three groups: structural, technical, and agrotechnical (Tab.1).

Table 1. Anti-erosion measures [5]

Structural	Technical	Agrotechnical
Increased afforestation and proportion of grassland in agricultural areas. Slopes with over 20% (12 °) inclination excluded from arable crop production. Imposition of anti erosion rules on land and farm planning, contouring of fields.	Construction of erosion prevention zones i.e. landslides, road gullies, quarries. Building of retention and anti-rubble reservoirs as well as gullies and suffusion trenches. Hardening roads and steep slopes. Slope levelling.	Crop rotation with winter crops, perennial crops and cover - crops. Soil mulching and organic manure application. Contour tillage and reduced or minimal tillage. Contour drilling and plant cultivation.

These anti erosion measures are not yet fully implemented both in land use planning and in practical farming. Thus, the situation is different in the five regions of the country most threatened by water and gully erosion. In the Carpathian and Sudety Mountain regions the actual erosion is markedly limited due to the large areas of forests (35 - 44% of total area), permanent grasslands (14 - 19%) and the high share of perennial crops in the arable crop rotations. Such land use structure is related to unfavourable climatic conditions and steep slopes. The worst situation is on the Loessial Plateau which is covered by highly productive loess soils which are highly susceptible to erosion runoff. This is a region of intensive land use with a very high density of rural population. The small extent of afforestation and grassland, and the domination of strip farming fields running downslope, all tend to increase the damage of erosion. Some areas of the territory show the highest gully densities in Europe. In the long run the prerequisite for diminishing the erosion risk here is the adoption of anti-erosion measures and land use planning in an integrated and rational way. A similar situation, with the exception of soils which are sandy loam and loamy sand, occurs in the

Beskidy Upland. In this region of small farms the field are arranged without regard to erosion prevention. In Baltic Lake Districts the soils originating from sandy and loamy deposits are moderately and weakly susceptible to erosion.

The structure of land utilisation is somewhat anti- erosion in character because forests cover 33% and grassland and lakes cover 16% of the total area. The problem in these regions used to be the prevalence of large, state farms which occupied 50% of agricultural lands. The fields were big and arranged without due consideration of anti-erosion measures in the hilly landscape. Today, the former state farms are being restructured and privatised, thus creating opportunities for rearranging the land structure with respect to erosion prevention.

3. Soil biological properties and crop rotations.

Good quality soil is characterised by an optimal content of organic matter, a deep plough layer and high biological activity. Measures of biological activity are gas exchange, enzyme activity, and the number and biodiversity of soil macro- and microbiota including plant pathogens and pests. The most important, soil biological property is the content of organic matter. The content of organic matter depends on climatic conditions, soil origin and granulometric composition, land utilisation, and the level of organic fertilisation. In mineral soils there is a close relation between the natural content of organic matter and the content of clay and silt (Tab. 2).

Table 2. The content of organic matter in mineral soils in Poland [13]

Soil textural group	Soils with respect to silt content		The share in agricultural land %
	not silty	Silty	
sands, slightly loamy sands	1.36	1.54	24.8
loamy sands	1.65	1.52	12.4
light loams	1.98	2.01	18.8
medium and heavy loams	2.84	2.75	13.2
Clays	3.56	2.89	1.0

The majority of soils in Poland are classified as light soils, i.e. belonging to sand and light loam textural groups. Consequently these soils show a low natural content of organic matter. As natural content one understands the content of organic matter in properly managed arable soils. This content can be increased by "good" crop rotation and by abundant farmyard manure application. The influence of management practices on organic matter content can be quantitatively evaluated in terms of organic matter reproduction coefficients (Tab. 3). The reproduction coefficient is the amount of organic matter, expressed in Mg/ha, by which the soil is enriched (+) or impoverished (-) as a consequence of one years cultivation of a given crop or application 1 Mg dry matter of organic manure [8].

Table 3. Organic matter reproduction coefficients (Mackowiak, 1992).

Crop or fertiliser	Coefficient in Mg organic matter / ha for the soils			
	very light, light	Medium heavy	heavy	black earths
Potato, sugar beets 1 ha	-1 26	- 1.40	- 1.54	- 1.02
Maize 1 ha	-1.12	- 1.15	- 1.22	- 0.91
Cereals and rape 1 ha	- 0.49	- 0.53	- 0.56	- 0.38
Pulses 1 ha	0.32	0.35	0.38	0.38
Grasses 1 ha	0.95	1.05	1.16	1.16
Leguminous crops 1 ha	1.89	1.96	2.10	2.10
FYM 1 Mg dry matter	0.35	0.35	0.35	0.35
Slurry 1 Mg dry matter	0.28	0.28	0.28	0.28
Surface wash 1 cm soil/ha	- 1.5	- 2.0	- 2.4	- 3.5

Cereals and oil seed crops (winter rape) impoverish the soil of organic matter by about 0.5 Mg/ year/ ha. The impoverishing power of row crops (potato, beet) and maize is twice or three time stronger. To counteract these negative influences crops that increase the organic matter content must be grown in the rotation or organic manure must be applied in ample quantities. In the last few years the balance of organic matter in Polish soils is worsening due to the unfavourable changes in crop structure and the decreasing number of animals (Tab. 4). These processes will, most probably, continue into the next millennium. There are management practices which, even with the given crop structure and animal number, can improve substantially the balance of organic matter. The most important are: growing cover crops, utilisation of industrial composts, and reduced or minimum tillage. Unlike some Western countries there is practically no import of organic matter to Poland via the concentrates for feeding the animals.

Table 4. The estimated balance of organic matter in Poland (arable soils) per 100 ha.

Crops or animals	1985/1990 years		1991/1996 years		2010/2015 years	
	ha or number per 100 ha	organic matter Mg/100 ha	ha or number per 100 ha	organic matter per 100 ha	ha or number per 100 ha	Organic matter per 100 ha
Row crops	16.4	- 22.9	15.3	- 21.4	10.2	- 14.3
Maize	2.4	- 2.76	2.8	- 3.19	3.1	- 3.56
Cereals and oil crops	61.7	- 32.7	66.9	- 35.5	70.9	- 37.6
Pulses	0.3	0.10	0.4	0.14	0.8	0.28
Perennial crops	12.7	15.2	10.6	12.7	10.8	13.0
Animal units	70.2	61.4	55.8	48.8	53.9	47.2
Total		18.3		6.65		5.02

Besides the balance of organic matter, "good " crop rotation is essential for increasing the number of useful soil micro- and macrobiota and decreasing the number of soil borne pests and pathogens.

Table 5. Antagonistic potential of soil against pathogens infecting winter wheat (*Gaeumannomyces graminis, Cercosporella herpotrichoides*)[10]

Crop rotation	Antagonistic potential of soil against		Yield of winter wheat Mg/ ha
	G. graminis	*C. herpotrichoides*	
wheat-rye-oats-wheat	133	133	4.3
wheat-rye-pulses-wheat	160	186	4.1
wheat-rye-wheat-wheat	106	38	3.0
wheat-wheat-wheat-wheat	54	36	2.0

The total mass of useful biota, particularly earthworms and larvae of insects in the soil under cereals monoculture is two to three times less in comparison to Norfolk rotation. A very important soil feature is so-called antagonistic potential against the soil- borne pathogens which damage some crops (Tab. 5).

A proper crop sequence and the dense canopy characteristic of a " good " crop rotation also contribute substantially to the diminution of soil infestation by seeds and rhizomes of annual and perennial weeds . The number of interventions and the doses of herbicides can be, hence, significantly reduced.

4. Soil fertility and fertilisation.

Soil fertility is defined as the capacity to provide plants with the necessary amount of mineral nutrients and water. Fertile soil should not be also contaminated by harmful substances like heavy metals or organic contaminants. As noted earlier , soil fertility is part of soil quality, while soil contamination falls into the category of soil degradation. Most of the soils in Poland show low natural fertility expressed by a high degree of acidity and a low content of available nutrients (Tab. 6).

Tabel 6. Soil fertility indices in Poland (Agrochemical Laboratories survey).

Soil category	Percentage of soils in the category in respect to			
	acidity, pH	phosphorus	potassium	magnesium
very acid/ very low	26	12	16	20
Acid/ low	32	27	27	22
Slightly acid/ medium	25	26	27	26
Neutral/ high	14	16	13	14
Alkaline/ very high	3	19	17	18

The main reasons for such low inherent soil fertility are the origin of soils from glacial, outwash, deposits and the predominant podzolization pedological process involving leaching of bases from the upper soil horizons. The process of soil acidification is hastened by anthropogenic influences like application of nitrogen fertilisers and the deposition of sulphur and nitrogen compounds from industrial and urban emission (Tab. 7).

Table 7. The extent of anthropogenic soil acidification [1]

Source of acidification	1985/90 years		1991/1996 years	
	Kg/ha	kmol H^+/ha	Kg/ha	kmol H^+/ha
Nitrogen fertilisers, N	72.5	3.01	41.3	1.69
Sulphur deposits, SO_2	115	3.58	89.4	2.79
Nitrogen deposits, NO_2	42.1	0.63	36.5	0.57
Total		7.22		5.05
Liming equivalent, CaO	202	7.22	141	5.05

Anthropogenic soil acidification processes are not counteracted sufficiently by limestone application. In spite of a State-subsidised liming programme the consumption of limestone is far below the demands evaluated on the basis of the current status of soil acidification and the expected course of acidification processes in the near future. The consumption of limestone was about 180 kg CaO per ha of agricultural land in 1986/90 and decreased to about 120 kg CaO per ha in 1991/1996. As reflected in Tab. 7, these amount do not even balance the losses of calcium due to anthropogenic processes.

The soil fertility indices for phosphorus and potassium showed a steady improvement until the late 1980s due to the high consumption of fertilisers and the positive balance of these nutrients. Since 1991, however the consumption of mineral fertilisers sharply decreased in consequence of removal the subsidies and the worsening economic situation of the farmers. In the years 1991-1996 the balance of all nutrients was negative and the same situation is expected until 2015, though the demand for fertilisers shows again an increasing trend (Tab. 8). The efficiency of fertilisers in Poland seems to be low in comparison to Western standards. The average input of nutrients in 15 EU countries in 1991/96 was estimated at 1.96 kg N, 0.82 kg P_2O_5 and 1.74 kg of K_2O per cereal unit of global crop production. These figures are about 30 % lower than the corresponding ones for Poland. They can be explained by worse soil and climatic conditions in Poland, which can only be compensated for by increased fertiliser applications, and by the accumulation of nutrients in the soils of Western countries in following long term high application of mineral and organic fertilizers. Good fertilisers practices , including optimal doses of NPKCaMg, are the main prerequisite for soil quality in Poland. The low consumption of limestone poses the threat of soil degradation through increasing acidification.

5. Soil physical properties and tillage practices.

It is much more difficult to quantify and to find threshold values for soil physical properties than for soil chemical properties. The two most commonly used soil physical properties are the water holding capacity and the soil compactness or density.

Water holding capacity depends mainly on the soil granulometric composition, i.e. soil textural class (Tab. 9). It has been already mentioned that most of the soils in Poland belong to the "light" category (see Tab.2) and therefore store small to moderate amounts of water available for plants. Crops commonly grown in the country (small grains, winter rape, potato) use water mainly found in the soil layer at a depth of 0 - 60 cm (Fig. 1). Easily available water stored to this depth suffices during the full vegetation period for only few days of unlimited evapotranspiration. Thus the yield of arable crops is highly variable and rainfall- dependent

Table 8. Fertiliser consumption and nutrient balances in Poland [2].

Characteristics	1985/90 years	1991/96 years	2010/2015 years
Nitrogen N kg/ha or kg/cereal unit of the global crop yield			
Uptake by crops	102.4	86.6	124
Input in organic fertilisers	45	36	35
Input in mineral fertilisers	72	41	72
Total input/ cereal unit	3.69	2.81	2.79
Phosphorus P_2O_5 kg/ha or kg/ cereal unit of the global crop yield			
Uptake by crops	37.6	32.1	46
Input in organic fertilisers	20	16	16
Input in mineral fertilisers	47	15	28
Total input/cereal unit	2.11	1.13	1.15
Potassium K_2O kg/ ha or kg/ cereal unit of the global crop yield			
Uptake by crops	116.3	95.8	131
Input in organic fertilizers	70	55	54
Input in mineral fertilizers	61	21	43
Total input/ cereal unit	4.13	2.27	2.53

The possibility of influencing soil water status by management practices is to a large extent limited to combating soil crusting. The most effective measures aim to shift evapotranspiration towards transpiration at the cost of evaporation. It is known that nitrogen fertilisation is effective in this respect and increases substantially the efficiency of water use by plants. It should be mentioned that irrigation, except for horticulture production, is practically non-existent in Poland. The reasons are high costs of installations, lack of water sources and the variable efficiency of this practice over the long run.

Table 9. Water holding capacity of different soils in Poland [14]

Soil textural group	Field water capacity mm/ 100 cm	Total available water mm/ 100 cm	Easily available water mm/ 100 cm
Loose sand	110	92	45
Slightly loamy sand	145	117	60
Loamy sand	180	145	75
Light loam	270	185	90
Medium loam	320	200	100
Heavy loam	400	240	120
Clay	460	220	110

Another very important physical parameter is soil compactness or density. The results of numerous vegetation and field experiments proved the close relationship between crop growth and development, including final yields, and the degree of soil compactness or soil density. Crop yield is particularly damaged by crusting and the high density of the upper soil layers (called sometimes tilth) because of the adverse effects on germinating seed and plant seedlings(Fig. 3) . Once the canopy is poorly established it can not fully recover in the latter stages of growth and development. Less but still quite negative is the influence of high soil density in the deeper layers most often know as a " plough sole ". This is a common phenomenon in Poland due to the low draught power of light tractors and ploughing from year to year at the same depth. Plough sole is one of the factors limiting the yield of sugar beet grown on silty soils. It is very difficult to quantify the threshold values for soil compactness or density, because they depend on the soil type, depth in the profile, soil water content, crop species (probably even crop variety), and the stage of crop growth and development . Recently the concept of expressing soil density in relative values , i.e.as a fraction of natural [15] or maximal [7] density offers new possibilities.

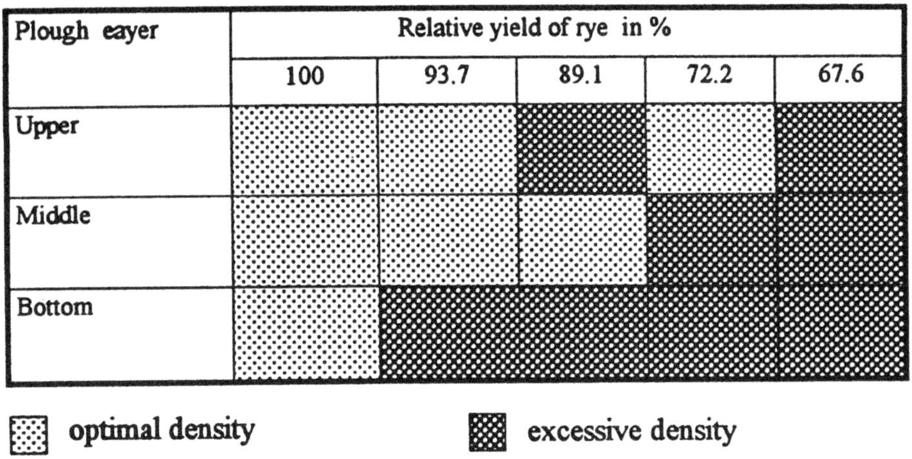

Figure 3.The influence of soil plough layer density on the relative rye yield [7]

6. Tillage practices

Soil physical parameters are influenced by all management practices such as crop rotation, and organic fertilisation but first of all by soil tillage. After the WW II the so called Polish system of soil tillage was established in the country, mainly due to the research and dissemination of the works of Prof. Swietochowski and his followers. In this system the whole process of soil tillage is split into several cultivation subsets executed in strict time order. The first is the post harvest cultivation set composed usually of skimming , disking and harrowing. The next is the summer-autumn pre sowing cultivation set or pre-winter cultivation set , depending on the crop sequence in the rotation. The main tool is the mouldboard plough supplemented by roller, cultivator or harrow. For spring crops the last one is the spring cultivation set including dragging, harrowing and/or disking. The last cultivation set was followed by drilling or planting the crop and then by pre- emerging or post- emerging cultivation.

In this system one cultivation set followed another and in each set the several cultivation processes were applied separately with the recommended time span between them. For instance ,the optimal time span between the summer-autumn plough and the drilling of winter cereals was to be 3 - 4 weeks which presumably allowed for the proper soil settlement. The system fits perfectly well to the contemporary horse draught power and to the very limited choice of agricultural machinery. The classical Polish system of soil tillage is still common among the small farmers poorly equipped with tractors and machines. The sight of a low-power tractor dragging a horse plough or harrow can still be seen in some regions of Poland.

Today , in medium sized, and the big farms a complex soil tillage system is gaining in popularity. In this system several tillage practices are executed simultaneously by using complex cultivation units i.e. ploughing-seed bed preparation unit, cultivation sowing, unit and others. The natural time-course of soil settlement processes are substituted by the proper application of several of kinds of rollers which are an indispensable part of all complex cultivation units. This system, coupled with the use high power but low pressure per area unit tractors, has a positive influence on physical soil parameters including soil density. The philosophy of classical and complex soil tillage systems is in fact the same and is based on the mouldboard plough.

Simplified soil tillage has been researched in the last decade and applied to some big farms. Results show that the elimination of the mouldboard plough and particularly direct drilling results in lower crop yields, but also results in an increase in the work and draught force efficiency (Tab. 10).

Table 10. The influence of tillage practices on crop yield [9]

Tillage system	Yield of successive crops in rotation Mg/ha			Crop production
	horse bean	winter wheat	spring barley	cereals units/ha
Classical	3.57	5.34	3.94	52.6
Simplified	3.51	4.70	3.54	48.9
Direct drilling	3.22	4.67	3.16	46.3

The yield decreases in simplified or direct-drilling tillage systems can be explained mainly by the stronger competition of weeds. Nevertheless, even after elimination of weeds by herbicide treatments, the yields were still lower, most likely due to the negative changes in soil physical properties manifested at least in the few first years after the elimination of mouldboard plough. This problem needs further investigation using large scale field experiments and by applying the more complex approach of sustainable crop production.

7. Conclusions

Due to pedological processes, unfavourable climatic conditions and generally poor management practices the soil quality in Poland is rather low. This, in turn has a negative impact on the nation's food self-sufficiency. Water and wind erosion threatens to some extent about a quarter of the territory but with the exception of the Loessial Plateau, it does not seem to seriously influence soil productivity. Anti-erosion measures are well known but not fully practiced , even in threatened areas, due to economic and structural reasons . The most important single factor limiting soil quality is the high degree of soil acidification. Low pH is an inherent feature of Polish soils and is exacerbated by anthropogenic processes. These processes are not counteracted sufficiently by limestone application , though the country has a huge deposits of carbonate and dolomite rocks. In the last few years another limiting factor has been the low use of organic and mineral fertilisers and the negative balance of nutrients. Tillage practices based on the use of mouldboard plough seem to fit local conditions. Elimination of the plough and direct drilling has had a negative impact on soil productivity.

8. References

1. Filpek, T and Kaczor, A. (1997) Dynamika antropogenicznej preji zakwaszenia rolniczej przestrzeni produkcynej w Polsce. *Ochrona I wykorzystanie rolniczej przestrzeni produkcyjnej Polski,* IUNG pp 35-44

2. Fotyma, M. and Krasowicz, S (1997) Polish agriculture, the present and the future. IFA Production and International Trade Committee Meeting, Warsaw, 14-15 October.

3. Józefaciuk, A and Józefaciuk, Cz. (1992) Struktura zagro e nia eroz wodn fizjograficznych krain Polski, *Pami tnikPu awski.* **Supplement 101,** 23-50

4. Józefaciuk, Cz and Józefaciuk, A. (1992) Struktura zagro enia erozj wodn fizjograficznych krain Polski, *Pami tnik Puawski* **Supplement 101,** 51-66.

5. Józefaciuk, A. and Józefaciuk, Cz. (1995) *Erozja agroekosystemów,* Biblioteka Monitoringu Środowiska, Warszawa.

6. Lal, R. (1994) Sustainable land use systems and soil resilience, in D.J. Greenland and I.Szabolcs (eds.), *Soil resilience and sustainable land use,* CAB International, pp. 41-76.

7. Lipiec, J. and Simona, C. (1994) Role of soil and climate factors in influencing crop responses to soil compaction in Central and Eastern Europe in.B.D. Soane and C.Van Ouwerkerk (eds.) *Soil compaction in crop production,* Elsewier, Amsterdam.

8. Maćkowiak, Cz. (1995) in Fotyma, M. and Mercik, S. *Chemia Rolna,* PWN Warszawa, pp.234.

9. Malicki, L. And Podstawka-Chmielewska, E. (1997) Soil cultivation method and yield of the crop rotation link: horse bean-winter wheat - spring barley, *Bibliotheca Fragmenta Agronomica.* **2B/97,** Puławy pp. 443 - 444.

10. Myśków, W. (1997) in Fotyma, M. and Kuś, J. Oddziaływanie rolnictwa na środowisko glebowe, *Ochrona i wykorzystanie rolniczej przestrzeni produkcyjnej Polski*, IUNG, Pulawy, pp. 155 - 171.

11. Oldeman, L.R. (1994) The global extent of soil degradation, in D.J. Greenland and I.Szabolcs (eds.), *Soil resilience and sustainable land use,* CAB International, pp. 99 - 127.

12. Papendick, R.I. and Parr, J.F. (1992) Soil quality-The key to a sustainable agriculture, *American Journal of Alternative Agriculture* **7,** 2-3.

13. Pietras, Cz. (1987), in Fotyma, M., Mercik, S., and Faber, A. *Chemiczne Podstawy Żyzności Gleb i Nawożenia,* PWRiL, Warszawa, pp. 57.

14. Ślusarczyk, E. in. Terelak, H. and Lekan, Sz. (1997) Zróżnicowanie środowiska glebowo-rolniczego Polski, *Ochrona i wykorzystanie rolniczej przestrzeni produkcyjnej Polski*, IUNG, Pulawy, pp.1 - 23.

15. Wojtasik, M. (1995) *Gęstość naturalna gleb mineralnych,* WSP, Bydgoszcz.

APPLICATION OF SLUDGES FOR REMEDIATION OF CONTAMINATED SOIL ENVIRONMENT

T. I. STUCZYNSKI
Institute of Soils Science and Palnt Cultivation
24-100 Pulawy, Poland

W.L. DANIELS
Deptartment of Crop and Soil Environmental Sciences, Virginia Tech
Blacksburg, VA 24061-0404, USA

F. PISTELOK
Center for Environmental Control and Protection
40-158 Katowice, Poland

K. PANTUCK
US EPA Region III,
Philadelphia, PA 19107, USA

R.L. CHANEY
Environmental Chemistry Laboratory, ARS-USDA,
Beltsville, MD 20705, USA

G. SIEBIELEC
Institute of Soils Science and Palnt Cultivation
24-100 Pulawy, Poland

1. Introduction

In the past ten years about 3000 waste-water treatment plants have been built in Poland. The result of this development has been a massive increase in overall sludge production. According to data supplied by the Ministry of Environmental Protection, 400,000 dry tons of sludge are generated every year in Poland [1]. Still it is difficult to estimate the amount of sludge which is used in agriculture or for reclamation purposes. However, it can be assumed that this comes to no more than 10 percent of all sludge produced in Poland on a yearly basis.

M.J. Wilson and B. Maliszewska-Kordybach (eds.),
Soil Quality, Sustainable Agriculture and Environmental Security in Central and Eastern Europe, 227–242.

The benefits of sludge-based fertilizer are widely known. In the 1970's and 80's a number of agencies and institutions conducted research regarding the use of sludge-based fertilizers and their effects on crop quality and soil productivity [2, 3, 4, 5]. During this same time period, a number of different laboratories were involved in the analysis of heavy metal and pathogen levels in sludge samples submitted by various municipal treatment plants around the country.

The results of this work should have been used to establish a data base for characterizing sludges on a regional and local basis. However the lack of coordination between these various institutions kept this opportunity from being effectively realized. According to the analytical work carried out by the leading center of agricultural research in Poland, the Institute of Soil Science and Plant Cultivation (IUNG), a number of sludges do not meet existing quality standards due to excessive concentrations of heavy metals (Table 1) or/and pathogens (Table 2) - the data does not include sludges from the Silesia region. Additionally, a number of sludges which would theoretically meet required standards are wasted in landfills, thus providing no benefit whatsoever to the farming community.

TABLE 1 Concentration of heavy metals in municipal sludges from 12 cities in Poland

Metal	Cd	Zn	Pb	Cu
Range (mg kg^{-1})	0.9-843	50-68880	11-881	7-2820
Average (mg kg^{-1})	7.70	1192	112	162
Maximum permissible level (MPL mg kg^{-1})	10	2000	500	1000
Sludges above MPL (%)	18	42.5	4.23	11.27

*IUNG unpublished data

TABLE 2 Sludges from 12 cieties in Poland that do not met the patogen control standards*

Pathogen	Rejection rate (%)
E. Coli	69
Clostridium perfingens	53

*IUNG unpublished data

Regarding the latter, the key factor behind the waste of this material basically comes from two sources. First, we should keep in mind the negative connotation of sewage sludge in every day society, as well as within the environmental community. People do not readily accept sludge-based projects in their own neighborhood simply because they have not been exposed to any sound information as to their benefits. Secondly, there are great bureaucratic problems surrounding the legal procedures regulating the use of sludge. The main problem here does not concern set criteria governing the maximum permissible levels of metals and pathogens since those thresholds in general are similar to acceptable standards in the European Union and in the United States. The problem originates in the lack of coherent standards, the multitude of licensing organizations and in the number of separate legal documents involved [6, 7, 8, 9, 10]. To date there is no comprehensive set of guidelines or references published in a single register. There has been encouraging news of late, that a register of this type is in the works [11], but we do not know what form this will take or whether or not it will be effectively put into use. Moreover, the application process requires approvals from three different institutions: the regional ecology department, regional inspectorate for environmental protection and the regional sanitary and epidemiology station.

However, even when taking into account the difficulties posed by sludge production and its disposal to agricultural land, it is relatively easy to obtain required approvals for using sludge in land reclamation work. This is particularly important in environmentally troubled regions such as the Silesia, which has suffered from smelting and mining activities.

Thus it became obvious that an urgent solution to this problem was needed. To their credit, local authorities in Silesia realized in the early 90's that a demonstration was necessary to encourage the proper utilization of sludge. This was the basis for establishing the Biosolids Subpart of the Silesia Project which addressed the use of biosolids in the reclamation of mining and smelting dumps [12]. Our project was deemed all the more necessary considering the fact that a significant portion of Silesia has been affected by these dump sites. We realized that a simple solution for stabilizing these sites would be to cover them with vegetation in order to reduce leaching of toxic elements, as well as to keep metallic fugitive dust from entering the environment.[1] One of the goals was to evaluate the possibility of adapting the rules of sludge management currently used in the United States to conditions commonly found here in industrial regions of Poland.

2. Characterization of Sensitive Areas in the Silesian District

Mining and smelting in Upper Silesia have a long tradition going back to medieval times. In the 13th century silver and lead smelters

---

[1] This project was inspired by work done in Palmerton, Pennsylvania, by Rufus Chaney and co-workers.

were established near Bytom and Tarnowskie Gory, and the mining and processing of zinc and lead ores has continued to date. Unfortunately, the non-ferrous industry has many adverse ecological effects associated with emissions of metallic dust. The extent of industrial contamination of all elements of the biosphere is enhanced by inadequate care of the environment in general.

The measures undertaken in recent years such as installing scrubbers, introducing closed-cycle industrial water systems and technologies designed to minimize waste production have contributed to a general reduction of air, water and soil contamination. However, these measures have not completely eliminated the impact of toxic waste piles on the environment. Water and wind erosion, which are present in varying degrees of intensity depending on local conditions in Silesia, lead to movement of toxic elements from dump sites to the biosphere. The most troubling risks which come from stockpiling smelter waste arise when dump sites are located in or near urban areas. These sites have a significant effect on nearby populations, and unfortunately this is frequently the case in Upper Silesia. According to data supplied by the Katowice Regional Statistics Department, the total area of smelter sites which require reclamation comes to approximately 300 hectares [13]. These sites are known to contain more than 87 million tons of waste. Each year this amount increases by approximately 400,000 tons. The reclamation of smelter waste sites which is currently being carried out within the framework of the Silesia Project is a joint effort of local government, industry and international research institutions/agencies, including the US Environmental Protection Agency (EPA), the Center for Research and Control of the Environment (OBIKS), Virginia Polytechnic Institute and IUNG. The main objective of the Silesia Project is the development of guidelines concerning all aspects of sludge use for the reclamation of degraded lands and waste sites.

3. Characterization of Sludges Produced in the Katowice Area

Available data indicate that the total yearly amount of sludge produced in the Katowice district comes to 26,000 dry tons. Stabilized sludge accounts for 96 percent of this amount [13]. A significant part of this sludge is used for land application - more than 50 percent. It should be taken into account that more than twenty large waster-water treatment facilities will be completed by the year 2000. Additionally, a number of existing facilities will be modernized which will enable them to increase capacity and overall performance capabilities. It is realistic to assume that by the year 2000 the volume of raw sewage will be doubled. Accordingly, the volume of sludge will also increase. On this basis, we can also state that there will be an increased demand for competent sludge management programs in this region

There are many factors which affect the composition and properties of sludge generated during biological sewage digestion [5, 13]. The main group of factors are as follows: 1) The size and type of a sewer system 2) The percentage of industrial sewage in a specific area 3) The technology used in sludge processing.

There may be significant differences not only between the properties of sludge coming from different plants, but also between different sludges generated at the same plant. As shown in Table 3, in average none of the sludges from the Silesia region met required standards for agricultural use. Most samples had elevated concentrations of zinc, and some also had elevated levels of lead and chromium.

These concentrations resulted from the use of zinc-coated piping in housing water systems. The mobilization of zinc into drinking water and then into the sewer system is enhanced by large doses of chlorine used for water purification. This is necessary to meet sanitary and bacterial standards. It is also evident that sewage from older housing centers tends to contain less zinc.

The main source of lead in sewage is a direct result of industry although lead also finds its way into sewage from street runoff. Due to the fact that this runoff is not separated from the sewer system, it naturally makes its way into municipal sewage.

Table 3 Characterization of sludges from Silesia Region*

Metal	Sludge sample	Range	Mean	S. D.
			$mg\ kg^{-1}$	
Cd	21	3-220	25	52
Zn	21	1350-13000	3551	2734
Pb	21	260-4000	521	951
Cu	21	41-4320	655	1108
Hg	21	126-2980	367	614
Cr	21	17-14030	10221	3213
Ni	21	15-403	74	104

*unpublished OBKIS data

When considering heavy metal sludge contamination we must define the maximum permissible levels of these elements. Obviously, in many cases excessive concentrations of metals in raw sewage have toxic effects on micro-organisms involved in treatment. This also contributes to the contamination of generated sewage sludge. This forces municipalities and industries to build sludge pre-treatment facilities.

The metal content in sludge may also be controlled somewhat by treatment process management in a particular plant. We observed a significant increase of heavy metals in activated sludge whenever its volume in an aerobic tank was too small. It is also noteworthy that deep mineralization of sewage sludge increases metal content.

4. An Approach to Wasteland Reclamation in Silesia

A traditional strategy for the reclamation of wastelands and degraded lands is based on top soiling methods followed by the intensive use of fertilizers and the planting of various grass mixtures. There were successful attempts to revegetate mining waste areas through the application of only mineral fertilizers and the direct planting of grasses [14]. In the end, however, these solutions were not found to be cost effective and sometimes were technically difficult. The primary limitation with topsoiling is the lack of quality soil material. Frequently, the soil material available for reclamation use in the Silesia region was much higher in contaminants than were some sewage sludges. Moreover, this soil material is of poor quality with respect to its content of nitrogen, phosphorous, potassium, other nutrients, organic matter and its adverse physical properties. Taking these arguments into account, it is reasonable to assume that sludge would be a feasible, and possibly quite effective, alternative to traditional topsoiling techniques. The main objective here is to develop and implement techniques for safe use of sludges which would meet all respective ecological, sanitary and hygienic standards [15]. There is a general consensus that clear and concise set of procedures and guidelines such as those which are currently in use in both the United States [16] and Western Europe [17, 18] need to be developed immediately. Such regulations and guidelines must take into consideration practical aspects of sludge disposal, not just thresholds. These would be sampling methodologies of an area designated for treatment, the extent of monitoring the site after treatment, techniques for treating slopes, safe distances from surface waters, drinking water supply facilities, etc.

Another important aspect in the safe use of biosolids takes the form of sludge application rates. It seems feasible that certain rules and practical solutions could be transferred from existing guidelines previously developed in other countries. For example, the implementation of certain standards and methods used for the disposal of sewage sludge as stated in the Federal Register [16] of the EPA would be attractive in this region.

5. A Pilot Biosolid Project

To validate biosolid use for the reclamation of wastelands a pilot project was established in the shutdown site of a Huta Warynski smelting plant. This site contained waste from two different smelting processes - Welz and Doerschel. The Welz process takes place in long, spinning furnaces which were used to enrich low-zinc and lead ores. In this process the load mix is reduced. The zinc oxide produced in this process, together with cadmium and lead compounds, are trapped in scrubbers. The remaining waste is transferred to the waste pile. The Doerschel process takes place in the furnace which simultaneously swings and rotates, and which is used specifically for the lead smelting process. The load consists of galena, lead oxide, lead sulfate, sodium carbonate, coke and recycled iron. In this process, lead compounds are reduced into pure metallic form while cadmium is volatilized and precipitates into dust - so called cadmium concentrate. The waste generated in these processes exhibits quite different properties and toxicity. The Doerschel waste as compared to Welz on the average contains more cadmium and lead. (Table 4). The Doerschel waste is characterized by an extremely high mobility of metals as measured by water extraction as well as by high salinity.

TABLE 4 Total metal content in waste materials sampled before treatment

Waste material	Zinc (g kg^{-1})		Cadmium (g kg^{-1})		Lead (g kg^{-1})	
	Mean	range	Mean	range	Mean	range
Welz	30.9	6.9-128	0.54	0.058-2.76	7.9	2.6-16.5
Doerschel	75.1	13.0-126	2.31	0.66-3.46	23.82	7.09-40.6

Both wastes, however, become an environmental hazard through leaching and wind erosion. The revegetation of such wastes is a challenging task for reasons of phytotoxicity. Since there was a lack of data concerning this subject, we needed a better understanding of the physical and chemical processes involved in revegetation. Therefore we conducted a number of field and pot experiments. The field experiments were designed to evaluate the impact of sludge and lime application rates on vegetation. In pot experiments we have tested a number of grass species regarding their adaptation to harsh environmental conditions. Our objective was to select the best performing species on the basis of low metal absorption and salinity resistance.

There were two specific experiments conducted in order to evaluate the potential for utilizing local sewage sludge in liming materials for revegetation at a lead and zinc smelter site in the Katowice area. The one-hectare plots contained either Doerschel or Welz material.

The sites were graded before treatment. The area was completely barren of vegetation. Before the experiments began more than 160 point samples were collected in a grid and analyzed for pH and total sulfur, zinc, cadmium and lead. Water-soluble forms of cadmium, zinc, lead, sodium and sulfates were then monitored during a three-year period. Electric conductivity (EC) was also measured in these water extracts.

The field experiment consisted of large demonstration plots on the smelter waste sites and replicated small-plot experiment in a randomized, complete-block design which was set up on both Welz and Doerschel types of waste in order to evaluate sludge-loading rates, lime-addition rates and grass species effects and interaction. A greenhouse pot experiment with Welz material was performed to characterize a number of grass and legume cultivars grown in a high-metal and high-salinity environment.

The extreme spatial variability regarding the chemical properties of wastes, however, did not allow us to draw any conclusions based on classical, statistical analysis. Therefore spatial analysis was used to interpret the data as shown in this paper. At the end of each growing season a monitoring program was conducted at each of the ten meter grid locations to measure above-ground biomass, metal content in plants and changes of waste-chemical properties.

In order to find an effective method for reducing metal mobility and toxicity in these smelter wastes several laboratory experiments were conducted using different amounts of calcium oxide and calcium carbornate as waste amendments.

Feeding studies were also conducted in order to look at the food-chain risk aspect associated with the revegetation of metal-wastes in Silesia.

5.1 RESULTS OF THE BIOSOLID PROJECT

The monitoring of chemical properties of wastes indicates that the primary reason behind phytotoxicity of some smelter wastes lies in the high mobility of zinc and cadmium as well as in low pH and high salinity levels as expressed by sodium and sulfate concentrations. Sludge application at the rate of 300 dry tons per hectare combined with the incorporation of lime in an oxide and carbonate form at the rate of up to 1 1/2 tons and 30 tons, respectively, per hectare ensure successful revegetation. The analysis of spatial distribution of vegetative cover and waste surface chemical properties has indicated that the method is productive if the following thresholds are not exceeded: soluble Cd (55 mg kg^{-1}), Zn(1000), Na (1600 mg kg^{-1}), SO_4^{2-} (20000 mg kg^{-1}) - Table 5. The pH threshold should not be lower than 6 - otherwise solubility of metals will increase dramatically, leading to phytotoxicity.

TABLE 5 Values (mg/kg)characterizing chemical properties of waste tolerated by grown on revegetated smelter waste.

Zinc		Cadmium		Lead		Soluble sodium	Sol. SO3	EC mS cm^{-1}
Total	Sol.	total	Sol.	total	Sol.			
100000	1000	1700	55	11000	3.7	1600	20000	5.4

The large spatial variability of properties responsible for phytotoxicity controls the spatial variability of biomass production. In order to achieve an equally-distributed ground cover, we would suggest that the reclamation work must first be preplanned with a detailed grid-based spatial analysis of basic waste chemical properties. This allows a site-specific treatment of the most problematic areas with specially-designed rates of biosolids and lime as well as appropriate grass mixes.

Regardless of the fact that the toxicity of Welz waste was very high, the treatment used allowed the establishment of ground cover over more than 80 percent of the area tested. This means that the adaptation capabilities of selected species were considerable. At the same time, when tested on Doerschel waste the same approach failed miserably because of high concentrations of soluble metals and salinity (Table 6).

TABLE 6 Chemical properties of waste material sampled before (1994) and after (1995) amendment with sewage sludge and lime

Waste material	Sampling time	Soluble zinc mg kg^{-1}	Soluble cadmium mg kg^{-1}	Soluble lead mg kg^{-1}	pH	EC
Welz	Before	343	17.6	1.8	7.0	7.3
	After	279	17.7	1.1	7.2	3.5
Doerschel	Before	1670	108	5.4	5.8	16
	After	983	57.4	2.9	6.0	9.0

* values reported reflect averages of 80 samples of each material

The results of the field experiments designed to determine the factors crucial to plant growth on reclaimed waste sites could not be statistically analyzed using classical analyses of variance and averages testing. This stemmed from the fact that the variability of the waste properties was much greater than the effects of treatment. The interpolation

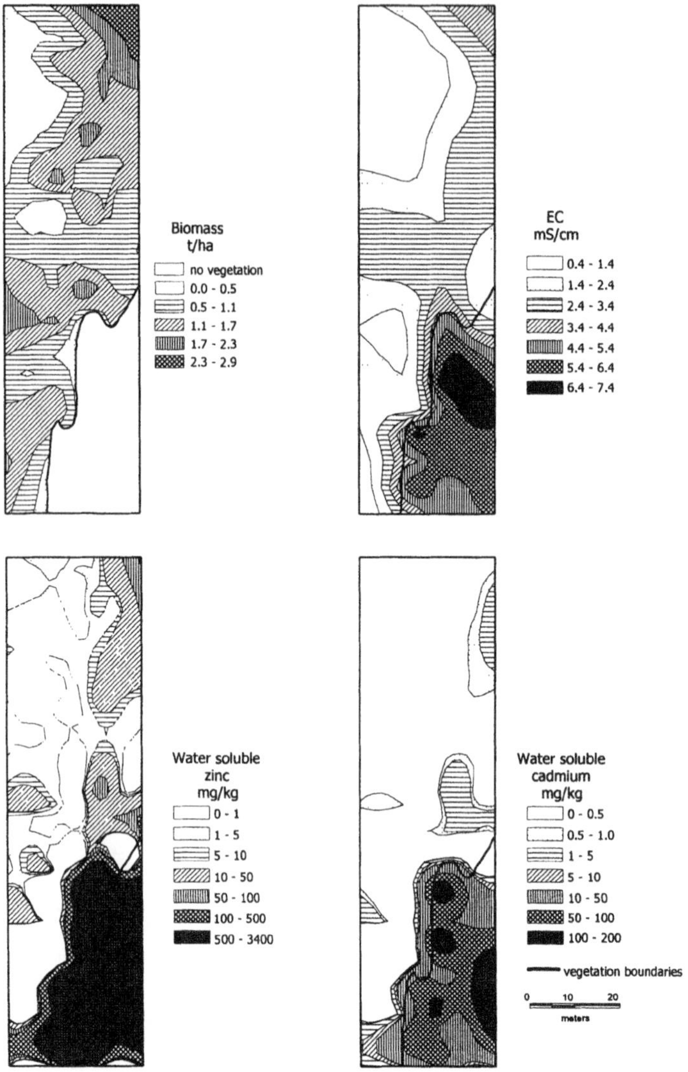

Figure 1. Standing biomass and WELZ waste chemical properties over the experimental area.

of waste properties overlaid with graphs of biomass performance has enabled us to identify and quantify factors responsible for phytotoxicity. It was assumed that areas with at least 80 percent of ground cover did not exhibit toxicity to plants even though the total metals, the water extractable metals and the salinity were quite high (Figure 1).

The Welz plots were covered by vegetation which consisted predominantly of grasses. A number of the twenty grass cultivars which we tested in pot experiments seemed to be useful for revegetation purposes and demonstrated different degrees of adaptation to chemical stress (Table 7). That allows us to propose a mixture of the most acid/salt-tolerant species which were selected from the list shown in Table 7. Such a selection of cultivars may be needed for different types of waste.

TABLE 7 Resistance of grass species and cultivars to metals and salinity

Grass species (cultivars)	Tolerance to metals	Tolerance to salinity
Lolium perenne (Solen)	+++*	+++
Lolium perenne (Argona)	+++	+++
Lolium multiflorum (Telga)	++	++
Lolium multiflorum (Koga)	++	++
Lolium x boucheanum Kunth. (Mega)	++	++
Calamagrostis (natural ecotype)	++	+
Poa pratensis (Alicja)	+++	-
Festuca rubra (Atra)	+	-
Festuca arundinacea (SZD 492)	++	+
Festuca ovina (Sima)	++	-

* degree of tolerance

The spatial variability present in the waste piles which were studied provided a unique opportunity to determine the extent of plant resistance to this adverse environment. From the results of our spatial analysis, it seems obvious that high salinity, and to a lesser extent soluble zinc and cadmium, are the most limiting factors which determine the effectiveness of revegetating these smelter waste sites with biosolids (Figure 1). However, it is also evident that these elements co-vary together and thus cannot be isolated as being singularly phytotoxic. These analyses thus enable us to conclude that the grass cultivars can adapt to the relatively harsh conditions that were seen in treated Welz material (Figure 1).

Adverse physical properties of the Doerschel material, particularly high compaction and sedimentation, also contributed to the total inhibition of plant growth. Lime and sludge at the rates used were not effective for the establishment of vegetation, although, their incorporation reduced metal solubility which will definitely decrease the potential of metal leaching from these piles (Table 6). We should emphasize that changes in pH and cadmium and zinc solubility in both Doerschel and Welz material - as affected by sludge and lime application - were smaller than expected. Evidently, liming was not effective for pH and metal solubility control with these materials - most likely due to limited solubility and/or occlusion with iron or other metal oxides that were present in solutions at very high levels. Our laboratory experiments demonstrated that heavy rates of calcium carbonate did not result in substantial increases of pH (Figure 2). Nor did they result in a reduction of cadmium and zinc solubility. On the other hand, calcium oxide reduced metal mobility to ppb levels although this effect may be temporary since the calcium oxide buffering system can change with time into calcium carbonate via CO_2 absorption. This seems very likely since the addition of calcium

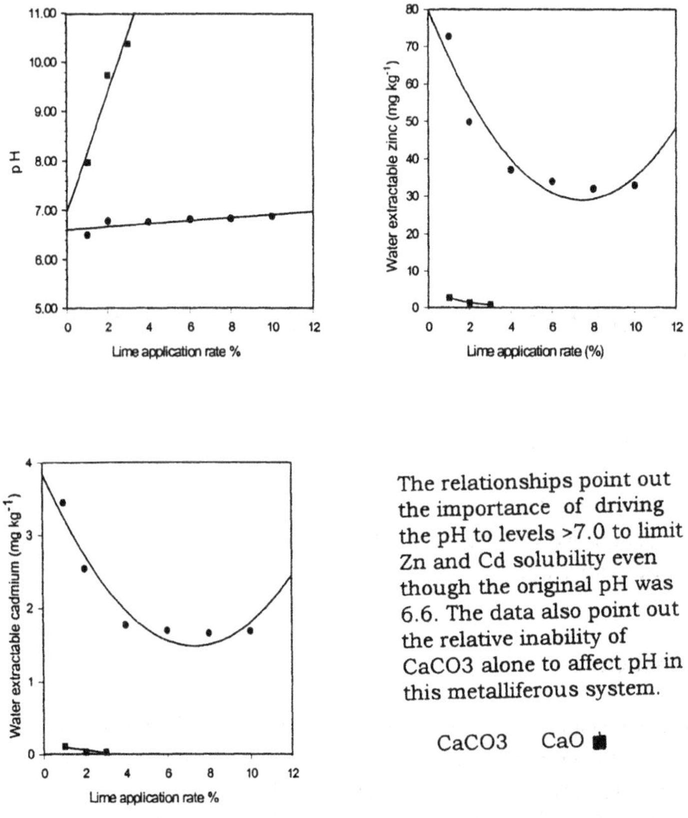

Figure 2. Influence of lime rate and type on pH and water soluble metal levels in WELZ waste.

oxide along with calcium carbonate to smelter plots did not affect the initial pH or metal solubility to any great extent after the first year of sludge application.

The materials similar to Doerschel waste may be treated with a less toxic waste cap as done in our experimental fields. A fifteen-centimeter cap of waste lime which subsequently received 300 tons of sludge created growing conditions for a tolerant grass seed mix. This type of treatment resulted in a 80-90 percent ground cover success rate with little of metal toxicity in the vegetation. On-site soil evaluation indicated that the roots penetrated to the lime/waste interface, but not more than two centimeters into the underlying toxic material. Furthermore, it was evident that even a minimal treatment enables vegetation to sustain itself through summer droughts. With time it can be noted that a number of perennial herbaceous and woody species invade the plots from the surrounding area. This supports the statement that the chemistry of toxic metal waste materials has thus been sufficiently stabilized by the use of lime and sludge in order to support long term plant growth.

In addition to the previous experiments, it is worth mentioning that a feeding study was conducted with young cattle which measured the extent of metal transfer from lead-, cadmium- and zinc-contaminated hay. This hay was harvested from smelter waste which had been reclaimed with lime and sewage sludge. We have found that forage crops grown on sludge-amended metal contaminated land are high in metals, but their bio-availability is greatly limited. There was no significant accumulation of lead, cadmium or zinc observed in muscle tissue. The growth of animals and their health conditons, as gauged by regular health inspection and blood testing, was not affected by a diet of these grasses. The metals absorbed from hay harvested from revegetated waste lands were found in both the liver and kidneys, but still the edibility thresholds were not exceeded after a 90-day feeding period. We have proven that crops grown on remediated soils do not pose any particular risk to animals, and thus do not affect the safety of the food chain. Still it should be noted that additional studies are needed to demonstrate the metal movement into organs of other mammals living under conditions similar to that of the previously conducted experiment.[2]

6. Conclusion

As we pointed out, the problem of sewage sludge disposal in Poland is difficult to solve in the short term. It is evident that the use of

---

[2] This information has been reported in a U.S.D.A. study [19]

sewage sludge for agricultural purposes and reclamation work is not as widespread as in the USA and in European countries. This is primarily because of sludge quality, but it is also important that there is no comprehensive and clear set of legal regulations to be used as guidelines for its use. It must be emphasized that the legal system in Poland allows the use of sludge for such purposes, although the criteria for use are not much different from criteria accepted in other countries. The primary difference lies in the bureaucracy surrounding approvals and administrative decision making.

Conflicts of interest, competency issues and duplication of relevant laws and regulations slow down the process considerably. These problems also tend to inhibit wider implementation for the use of sludge by thwarting useful initiatives before they get under way.

From a practical point of view, the poor quality of sewage sludge often found in industrial areas in the Katowice region automatically eliminates them from agricultural use. It seems that the only valid solution for the intelligent management of sewage sludge in such regions would be to use them for the stabilization and revegetation of industrial waste lands. The results of our research indicate that sewage sludges can be successfully used for the reclamation of toxic smelter waste as an alternative to traditional methods such as topsoiling. Sewage sludge has many benefits, one of the most important being that it is cost effective as compared to traditional methods. This is primarily because transportation costs are reduced along with the cost of revegetation. At the same time, the problem of sludge land filling is also solved.

Regarding metal waste, the main roadblocks we are facing come from salinity and the high content of heavy metals in a mobile form. This problem is further enhanced by spatial variability which forces us to decide the application rates of sludge and lime specifically for each given site, and here even within one pile. GIS methods and grid sampling are useful tools for achieving this. A crucial factor for successful revegetation of zinc- and lead-contaminated smelter waste sites in Silesia is their salinity. High concentrations of metals in their soluble form are only of secondary importance because their mobility can be reduced by appropriate forms and doses of lime.

For waste characterized by medium salinity such as Welz waste the recommended rate of sludge should not be higher than 300 dry tons per hectare under average conditions. Waste which demonstrates higher salinity, such as Doerschel waste, must be treated differently. Capping such waste with less toxic waste - ideally lime waste - followed by sludge application and grass seeding has great potential.

Similar methods can be used for other types of waste treatments such as mining spoil which is addressed in other publications [20, 21].

An integral part of a biosolid reclamation project is the selection of grass species and cultivars that are resistant to toxicity. The appropriate selection creates conditions for good coverage of an area and limits the movement of toxic elements into the terrestrial ecosystem. The metal content in the biomass of selected species also inhibits the impact of metals on the health conditions of organisms/animals returning to reclaimed areas.

It is encouraging that the results of this project seem to have implementation potential and have been well recieved by both local authorities and industry in Silesia. This technology similarly seems to have potential for the entire Central and Eastern European region where lands have been degraded. As a byproduct of this study, a set of guidelines and procedures is in the process of being developed which will address not only current laws and regulations, but appropriate management practices and solutions as applied to land reclamation with sewage sludge.

It is noteworthy that a number of other countries such as the Ukraine, Hungary and South Africa have shown interest in biosolid applications which follow the approach presented. This could be taken as an optimistic preview of what is to come, or at the least, a step in the right direction when considering the environmental and waste-management problems now faced by countries in transition.

7. References

1. Koblak-Kalinska E. (1996) Wykorzystanie osadow sciekowych w swietle przepisow prawa, in J. B. Bien (edt.), *Wykorzystanie Osadow Sciekowych - Techniczne i Prawne Uwarunkowania, Wydawnictwo Politechniki Czestochowskiej,* ISSN 1234-9895, Czestochowa, pp. 17-20.
2. Cebula, J. (1980) Kryteria przydatnosci osadow sciekowych do celow rolniczych, *Ochrona Srodowiska* **6,** 8-12.
3. Cebula, J., Jodkowska T. (1980) Technologiczne aspekty rolniczej przydatnosci osadow w wybranych oczyszczalniach œciekow w Polsce, *Ochrona Srodowiska* **6**, 13-18.
4. Piotrowska M., Dudka S. (1987) Metale sladowe w osadach sciekowych a mozliwosc nawozenia, *Archiwum Ochrony Srodowiska* **1-2**, 65-72.
5. Roszyk E., Roszyk S., Spiak Z. (1987) Wartosc nawozowa osadow sciekowych z niektorych oczyszczalni poludniowo-zachodniej Polski. Cz. I. Sklad Chemiczny osadow, *Rocz. Gleb.* 38 **3**, 93-102.
6. Tymczasowe wytyczne w sprawie klasyfikowania odpadow (1980). Ministerstwo Administracji i Ochrony Srodowiska, Warszawa. (Ministry of Environmental Protection letter)
7. Pismo Ministerstwa Zdrowia i Opieki Spolecznej z dnia 16.02.1984 w sprawie przyrodniczego wykorzystania osadow sciekowych z oczyszczalni komunalnych metod[1] hydroobsiewu (1984). Ministry of Health letter)
8. Metody kontroli stanu sanitarnego osadow sciekowych (1985) Opracowanie na potrzeby PIS. Instytut Medycyny Wsi, Lublin. (preliminary sludge control guidelines)
9. Zarz[1]dzenie Ministra Ochrony Srodowiska, Zasobow Naturalnych i Lesnictwa z dnia 7.07.1986 w sprawie rolniczego wykorzystania sciekow. (1986) *Monitor Polski* nr **23**, poz. 170. (Ministry regulations)

10. Ustawa z dnia 31.01.1980 o ochronie i ksztaltowaniu srodowiska (1994) Dz.U. z 1994 roku nr **49**, poz. 196. (Environment Protection Act)
11. Bernacka J., Siuta J., Wasiak G., Pawlowska L. (1996) Wytyczne przyrodniczego uzytkowania osadow z miejskich oczyszczlni sciekow. Etap I IOS, Warszawa (report)
12. Stuczynski T.I., Daniels W.L., Pantuck K., Pistelok F. (1986) Stabilization and revegetation of metal smelter wastes in Poland with lime and sewage sludge applications. *Symposium Proceedings. Third International Symposium and Exhibition on Environmental Protection in Central and Eastern Europe.* Warsaw September 10-13 1996, pp. 402-404.
13. Pistelok F., Stuczynski T.I., Daniels W.L., Pantuck K (1995) Aspekty prawne wykorzystania osadow sciekowych, *Materialy na XXIX Konferencje Postep Techniczny w Dziedzinie Oczyszczania Sciekow,* Katowice, 12-14 Pazdziernika 1995 pp. 221-223. (symposium proceedings)
14. Patrzalek A, Strzyszcz Z. (1980) Badania terenowe nad mozliwoscia biologicznej obudowy osadnikow poflotacyjnych zielenia niska, *Archiwum Ochrony Srodowiska* **3-4**, 165-180
15. Pantuck K , Daniels W.L., Stuczynski T.I., Pistelok F. (1986) Development of guidelines for the reclamation of mining wastes with sewage sludge in Upper Silesia, Poland. *Symposium Proceedings. Third International Symposium and Exhibition on Environmental Protection in Central and Eastern Europe.* Warsaw September 10-13 1996, pp. 405-407.
16. USEPA (1992) *Standards for The Use and Disposal of Sewage Sludge, Final Rule.* 40 CFR (Code of Federal Regulations), Parts 257, 403 and 503.
17. Davis R.D. and Hall J.E. (1997) Production, treatment and disposal of wastewater sludge in Europe from a UK perspective, *European Water Pollution Control* **7**, 9-17.
18. Bergs C.G., Linder K.H. (1997) Sewage sludge use in the Federal Republic of Germany, *European Water Pollution Control* **7**, 47-52.
19. Stuczynski T.I., Chaney R.L. (1997) Improving methods for remediation of metal contaminated soils, USDA Grant Report No. FG-Po-388.
20. Sopper W.E. (1992) Reclamation of mined land using municipal sludge, *Adv. Soil Sci.* **17**, 351-431
21. Daniels W.L., Haering K.C. (1994) Use of sewage sludge for land reclamation in central appallachians, in C.E. Clapp, W.E. Larson, R.H. Dowdy (eds.), *Sewage Sludge: Land Utilization and the Environment,* SSSA Miscellaneous Publication, Madison, WI, USA.

PAST, PRESENT AND FUTURE STATUS OF N-FERTILIZATION POLICIES IN HUNGARY

T. NÉMETH
Director
Research Institute for Soil Science and Agricultural Chemistry of the Hungarian Academy of Sciences
1022 Budapest, Herman Ottó u. 15.

1. INTRODUCTION

Soils are the most important factors of production in agriculture and the most important, renewable, natural resource in Hungary (1). They are also important in the element cycles where geological and biological processes intersect each other. The circulation of plant-available nutrients slows down in the soil for a "moment" (compared with geological time) and nutrient compounds are able to be retained. This property of the soil brings about a relative accumulation of available forms of those elements most essential for plants and this is the circumstance which enables plants and other organisms to live on Earth. It is well known that the soil can be regarded as a heterogenous system, the properties of which vary in horizontal as well as in vertical directions and some of its parameters may even change with time. During the application of a recommendation system, the magnitudes and proportions of the causes of errors depend on the soil properties, on the methods used for sampling and analysis, but they may also vary according to the nature of nutrient elements (2). The soil properties, particularly the content and distribution of plant nutrients, are greatly influenced by the agrotechnology used, the amount of fertilizers applied, and the procedure of fertilization.

The total element content of an average chernozem soil in the upper 100 cm soil layer (3) is the following (kg/ha):

Nitrogen	5000	Phosphorus (P_2O_5)	1500
Potassium (K_2O)	12000	Calcium	15000
Magnesium	800	Iron	30000
Zinc	150	Copper	60
Manganese	400	Boron	30
Molybdenum	9		

Only a part of the total element content of the soil is available for the plants, depending on the characteristics of the element (solubility, ratio between the total and soluble amounts) and on the environmental conditions (soil physical and chemical properties). This is why recommendations for farmers are needed for reasonable use of their land.

M.J. Wilson and B. Maliszewska-Kordybach (eds.),
Soil Quality, Sustainable Agriculture and Environmental Security in Central and Eastern Europe, 243–252.

In Hungary, during the last few decades, the farmers applied more NPK than the crops needed and from the middle 50's till the end of 80's there was a sharp increase in the total fertilizer consumption (*Table 1*).
Nitrogen accumulates in the soil organic matter, which contains about 92% of the total N in the soil. Nitrogen in mineral forms is usually under 2%, when fertilization is based on the balance sheet method, but this is greatly affected by certain climatic and soil conditions, as well as by the population of soil microorganisms and land use. Although the ratio between the organic and mineral forms can only be modified to a small degree, under certain circumstances a great part of the surplus nitrogen can be accumulated in the form of nitrate, which often occurs in Hungarian soils (4, 5, 6, 7). Nitrate is of course able to be taken up by plants very easily.

TABLE 1. Farmyard manure and fertilizer use in Hungary, 1931-1996.

Year	Farmyard Manure million t/year	Fertiliser active ingredient (1000 t/year)				Arable land (kg/ha/y)
		N	P	K	Total	
1931-40	22.4	1	7	1	9	2
1951-60	21.2	33	33	17	83	15
1961-65	20.6	143	100	56	299	57
1966-70	22.2	293	170	150	613	109
1971-75	14.8	479	326	400	1205	218
1976-80	14.3	556	401	511	1468	250
1981-85	15.4	604	394	495	1493	282
1986-90	13.2	559	280	374	1213	230
1991	8.0	140	23	33	196	37
1992	7.2	148	21	20	189	36
1993	5.0	160	25	21	206	38
1994	4.9	222	27	31	280	50
1995	4.8	191	29	27	247	47
1996	4.0	203	29	27	270	54

2. NITROGEN MANAGEMENT IN HUNGARY TILL THE 1950s

From early times most Hungarian farmers, living on fertile soils, were able to produce considerable crops using traditional crop production practices, i.e. fallows, crop rotation. Organic manure was almost the only added source to make up for nutrient deficiencies, but sufficient quantities were not available. This was because of the low animal density due to the environmental circumstances (unfavourable meteorological conditions - poor grasslands and pastures). This period of the plant cultivation may be described as „exhausting", as can be seen from the nitrogen balance data from the 1930s (*Table 2*).

TABLE 2. Nitrogen balances of the agricultural area of Hungary (kg/ha) [7,8]

Items of balance	1932-36	1960-64	1971	1975	1985	1990	1991
Taken up with yield	40	47	6	80	96	80	103
Supplied							
with farmyard manure	7	7	8	9	8	6	6
with fertilizer	-	16	57	79	96	55	23
with by-products	-	-	6	8	12	10	14
Total supplied	7	23	71	96	116	71	43
Balance	-33	-24	7	16	20	-9	-60
Intensity of balance, %[1]	18	49	111	120	121	89	42

1 Quotient, which expresses how much N (in percentage) taken up by the yield was replaced on the whole.

The yearly report for Hungarian agriculture in 1891 by the Minister of Agriculture presented the average yields of the main crops, as follows:

wheat	1.14 t/ha
rye	0.86 t/ha
barley	1.06 t/ha
oats	0.89 t/ha
oilseed rape	0.73 t/ha.

In the opinion of the Minister these yields were too low taking into account the favourable natural endowments (soil and climatic conditions) and were due to the system of extensive agriculture practice (10). At the end of the last century and the beginning of the 1900s it was thought, that under the usual environmental circumstances the nitrogen and potassium supplies of Hungarian soils were able to provide the cultivated crops with enough nutrients. At that time almost the only mineral fertiliser source applied to the fields was phosphorus, at the beginning of the fertilisation known as Thomas phosphate, and later as superphosphate (11). The overall opinion of the advisers was that nitrogen fertilisation (mainly in the form of Chile saltpetre) was needed only if the farmer had applied organic manure a long time ago and apart from this situation it was considered that only phosphorus fertilisation was required to reach average yield value.

The World's fertiliser consumption and the future trends was discussed by Schneider (12). At that time that time the leading countries were Belgium with 10.45 kg N/ha, 24.55 kg P_2O_5/ha and 4.45 kg K_2O/ha, and Germany with 3.97 kg N/ha, 12.16 kg P_2O_5/ha, and 8.2 kg K_2O/ha application. Hungarian consumption was only 0.05 kg N/ha, 1.17 kg P_2O_5/ha and 0.12 kg K_2O/ha active ingredients, respectively, for arable land (13).

3. NITROGEN MANAGEMENT FROM THE EARLY 1950s TILL THE LATE 1980s

As mentioned above, from the turn of the century to the late 50's, N-balances for Hungarian agriculture were strongly negative. The nitrogen given to the fields in different forms (mainly farmyard manure) was less than the amount removed by the harvested yields (20-30 kg N/ha/year negative balance). [The total amount of the farmyard manure applied to the fields did not increase from the beginning of the 30s, because of a shortage in animal production, due to unfavourable grazing conditions.] With the wide introduction of fertilisation in the late 60s and early 70s nitrogen balances became positive, with an average of 7 kg/ha. For the next 20 years until the end of the 80s, the yearly nitrogen balance remained positive, varying between 16-27 kg N/ha/year in this period. From the early 90s because of economic and political reasons, the use of fertiliser dramatically decreased in Hungary and this situation has resulted in a big change in nutrient balances (*Table 2*). Thus, in 1991 the nitrogen balance was strongly negative (-60 kg N/ha/year), for the first time since the early 70s and this trend has continued (4, 8, 9, 14).

The long-term positive nutrient balances during the two decades between 1970-1990 caused an enrichment of arable soils. Under Hungarian environmental conditions the residual P and K are found mainly in the ploughed layer (15). Some surplus nitrogen enters soil organic matter via biological pathways, some may accumulate in the form of nitrate in the profile and some may be leached to deeper soil layers or to the groundwater (4). The fate of nitrate in the vadose zone depends on the physical characteristics of the soil and on environmental conditions.

4. NITROGEN FERTILISER RECOMMENDATIONS

4.1. Early times

Questions of the fertiliser recommendation system were widely discussed in Hungary during the first decade of this century (16). The main topics of these discussions being like now: how much fertilizer can be applied, when and where. Most of the communications from this time underlined the importance of the field experiments and according to expert opinion, soil analyses alone (without local experimental background) were not enough to determine the amount of the fertiliser needed. Hungarian soil fertility and plant nutrition research has been, and still is, rich in field experiments. Journals and periodicals from the end of the last century contain numerous data on the uses and experimental applications of nutrient-containing materials, so as to help the farmers in fertilization practice.

At that time the Journal "*Köztelek*" was the basic journal for farmers, giving information about a wide range of agricultural topics. A special column of this Journal gave information on soil fertility and plant nutrition and, for example, in 1900 the main topics of this column were nitrogen in soils by Treitz (17) and liming by Spiegel (18). The Journal also contains informations on plant nutrition and soil fertility, giving recommendations for different cultivated crops. The basic nutrition advice for wintercereals at that time was that wheat should follow rape, barley, clover or silage maize which have received farmyard manure before sowing, or fallow which has received farmyard manure.

On fertile soils, wheat was sometimes sowed without farmyard manure application, and sometimes it was the first cultivated crop after farmyard manure application. Advice was also given about fertilisation; for example, after good results had been obtained from field experiments, phosphorus fertilisers were recommended in autumn before sowing (19). In the first decade of this century, the nutrition advice for winter cereals was modified. Fertiliser application had gained ground and phosphate was recommended in autumn with 1/3 of the recommended nitrogen, while the remaining 2/3 was advised for spring application (11). In addition there was an increasing demand for quality control of fertilizers (20).

4.2. During intensive agricultural development

During the development of a fertiliser recommendation system, the greatest difficulty was to determine the amount of nitrogen to be applied. In Hungary the widely used N-fertilisation recommendations in the past were based on the organic matter (humus) content of the ploughed layer. During the period between the early sixties and middle eighties, when intensive systems of agriculture were practiced, quantities of nutrients in excess of the crop's demands were often applied in the form of fertilisers. Because of this, the nutrient (NPK) balance of many fields became positive and their nutrient supplying capacity improved. This was favourable for soils with a low nutrient content and a weak nutrient supply and under these conditions, fertilisation played a significant role in the sustainability of the yields. In contrast, there were many fields where after the plant's nutrient demand had been satisfied, the surplus nutrients could easily be detected in the ploughed layer, particularly for P and K using the standard AL-method (15).

TABLE 3. Nitrogen supplying categories in the six soil group (OM %)

Soil group	K_A*	Organic matter (%) very poor	poor	medium	good	v.good
I.	>42	<2,00	2,01-2,40	2,41-3,00	3,01-4,00	>4,01
	<42	<1,50	1,51-1,90	1,91-2,50	2,51-3,50	>3,51
II.	>38	<1,50	1,51-1,90	1,91-2,50	2,51-3,50	>3,51
	<38	<1.20	1,21-1,50	1,51-2,00	2,01-3,00	>3,01
III.	>50	<2,00	2,01-2,50	2,51-3,30	3,31-4,50	>4,51
	<50	<1,60	1,61-2,00	2,01-2,80	2,81-4,00	>4,01
IV.	30-38	<0,70	0,71-1,00	1,01-1,50	1,51-2,50	>2,51
	<30	<0,40	0,41-0,70	0,71-1,20	1,21-2,00	>2,01
V.	>50	<1,80	1,81-2,30	2,31-3,10	3,11-4,00	>4,01
	<50	<1,40	1,41-1,80	1,81-2,60	2,61-3,50	>3,51
VI.	>42	<1,30	1,31-1,70	1,71-2,40	2,41-3,30	>3,31
	<42	<0,80	0,81-1,20	1,21-1,90	1,91-2,80	>2,81

K_A* = plasticity index

For developing a better nitrogen recommendation system in Hungary, the results of long-term field experiments gave the basis of the calculations. Two systems were widely used from middle of the seventies in Hungarian agriculture, both based on the organic matter content of the ploughed layers of arable soils (4).

A systematic program (sampling, analysis, recommendation) for controlling the plant nutrient status of cultivated soils was organized in the mid-1970's in Hungary in a 3-year cycle. Soils were grouped into six main categories according to their characteristics, as follows:

I.	Chernozem soils
II.	Brown forest soils
III.	Meadow soils
IV.	Sandy soils
V.	Salt affected soils
VI.	Soils with shallow depth

To characterize the N supplying capacity of these soils, they were classified as very poor, poor, medium, good and very good in each of the six main soil groups, and the texture of soil (K_A = *plasticity index*) was considered as another basic parameter (*Table 3*).

During the mid-80s it seemed appropriate to develop a new, user friendly, computerized fertiliser recommendation system in RISSAC. The differences between the two systems were in the grouping of soils. During the elaboration process, every soil on agricultural land was grouped into a 4-digit-code system according to its characteristics (4, 21). The first digit of this new land-site category system refers to the soil water regime and climate for soils with no fertility limitation processes such as

1 for chernozem-like soils, migration type,
2 for brown forest-like soils, leaching type,
3 for meadow soils, hydromorphic type.

(There are 5 other land-site categories reflecting fertility limitation characteristics like salinization, skeletal soils, etc.).Various categories were established within individual soil types to indicate the soil fertility level. The second digit for the 1-3 land-site groups refers to soil texture, the third to soil reaction (pH) and $CaCO_3$-content, and the fourth to organic matter status.

Altogether 280 existing soil mosaics were distinguished in Hungary on the basis of this classification system. In the next step the existing soil mosaics were classified into aggregated land-site groups, using a matrix table compiled by specialists on issues regarding the major plant nutrients. Six aggregated groups were elaborated for N, P, K and Mg, and three for Ca and within each group the nutrient status of the soil was characterized by nutrient supply categories. For nitrogen this was based on the organic matter content of the soil, while for P and K it was based on the AL-soluble P_2O_5 and K_2O contents, respectively. The system has five nutrient supply categories for each of the elements (NPK), from 1 (low) to 5 (high) (*Table 4*).

TABLE 4. Nitrogen supplying categories according to the organic matter content of the ploughed layer (4,21)

Group	Supplying categories				
	1	2	3	4	5
	Organic matter (%)				
I.	<1,0	1,0-1,7	1,8-2,4	2,5- 3,0	3,0<
II.	<1,5	1,5-2,0	2,1-3,0	3,1- 3,5	3,5<
III.	<2,0	2,0-2,5	2,6-3,5	3,6- 4,0	4,0<
IV.	<2,5	2,5-3,0	3,1-4,0	4,1- 5,0	5,0<
V.					5,0-15,0
VI.					15,0>

The fertiliser requirement of the crops is also taken into account on the basis of their specific nutrient contents [specific nutrient contents = N, P_2O_5 and K_2O (kg/ha) for the production of 1 ton yield] - according to the 5 soil nutrient supply categories. The system was developed so that these specific nutrient contents, could be modified by nutrient supply categories, in relation to the expected yield (i.e. for 5.5 t/ha winter wheat this amount has to be multiplied with 5.5, respectively, for N, P and K

4.3. Future trends and scenarios concerning N-fertilisation

A positive nutrient balance with respect to soil N content, as in the case of phosphorus and potassium, may result overfertilisation, the harmful effect is greatest in the case of N. The nitrogen overfertilisation cannot be followed by analysis of soil organic matter, and measurement of soil organic-C content is not the best tool to detect changes in the easily available soil N fraction. It is well known that the effect of N-fertilisation on the N content of the soil can be characterized by measuring the mineral-N (exchangeable ammonium-N and nitrate-N) content of the soil (22) and a nitrogen fertilisation advisory system was also developed in Hungary on this basis (5). The basic idea of the newly introduced system is that proper use of the organic and mineral N-fertilizers to arable crops means that little or no mineral nitrogen should remain in the soil at the end of the growing season. Depending on the environmental (mainly meteorological) conditions that occur until the sowing of the next crop, N-mineralization will occur in the soil. The mineral-N fertilisation system takes this process also into account and it is based on the measurements and calibration of residual and mineralized nitrogen.

In previous studies (4, 5, 6, 7) it was found that under Hungarian environmental conditions the exchangeable ammonium concentration did not reflect the different rate of the nitrogen application, but was characteristic of soil type rather than treatments. However, the nitrate content measured in the profiles strongly depends on the rate of fertilisation. For this reason the newly introduced N_{min}-method contains measurements of both mineral N forms, but for calibration uses only the measured nitrate-N.

5. ENVIRONMENTAL ASPECTS

As mentioned above, during the phase of intensive agricultural development, Hungarian farmers applied more nitrogen fertiliser than the plant's nitrogen demand, as shown by the budgetary approaches (Table 2). The residual nitrate-N at the end of a cropping cycle is subject to different transformation and transport processes. One of these processes is that the nitrate can be leached out from the rooting zone under unfavourable environmental conditions.

The fate and behaviour of the surplus nitrogen in the soil-plant system was studied, in several long-term field experiments (4, 5, 6, 7). In one, on a calcareous chernozem soil, the mineral-N content of the upper 60 cm soil layer and the seasonal dynamics of the N forms were investigated, after an application at four different N-fertilisation rates (15, 23), over an eleven year period as well as after 12 years. The movement, accumulation and distribution of the NO_3-N in the profiles down to six meters, after 12 years is illustrated on Figure 1 (4, 5, 7). Significant nitrate-N accumulation was found between 60 and 200 cm in the intensively fertilised treatments (i.e. having received 200 and 300 kg N/ha/year). There was practically no difference in the nitrate-N distribution of unfertilised plots and of the plots which had received the lower application rate (100 kg N/ha/year). The amounts of nitrogen accumulated in the profiles of the plots fertilised yearly with 300 kg N/ha were ten times higher than those measured in the unfertilised plots. The effect of the overfertilisation in the residual nitrate-N form could be detected down to 350-400 cm (7).

FIGURE 1. Nitrate-N distribution in soil profiles of a 12-years-old field experiment

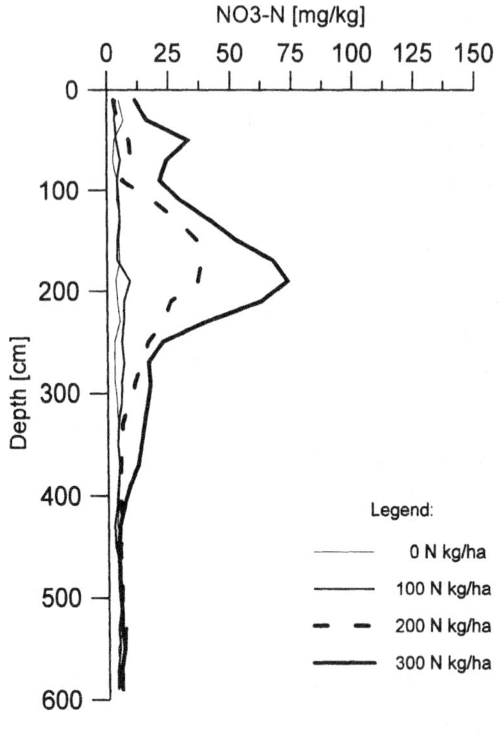

The nitrogen (mostly nitrate) arriving in the deeper soil layers and/or to ground water originates from various sources, one of them being the non-point load from agricultural land use. Contamination of shallow ground water is thus partly caused by intensive agricultural activities, via increasing nitrate leaching.

6. CONCLUSIONS

The goal of efficient N management is to minimize or eliminate nitrate leaching losses which involves controlling the excess of nitrogen (in the form of nitrate) in the root zone at the end of the growing season as well as at times when intensive rainfall is probable. This goal can only be realised through well-planned field experiments. Evaluation of Hungarian experimental data-sets shows that it is possible to improve the nitrogen management practices so as to secure stable and sufficient yields and to protect the environment at the same time.

The different nitrogen budgetary approaches at different scales (country, county, farm, field, experiment) and the results concerning the fate and behaviour of nitrogen in the long-term field experiments proved that with appropriate use of fertiliser N the amount of the residual nitrogen could be as low as in the unfertilised plots. In addition, no nitrate-N accumulation therefore was detected under the root zone even for the migration type soils. It is concluded that the harmful effect of fertilization can be avoided if N fertilisers are applied at the recommended rates, taking into account the N demand of the crops for the expected yield, and the specific environmental conditions.

7. REFERENCES

1. Várallyay Gy. (1990) Soil quality and land use, in D. Hinrichsen and Gy. Enyedi (eds.), *State of Hungarian Environment*, Statistical Publishing House, Budapest, pp. 91-123.

2. Kádár, I. (1986) A tápanyagvizsgálatokat célzó talajmintavétel problémái hazánkban. (Problems in sampling for determining the nutrient supply of soils.) *Agrokémia és Talajtan* **35**, 405-414.

3. Németh T. and Kádár I. (1991) Macro- and micronutrients in Hungarian soils., in I. Pais (ed) Proc. "*Cycling of nutritive element in geo- and biosphere*" IGBP Symp. of the Hung. Acad. of Sciences. Univ. of Hortic. and Food Ind., Budapest, pp. 19-52.

4. Németh, T. (1995) Nitrogen in Hungarian soils - nitrogen management relation to groundwater protection, *J. of Contaminant Hidrology* **20**, 185-208.

5. Németh, T. (1996a) *Talajaink szervesanyag-tartalma és nitrogénforgalma*. (Organic matter and nitrogen content of soils), MTA TAKI, Budapest.

6. Németh, T. (1996b) Nitrogen balances in long-term field experiments. *Fertilizer Research* **43**, 13-19

7. Németh T., Kovács G. and Kádár I. (1987-1988) A nitrát, szulfát és a sóbemosódás vizsgálata m trágyázási tartamkísérletben. (Nitrate, sulphate and "Water soluble salts" accumulation in the soil profile of long-term fertilization experiment*) Agrokémia és Talajtan* **36-37**, 109-126.

8. Kádár, I. (1979) Födm veléünk nitrogén, foszfor és kalium mérlege. (Nitrogen, phosphorus and potassium balance studies in agriculture) *Agrokémia és Talajtan* **28**, 527 – 545

9. Csathó, P. (1994) A magyországi talajok NPK-mérlege 1990 – ben. (NPK balances of the Hungarian soils in 1990 and 1991). Növénytermelés **43**, 551 - 561

10. Baross, K. (1892) A földmívelési miniszter jelentése 1891 évi müködésér 1. (Minister of Agriculture's report for 1891), *Köztelek* 10 December, 1892. pp. 1-7.

11. Cserháti, S. (1908) szibúza trágyázása. (Fertilization of winter wheat), *Köztelek* **18**, 1840-1841.

12. Schneider, W. (1909) *Mineralische Düngermittel und Ernteertrage*. Verlag von N. Kymmel, Riga.

13. Konkoly, T.S. (1911) A világ m trágyafogyasztása. (Fertilizer consumption of the World),*Köztelek* **21**, 3187-3189.

14. Csathó, P. and Árendás, T. (1997) The effect of organic matter content on crop responses to N given in mineral or organic forms. *Agrokémia és Talajtan* **46**, 63-76.

15. Sarkadi J., Németh T. and Kádár I. (1986) A talaj könnyen oldható tápanyag-tartalmának heterogenitása. (Heterogenity of the readily soluble nutrient content of soil.) *Agrokémia és Talajtan* **35**, 295-306.

16. Gyárfás, J. (1908) M trágyázási kérdések (Questions concerning to fertilizer applications)*Köztelek* **18**, 2570-2571.

17. Treitz, P. (1900) Nitrogén a term talajokban. (Nitrogen in soils), *Köztelek* **10**, 1182-1183.

18. Spiegel, J. (1900) A mésztrágyázásról. (Liming), *Köztelek* **10**, 1370-1371.

19. Rovara, F. (1892) Az szi gabona trágyázása. (Fertilization of winter cereals), *Köztelek* **2**, 1416.

20. Faltin, A. (1911) A Thomas-salakok forgalomba hozásának feltételei. (Conditions of Thomas phosphate trade), *Köztelek* **21**, 1803-1804.

21. Várallyay Gy., Buzás I., Kádár I. and Németh T. (1992) New plant nutrition advisory system in Hungary, *Commun. Soil Sci. Plant Anal.* **23**, 2053-2073.

22. Wehrmann, J. und Scharpf, H.C. (1979) Der mineralstickstoffgehalt des Bodens als Mass-stab für den Stickstoffdüngerbedarf (N_{min}-methode), *Plant and Soil* **52**, 109-126.

23. Németh, T. and Buzás, I. (1985) Characterization of the mineral nitrogen content of soils for fertilization advices, in E. Welte and I. Szabolcs (eds.) *Fight againts hunger through improved plant nutrition*. Proc. 9th World Fertilizer Congress held in Budapest, Hungary. June 11-16., 1984. CIEC Goltze-Druck, Goettingen, Volume 2, pp. 220-224.

SOIL TILLAGE SYSTEMS APPLIED IN RUSSIA TO PREVENT SOIL EROSION AND DEGRADATION

A.G. Bondarev

V.V.Dokuchaev Soil Science Institute. Moscow, Russia

1. Introduction

As in many other countries, the soil tillage systems applied in Russia have undergone great change during the past 20 to 30 years. Systems based on deep mouldboard ploughing, which have dominated agricultural practice for a long period of time, are making way for new systems oriented towards minimum tillage. However, as there are positive and negative effects of both traditional ploughing and minimum tillage, systems have been developed which aim to combine the advantages of mouldboard and sub-surface treatment of soils at different depths, at the same time reducing the disadvantages of these systems. A great diversity of these combined tillage systems exist in Russia. They have been widely adopted over vast areas of the country, depending upon the natural soil-climate zones, the susceptibility of soils to erosion and the extent to which this erosion has already taken place or to which the soil properties have been degraded.

2. Soil degradation and soil type

According to data obtained during a cadastral survey completed in January, 1996, the area of agricultural land in Russia that is vulnerable to water and wind erosion amounts to more than 117 million ha. including 51 million ha. that are actually eroded. In a number of regions, susceptible eroded soils account for 50 to 90% of the total area of agricultural land. Water erosion of soils is widespread in almost all regions, with the exception of the North Caucasus and Siberia. The total area of such lands is estimated at 35.1 million ha., including 25 million ha. used for arable crops. Soils with a light granulometric composition are particularly prone to wind erosion and this is a serious problem in the steppe and semi-desert zones of the country. Degradation of soil physical properties, leading ultimately to erosion, appears to be due mainly to two inter-related processes, namely disaggregation of the soil mass resulting from decreasing organic matter contents and soil over-compaction due to heavy agricultural machinery. Because soil degradation and erosion in Russia is caused by human activities and varies according to soil type, it is evident that tillage systems should be matched to soil type and much research has been devoted to this end.

According to Yakovlev *et al,* (15), chernozems account for 52.6% of agricultural land used for arable crops in Russia, amounting in total to 69.43 million ha. Other soil types, such as grey forest soils, podzolic soils and sod-podzolic soils occupy 14 to 15% of the total agricultural area and chesnut soils about 11%. The physical properties of all these soil types have drastically deteriorated.

M.J. Wilson and B. Maliszewska-Kordybach (eds.),
Soil Quality, Sustainable Agriculture and Environmental Security in Central and Eastern Europe, 253–257.

The chernozems possess the most favourable physical properties, but these properties vary to a considerable extent according to the particular genetic features of the soil and the degree of human-induced degradation. Thus, most (75-80%) of the typical leached, thick, chernozems retain physical properties close to what may be considered as optimal. However, the shallow podzolized and low-humus varieties of chernozem, particularly those occurring in the southern part of the country, are characterized by large blocky structures, dense consistence and low water permeability. Such chernozem sub-types occupy about 50% of the total area covered by chernozem soils in Russia.

Only about 8 to 10% of the sod-podzolic and grey forest soils possess physical properties close to those regarded as optimal for these soils. Weak or moderate degradation is characteristic, a trend that seems to be increasing. It has been forecast that one third of all arable sod-podzolic soils will be highly degraded in the near future (3).

The loamy and heavy loamy chesnut soils vary greatly in their physical properties but for the most part they are weakly to moderately degraded. Chesnut soils with optimal physical properties are rare and are found only in areas used for horticultural purposes or for controlled experimental tests. Some chesnut soils have become highly degraded because of poor irrigation techniques.

3. Soil tillage systems involving sub-surface cultivation

As noted above, the search for different ways of changing the practical basis of soil tillage systems in Russia and countries of the former Soviet Union has arisen, to a large extent, because intensified mechanization of agricultural production has led to diminishing contents of organic matter, soil compaction, loss of structure and ultimately degradation and erosion. (4, 5). Based on the ideas of Maltsev (11) over forty years ago, sub-surface tillage is now seen as a means of combating soil degradation and is increasing in importance. Maltsev's research was carried out on chernozems in Siberia and is now accepted as the basis for the development of systems involving minimal soil treatment. Essentially, the system involves cultivation of the surface with a stubble plough and repeated loosening of the deep sub-surface. Baraev (14) further developed this system, taking into consideration Canadian experience in the use of different sub-surface implements. The system effectively decreased wind and water erosion, helped conservation of soil moisture through the slow accumulation of stubble on the soil surface, and decreased the rate of loss of soil humus through mineralization. The system, therefore, emphasizes soil protection. The results of Kiryushin and Lebedeva (10) show how loss of humus is slowed down in southern carbonate chernozems as a result of sub-surface cultivation. (Table 1).

Sub-surface cultivation is being effectively used as a means of soil protection, particularly in areas like Siberia and Kazakhstan which have seen a programme of intensive development of virgin lands since the 1950s.

Table 1. Humus content (%) in southern carbonate chernozems following mouldboard ploughing and sub-surface tillage systems in grain crop-fallow rotations.

Soil Tillage System	Soil Layer (cm)			
	0-5	5-10	10-15	15-50
Mouldboard Plough	4.43	4.54	4.44	4.42
Sub-Surface Cultiv.	4.91	4.84	4.71	4.52
Difference	0.48	0.29	0.27	0.10

There are, however, some shortcomings in the implementation of sub-surface cultivation systems, the main one being the increase in weeds. The necessity of tackling this problem in many steppe regions of the Urals and Siberia has meant that there may be a delay in sowing grain crops, wheat in particular. In using a sub-surface cultivation system, the use of bare fallow in rotations in conjunction with herbicides plays an essential role in weed control.

4. Recommendations for using minimal soil tillage in Russia

In order to reduce energy expenditure for soil tillage, as well as soil compaction, the relevant research centres of the Ministry of Agriculture and the Academy of Agricultural Sciences published recommendations to minimize soil tillage (13) and the impact of agricultural machinery on soils (7). Some of the main recommendations were as follows:

(a). Soils should be prepared for growth of winter crops by surface and shallow cultivation rather than by deep mouldboard ploughing. Cultivation by machines with a wide span, disc harrowing, stubble ploughing and rotary cultivation were recommended.

(b). The number and depth of soil tillage operations, pre-sowing and inter-row cultivations, together with herbicide applications should be reduced.

(c). Several treatments could be combined into one process using suitable machines, thus reducing the number of machine passes.

(d). The area of fields could be made smaller by the adoption of strip tillage.

(e). Minimum tillage is recommended for soils with favourable physical properties, such as non-degraded grey forest soils, chernozems, chesnut soils etc, and where weed growth does not cause problems.

The State Standards body has also made recommendations about the permissible pressure on soils by different machines (8). The maximum permissible pressure should not exceed 80 kPa in spring and 100-210 kPa in the summer-autumn period. The shrinkage stress at a depth of 0.5 m should not be more than 25-35 kPa in spring and about 30-50 kPa in summer and autumn.

As a rule deep mouldboard ploughing is used for tilled crops requiring incorporation of NPK fertilisers. Sub-surface cultivation mostly applies to soils under winter and spring grain crops that are grown after tilled crops. The choice of combined operations for soil treatment is determined by the requirements of the crop, climatic conditions and soil cover structure. The above measures for minimum tillage effectively decrease the problem of soil compaction but do not solve the problem completely.

5. Concluding remarks

The most important function of soil tillage is to provide control over soil physical properties and regimes (structural state, water, air, heat and nutrition regimes). For this reason, it is evident that cultivation should take into account the soil physical properties, or strictly speaking the difference between actual and optimal parameters of the soil physical properties (1, 12, 6). For example, optimal equilibrium density has been determined for the main soil types and cultivated crops in Russia and are expressed in the following ranges (gr cm3).

1.0 - 1.3	loamy and clayey soils
1.1 - 1.4	light, loamy soils
1.3 - 1.5	sandy loamy and sandy soils

Within this range, the lower value of the soil equilibrium density is considered as the optimum for tilled crops and vegetables which are sensitive to soil density and excessive soil wetness. The higher value of the range is the optimum for grain crops which are not so sensitive to soil density or soil wetness

Optimal parameters for other soil physical properties have also been determined including soil structure, porosity, water permeablity and water availability. However, actual soil physical properties in any given field are usually unknown. Thus, the Russian farmer, as in other countries, does not know the actual state of the soil in terms of its physical properties at the moment when it is necessary to choose soil tillage type. It is estimated that about 40% of the energy expenditure used for soil tillage is not required from a scientific point of view. It is important, therefore, that a Soil Advisory Service should be able to provide appropriate information to farmers with regard to the physical properties of their soils with respect to both actual and optimal parameters. Rapid methods for determining soil physical properties are evidently needed. Expanded knowledge of this sort would place the choice of a soil tillage system on a more scientific basis and would ensure that the soil was protected against degradation and erosion and, at the same time, would enhance soil fertility.

6. References

1. Bondarev, A.G., Bakhtin, P.U. and Voronin, A.D. 1986. Fizicheskie i fiziko-tekhnologicheskie osnovy plodorodia pochv. (The physical and physico technological basis for soil fertility). *100th Anniversary of Genetic Pedology*. Nauka, pp 178-184

2. Bondarev, A.G. , 1988. Problemy regulirovania fizicheskikh svoistv poch v intensivnom zemledelnii. (Problems of controlling soil physical properties in intensive farming). *J. Pochvovedenie,* No. 9. pp 64-70.

3. Bondarev, A.G. , 1988. Prognoz otsenki uplotnenia pochv dvizhitelyami. (Predicted estimation of soil compaction by agricultural machinery). *J. Agricultural Engineering* No. 5, p. 13

4. Bondarev, A.G. , 1990. Problema uplotneniya pochv selskokh ozyaistvennoi teknikoi i puti ee resheniya (Problems of soil compaction by agricultural machines and ways of solving them). *J. Pochvovedenie.* No. 5. pp. 31-37

5. Bondarev, A.G., 1994. Sapozhnikov. Pereuplotnenie pochv selskochozyastvennoi tekhnikoi, prognoz yavlenia i protsessy rasuplotneniya (Soil overcompaction by agricultural machines and processes of decompaction). *J. Pochvovedenie.* No. 4. pp. 58-64.

6. Bondarev, A.G., 1994. Teoreticheskie osnovy i praktika optimizatsii fizicheskikh uslovj plodorodia pochv (Theoretical basis and practice of optimizing physical conditions for soil fertility). *J. Pochvovedenie.* No. 11. pp. 10-15.

7. Anon. 1985. Vremennye rekommendarsii po ogranicheniyu urovnya vozdeistvia dvizhitelei selskokhozyastvennoi tekhniki na pochvu (Provisional recommendations to eliminate the impact on soil by agricultural machines). *Agropromizdat.* 14pp.

8. GOST - 26955-86. 1986. Tekhnika selskokhozyastvennaia mobilnaya. Normy vozdeistviya dvizhitelei na pochvu (State Standard 26955-86- Mobile agricultural machines.Permissible pressure on soil of agricultural machines.) Moscow Publish. House of Standards. pp. 1-7.

9. Gosudarstvennyi doklad o sostoyanii i ispolzovanii zemel Rossii za 1995 (State report on the status and usage of lands in Russia for 1996). 1996. Moscow Russolit. 120 pp.

10. Kiryushin, V.I., 1996. Ekologicheskie osnovy zemledelia.(The ecological basis of farming). Moscow, Kolos, 366pp.

11. Maltsev, T.S., 1955. Voprosy zemledelia. (Problems of farming). Moscow. Selkhozgiz.

12. Medvedev, V.V., 1988. Optimizatsiya agrofizicheskikh svoistv chernozemov. (Optimization of agrophysical properties in chernozems). Moscow. *Agropromizdat* 157pp.

13. Anon. 1985. Minimalizatsiya obrabotki pochv (recomendatsii) (Minimum soil tillage, recommendations). Moscow. *Agropromizdat.* 14 pp.

14. Baraev, A.I., (editor) 1975. Pochvozatschitnoe zemledelie (Soil protection farming). Moscow, Kolos.

15. Yakovlev, A., *et al.,* 1994. Sovremennoe sostoyannie pochvennogo pokrova Rossii i osnovnye prichiny ego degradatsii. (The present day status of the soil cover in Russia and the main causes for its degradation). *Intern, Agric. J.,* No.2. pp. 13-21.

A MECHANISM FOR THE SELF-REGULATION OF FERTILITY IN UKRAINIAN CHERNOZEMS

M.K. Shikula
National Agricultural University of Ukraine, Kiev

1. Introduction

For many centuries, agriculture in Ukraine has been associated with decreasing soil fertility. Effectively, crops were grown at the expense of losses in potential soil fertility leading also to the degradation of soil physical properties of agricultural importance. Nutrient losses were partly compensated for by the application of organic and mineral fertilizers so that a gradual decline has occurred in soil nutrient status. As a result the biological cycling of many elements in the soil has been disrupted, leading to ecological changes on a landscape scale, enhanced wind and water erosion and ultimately to even greater losses in soil fertility [4,5].
The development of ways of maintaining and enhancing soil fertility - in other words of extended reproduction of soil fertility - has been a component of the "green" revolutions that have occurred in the majority of the technologically advanced countries. This has ensured that the conditions exist for the realization of the biological possibilities for new varieties and cultivars of plants. In fact, sustainable soil fertility has become the key to the realization of the "green" revolutions.

2. Dehumification of Ukrainian Chernozems.

Ukraine has inherited from Nature some of the richest chernozemic soils in the world. These soils make up about 68% of the total area of Ukraine's soil cover. But, over the centuries the organic matter status of the soils has declined considerably as a result of the disruption of the biological cycling of elements and the violation of the law of return in farming. Plant nutrient elements have been steadily depleted by crop growth and this is associated with a diminished content of soil organic matter. In many ways soil organic matter may be regarded as the energy source for soil formation and its diminution is considered to effectively slow down the natural pedogenic processes involved in the reproduction of soil fertility [1,5]. In Table 1 it can be seen that the top layer of a Ukrainian chernozem contains 10 to 11% of soil organic matter in the virgin state, but that after 54 years of agricultural use this has declined to 5 to 6%.
These results are quite typical for other chernozems. Thus, in the example shown in Table 1, in 54 years total stocks of humus have declined by some 51.3 tons/ha, on average about 1 ton/ha/yr [4,5]. Even steeper rates of decline are observed where there is an inappropriate balance between organic and mineral fertilizers. For centuries in the southern forest steppe zone and the northern steppe zones of Ukraine, there has existed a belt of typical, medium humic chernozems. With the intensive application of mineral fertilizers and the neglect of organic fertilizers these soils have been transformed into chernozems of low humic status. (Table 2)

M.J. Wilson and B. Maliszewska-Kordybach (eds.),
***Soil Quality, Sustainable Agriculture and Environmental Security in Central and Eastern Europe*, 259–266.**

TABLE 1. Contents and stocks of total organic matter in medium-humic typical Chernozem of Mikhailovskaya virgin land and in the 54-year-old plough land

Depth, cm	Virgin-land soil			Plough land, 54-year-old			Changes in humus content and stocks in 54 years as compared with virgin soil	
	humus, %	bulk density, g/cm^3	stocks of humus, t/ha	Humus, %	bulk density, g/cm^3	stocks of humus, t/ha	%	t/ha
0–10	10,39	0,77	80,0	5,61	1,00	56,1	–4,78	–29,9
10–20	7,82	1,06	82,9	5,71	1,22	69,7	–2,11	–13,2
20–30	6,46	1,06	68,5	5,70	1,22	69.5	–0,76	+1,0
30–40	5,54	1,10	60,9	5,14	1,19	61,2	–0,40	+0,3
40–60	5,25	1,07	112,4	4,47	1,13	101,0	–0,78	–11,4
60–80	4,58	1,07	98,0	4,30	1,13	97,2	–0,28	–0,8
80–100	3,96	1,09	86,3	3,74	,11	83,0	–0,22	–3,3
0–100	5,78	1,04	589,0	4,72	1,14	537,7	–1,06	–52,3

TABLE 2. Change in the content of humus in a deep ordinary clay-loam Chernozem of Kirovograd agricultural station (1958–1978)

Depth, cm	Humus content, %					
	on watershed			on gentle slopes		
	1958	1978	Decreased in 20 years	1958	1978	Decreased in 20 years
0–20	6,5	4,9	–1,6	6,7	4,2	2,5
20–30	5,3	4,8	–0,5	5,3	3,5	1,8
30–40	5,2	3,9	–1,3	4,4	2,9	1,5
40–50	4,5	3,3	–1,2	4,3	2,4	1,9
50–60	4,1	2,8	–1,3	3,4	2,0	1,4

Data such as these have stimulated research into how this potentially catastrophic dehumification of Ukrainian chernozem soils can be stopped and have provided a starting point for the discovery of a biochemical mechanism involved in the reproduction of humus and self-regulation of soil fertility.

3. Annual humus cycles in agroecosystems.

The extended reproduction of humus in the soil can be illustrated by a field experiment conducted at the Novoodessky Plant Breeding Station in the Nicolaev area. (Table 3)

During the 20 years of the experiment, using two types of soil tillage (ploughing and non-plough tillage) the following rates of fertilizer were applied every year; farm manure 2 tons/ha, winter wheat straw and sunflower stalks 24 tons/ha, mineral fertilizers (NPK) 180 kg/ha. In the experiment with ploughing, where the organic fertilizers were ploughed down to a depth 22 to 27cm, what may be termed a simple reproduction of humus was observed, but in the non-plough variant where organic fertilizers were incorporated into the soil by a heavy disk harrow to a depth of 8 to 10 cm an elevated reproduction of humus was observed [4].

TABLE 3.Change of the contents and stocks of humus in southern loam-clay chernozem on various system of soil tillage for 1961-1980

Depth, cm	Ploughing to the depth 22-27 cm		Non-plough tillage to the depth 22-27 cm		± compared to ploughing after a period of 20 years	
	content of humus, %	stocks of humus, t/ha	content of humus, %	stocks of humus, t/ha	content of humus, %	Stocks of humus, t/ha
	Initial content of humus in 1961					
0–10	3,79	–	3,79	–	–	–
	Content of humus in 1980					
0–10	3,81 ±0,04	40,0	4,14 ±0,06	42,6	+0,33	+2,6
10–20	3,84 ±0,04	45,3	4,05 ±0,06	48,6	+0,22	+3,3
20–30	3,80 ±0,03	46,4	3,95 ±0,06	47,8	+0,15	+1,4
30–40	3,49 ±0,05	43,3	3,57 ±0,09	43,9	+0,08	+0,6
40–60	2,55 ±0,06	67,8	2,52 ±0,06	67,0	–0,03	–0,8
0–60	3,34 ±0,05	242,8	3,46 ±0,07	249,9	+0,12	+7,1

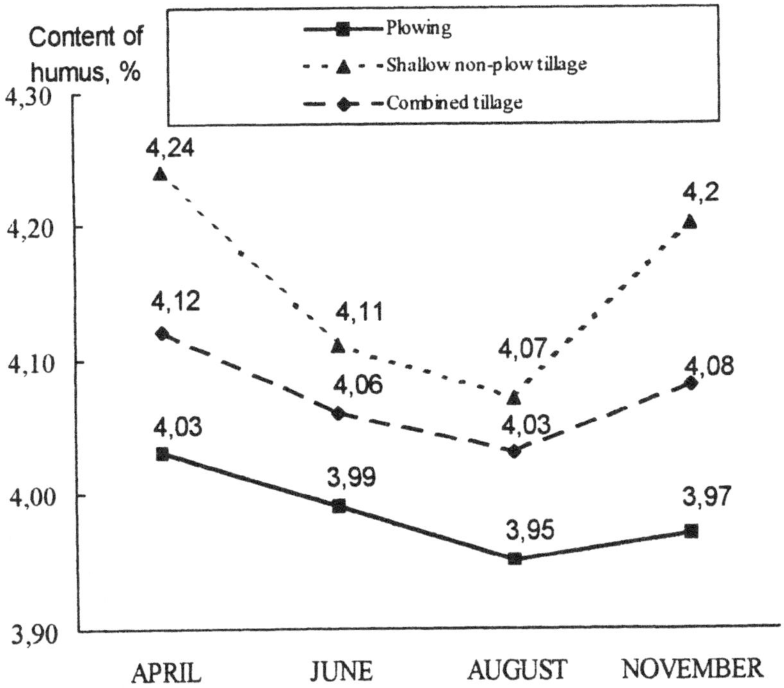

Figure 1. Annual cycles of organic matter content in southern clay loam chernozem under various systems of soil tillage explored for 18 years.

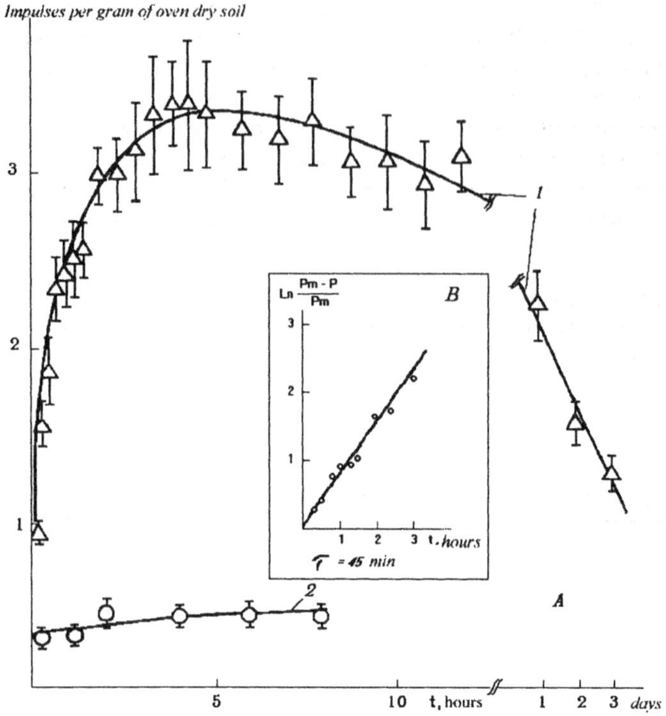

Figure 2. Dynamics of ^{14}C inclusion from $NaH^{14}CO_3$ into soil organic matter:
A – in fresh soil samples;
B – after their thermal processing.

Monitoring of humus content during the period of vegetation growth led to the discovery that there were annual cycles of humus content in agroecosystems. Figure 1 shows that the cycle is most clearly displayed when the top layer of the soil is not inverted by tillage.

When the top layer is inverted the soil finds itself in a condition of "shock" causing the death of soil biota whereby anaerobic forms die from exposure to oxygen and aerobic forms die from lack of oxygen. In these circumstances the cycle in the humus is barely discernible. The higher the range of the annual cycle of the humus content, the greater the ability of the soil to self-regulate its fertility. From spring until the end of summer (in the descending part of the cycle) a significant amount of humus (6 to 10 tons/ha) in the plough layer is spent for the maintenance of plants during their intensive growth.

4. Role of ATP

The question arises as to the storage of an appropriate energy source for the operation of the cycle in the autumn and the early spring. ATP is an accumulator of biological energy and this substance was detected in the soil. In the soil subjected to the non-plough tillage there was 25 to 200% more ATP than in the ploughed soil. ATP is dependent upon the biological activity of the soil micro-organisms and in the 0-15 cm layer of the non-ploughed soil, this activity was 1.5 to 2 times greater than in the ploughed soil. This difference was especially noticeable in variants where organic fertilizers were applied or where residues were left at the surface. Further details of the processes involved in the reproduction of humus have been clarified using radioisotopes (14C). In the soil, heterotrophic organisms assimilate carbon from carbon dioxide in the soil air (Figure 2) resulting in the formation of new organic matter with a protein content of 30 to 42% and a lipid content of 1 to 5%.
The carbon source for the micro-organisms would be dead organic residues, farmyard manure, straw and various vegetative residues. After the destruction of the micro-organisms the protoplasm leaves the cell for the soil solution and interacts with the organic acids of the soil. In the first days the new substances prevail in the fulvic acid fraction at a ratio of 4:1 but after 30 days this situation is reversed.

5. Overview of the cycle.

In general terms, it seems that the reproduction of humus in chernozem soils, which is regarded as the essence of the self regulating process of soil fertility, involves the assimilation of carbonic acid from the soil solution and of carbon dioxide from the soil air by heterotrophic micro-organisms followed by the incorporation of newly formed organic substances from these micro-organisms into the humic fraction. The process ultimately ensures that the soil-plant relationship is sustainable at an optimum level over the long term. A general scheme for this self regulatory process of soil fertility is shown in Figure 3.

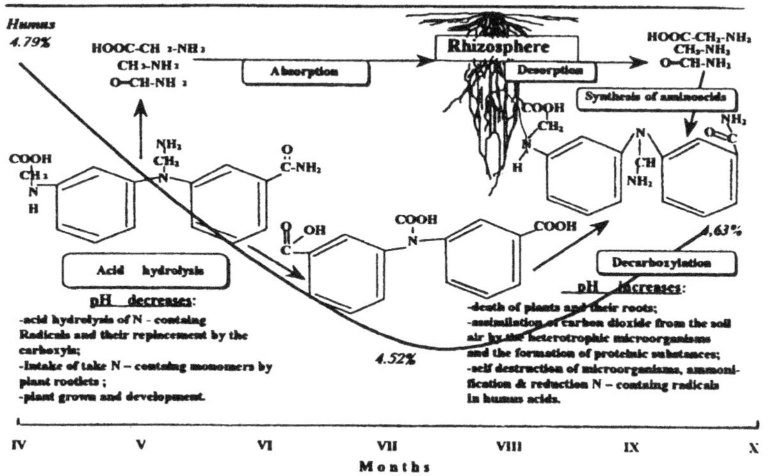

Figure 3. Mechanizm of soil fertility self - regulation according to N.K.Shikula

At the start of plant growth, carbon dioxide is released from plant rootlets, as well as other organic acids, resulting in a drop of soil pH by about 0.6 to 1.2 units. There then the pass through the membranes of the rootlets and are absorbed in the plants and used in crop growth. At the end of the growing season acids are no longer released from the plant rootlets to the soil solution and pH is now restored to its initial level. Heterotrophic micro-organisms now feed off the dead organic residues and assimilate carbon dioxide and create new protein substances. With the death and decomposition of these micro-organisms, their protoplasm enters the soil solution where it reacts with the humus acids causing their ammonification and replacement of carboxyls by N-containing radicals. The annual cycle regulating soil fertility is now complete and the humus acids have obtained the structure and composition that they had in the in the spring prior to the start of plant growth [2,3].

6. Relevance to Soil Management.

This self-regulating cycle is part of the natural process of soil formation and is highly relevant to soil management in agroecosystems. Thus, the intensity of carbonic acid assimilation by heterotrophic micro-organisms is influenced by the amount and quality of organic substances (organic fertilizers), the depth of their incorporation into the soil (aerobiosis and anaerobiosis), reaction of the soil solution, presence of di- or tri-valent cations for the irreversible coagulation of organic colloids, soil moisture and temperature. Most of these factors can be adjusted by soil tillage and mulching of the soil surface by plant residues, application of organic fertilizers and chemical amendments.

Methods of soil tillage have the most significant effect upon the intensity of carbonic acid assimilation by soil mocro-organisms. Conservation non-plough tillage incorporates residues and organic fertilizers into the top 8 to 10 cm of the soil and results in factors of organic matter humification that are up to one third higher compared with conventional tillage. This enables a non-deficit and extended balance of soil humus to be reached using lower rates of organic fertilizers.

There is a clear correlation between the status of soil organic matter and overall soil properties and behaviour. The annual cycle of humus content is matched by similar cycles of other soil properties, suggesting that soil organic matter plays a key role. In this connection, monitoring the content of soil humus can help to assess general trends in soil fertility status. If humus is decreasing over time then it is concluded that the potential productivity is also decreasing. If humus content is maintained at the same level then the situation is one of simple reproduction of soil fertility, whereas if humus is increasing then this is a case of extended reproduction of soil fertility.

Table 4 shows the influence of various soil management practices (tillage, mulching, fertilization etc,) on the intensity of the soil forming process.

It can be seen that farmyard manure added at the rate of 12 tons/ha ensures a non-deficit balance of humus. Use of conservation tillage significantly increases humus content (+0.09%) and addition of mineral fertilizers to farm manure at a rate of 235 kg/ha also tends to increase the humus content within the limits of experimental error (+0.03%). The same rate of fertilizers with conservation tillage brings about an 8 times increase in humus content compared with conventional ploughing. A further increase in the rate of mineral fertilization (35 kg/ha) results in a slowing down in the process of humus formation.

TABLE 4. Influence of tillage mulching and fertilizing on the contents of humus in an arable layer of a typical Chernozems (1980-1988), %

System of fertilizing per 1 ha of crop rotation	System of soil tillage					
	Conventional		Combined		Conservation minimum	
	content	± to control	content	± to control	content	± to control
	Initial content of humus in 1980					
No fertilizers (control)	2,95	–	2,94	–	3,07	–
	Initial content of humus in 1988					
No fertilizers	2,88	–0,07	2,91	–0,03	3,6	–0,01
Farm manure 12t-Background (B1)	2,95	0	2,96	+0,02	3,16	+0,09
B1 + $N_{78}P_{70}K_{78}$	2,99	+0,03	3,02	+0,08	3,31	+0,24
B1 + $N_{127}P_{115}K_{125}$	2,97	+0,02	3,00	+0,06	3,12	+0,5
Straw 2,4t+N_{24}— Background (B2)	2,96	+0,01	2,96	+0,02	3,24	+0,17
B2 + $N_{78}P_{70}K_{78}$	2,98	+0,03	2,99	+0,05	3,41	+0,34
B2 + $N_{127}P_{115}K_{125}$	3,02	+0,07	3,09	+0,15	3,36	+0,29
	LSD_{05} for fertilizing = 0,04 %, for tillage = 0,06 %					

Table 4 also shows that the application of winter wheat straw at a rate of 2.4 tons/ha, with the addition of 10 kg N for every ton of straw, provides the same reproduction of humus as 10 tons/ha of farmyard manure for both conventional and combined tillage. With conservation tillage, however, the same amendments result in a much higher reproduction of humus (+0.17%) and with further addition of mineral fertilizers humus accumulation increases to + 0.34%.Thus, in terms of humus accumulation 1 ton of straw (with appropriate addition of N) is equivalent to 5 tons of semi-decomposed farm manure.

It may be concluded that it is difficult or impossible to ensure an extended reproduction of humus (and soil fertility) with conventional tillage. Conservation tillage, using the same rates of fertilizers, did lead to extended reproduction of humus, whereas the variant with combined tillage, that is ploughing on row crops and conservation tillage on grains, was characterized by intermediate levels of humus accumulation and crop productivity.

7. Conclusions.

1. For Ukrainian chernozems, mouldboard ploughing is an inappropriate method of soil tillage. It effectively disrupts biological cycling of major nutrients, reduces the humification coefficients of organic fertilizers and results ultimately in extensive dehumification with loss of soil fertility. Essentially, the practice runs contrary to the natural processes of soil formation.

2. Soil tillage that does not involve inversion of the furrow slice is much more in keeping with natural soil formation processes. In Ukrainian chernozems it maintains or extends the process of humification in crop production systems.
3. The self-regulating processes that promote fertility in Ukrainian chernozems involve the assimilation of carbon dioxide and carbonic acid from the soil by heterotrophic micro-organisms and creation of new protein-like substances. These substances restore the nutritive properties of soil humus, which decline during crop growth, and also improve other properties and behavior of soils.
4. A better understanding of the self-regulatory processes involved in maintaining soil fertility allows control and management of soil formation processes in agroecosystems through consideration of tillage systems, fertilization practices and land improvement, with considerable savings in money and labour.

8.References

1. Shykula M.K.(1997). Biochemical mechanism for soil fertility self-regulation. – *Scientific proceedings of Nat. Agric. Univ.* - **1**, 163-171.
2. Shykula M.K., Melnychuk D.O. et al. (1995) Heterotrophic fixation of carbon dioxide and it role in chernozem`s soil organic matter formation. *Proceedings of agricultural science*-8, 23-28.
3. Shykula M.K., Ridei N.M.. et al. (1996) Intensity of carbon dioxide assimilation by varios soil units of Ukraine. *Proceedings of agricultural science* – **8**, 15-19.
4. Shykula M.K., Nazarenko G.V.(1990) *Minimum tillage of Chernozems and Reproduction of their fertility*, Moscow, Agropromizdat.
5. Shykula M.K., Antonets C.C., Kapshtyk M.V.(1998) et al. *Soil fertility Reproduction in Conservation Farming*, Kyiv, Oranta.

CONSERVATION NON-PLOUGH SYSTEMS OF CROP PRODUCTION IN UKRAINE WITH INCREASED REPRODUCTION OF SOIL FERTILITY

M.V. Kapshtyk, M.K. Shikula and L.R. Petrenko
National Agricultural University of Ukraine, Kiev

1. Introduction.

Ukraine has a great potential for the development of agriculture with 43 mln.ha of farm land mainly in chernozems (base-rich Mollisols) reaching 68% of the total area of the country. Ukraine could have become a leading producer and exporter of grain and other agricultural products, but an imperfect system of economic management and the adoption of irrational technologies based upon mouldboard ploughing has caused considerable soil degradation by water and wind erosion, mainly through the loss of organic matter and increasing acidity. In addition, widespread pollution by nuclear wastes, heavy metals and various chemicals has occurred, affecting the health and genetic stock of the Ukrainian people [3].

The chernozems of Ukraine are her most valuable resource, representing a legacy bequeathed to generations of Ukrainian people. These soils are very fertile but they have become considerably impoverished and degraded due to centuries of mismanagement. Soil organic matter is particularly vulnerable to severe degradation and there is a good correlation between the content of humus and other soil properties that relate to soil fertility [1]. Centuries of soil dehumification have been caused by the violation of the principle of returns in farming [4]. Crop yields have depleted nutrients from the soil and the decomposing organic matter does not become a source of humus.

2. Soil organic matter content and conventional tillage.

A decline in organic residues leads to a lower intensity of soil formation processes, principally those relating to biogeochemical cycling of nutrient elements and energy which are vital for the reproduction of soil fertility. However, the lower fertility and bioproductivity of cultivated chernozems compared with virgin and long-fallow soils is caused not only by losses of organic matter and nutrients but also by changes in the composition of organic matter and its inability to participate in the seasonal cycles of transformation [5]. Changes in the composition of humus in cultivated chernozems compared with uncultivated virgin soils (in the Michailivska Tsilyna, Sumy region) show lower amounts of humus substances, including both humic and fulvic acids, although the total amount of these substances us still high in cultivated soils especially in the 0-10 cm layer. (Table 1).

However, the Ch:Cf ratio has increased significantly, especially in the 0-10cm layer of the soil. In considering the amount of humic and fulvic acids as percentages of total organic carbon one may come to a different conclusion about the composition of soil humus. Thus, cultivation increases the content of soluble substances in humus (free and Ca-bound) including humic acids and reduces the content of insoluble residue in the 0-20 cm layer. Therefore, the quality of soil organic matter may be regarded as degraded

M.J. Wilson and B. Maliszewska-Kordybach (eds.),
Soil Quality, Sustainable Agriculture and Environmental Security in Central and Eastern Europe, 267–276.

by the influence of cultivation. In this respect, the organic matter quality in virgin soils may be considered as a model for the reproduction of organic matter in cultivated soils

TABLE 1. Quality of SOM in typical Chernozem under influence of various systems of land use (Preserve "Mykhailivska tsilina", Sumy region)

Variant	Layers, cm	Total C		$N_{extractable}$		N_H		N_F		$N_{nonextractable}$		N_H: N_F
		%	% of C	%	% of C	%	% of C	%	% of C	%	% of C	
Virgin soil	0-10	5,90	100	2,72	46,1	2,00	33,8	0,822	13,9	3,08	52,2	2,43
	10-20	4,70	100	2,36	50,3	1,73	36,8	0,63	13,4	2,34	49,7	2,74
	20-30	3,84	100	2,02	52,6	1,49	38,7	0,536	13,9	18,20	47,4	2,77
	30-40	3,88	100	1,87	48,3	1,29	33,3	0,582	15,0	2,01	51,7	2,23
Cultivated soil	0-10	3,71	100	2,02	54,5	1,49	40,3	0,527	14,2	1,69	45,5	2,83
	10-20	3,72	100	1,91	51,4	1,41	38,1	0,499	13,4	1,80	48,6	2,82
	20-30	3,57	100	1,83	51,4	1,37	38,3	0,466	13,1	1,74	48,6	2,69
	30-40	3,49	100	1,71	49,0	1,19	33,2	0,523	15,0	1,71	49,0	2,27

Nevertheless, some authors claim that losses of SOM caused by cultivation are compensated by an improvement of its quality [6]. Evidence cited for this includes an increasing Ch:Cf ratio, an increase in the insoluble residue content which is more firmly bound to the mineral part of the soil, and an increase in the active part of the humus molecule and the overall chemical reactivity of humus. But in our opinion, changes in the quality of humus and soil fertility in general have a negative direction.

TABLE 2.Cation exchange capacity of typical clay loamy Chernozem under influence of various land management for 17 year, meq/100g^{-1} of soil

Layers, Sm	System of land management					
	Long-term grassland		Cultivated soil, non-fertilized		Cultivated soil, fertilized*	
	Meq/100g^{-1}	%	meq/100g^{-1}	%	meq/100g^{-1}	%
0-10	37,8	100	31,5	83,3	35,9	95,0
10-20	35,6	100	31,2	87,6	34,5	96,9
20-30	34,5	100	30,7	89,0	33,0	95,7
30-40	31,2	100	27,3	87,5	30,5	97,8
40-60	28,5	100	25,2	88,4	30,1	106
60-80	19,7	100	21,5	109	21,4	109
80-100	17,1	100	17,8	104	17,4	102

*Fertilizers were used in rate of 9 t of manure + $N_{28}P_{28}K_{28}$ per 1 ha of crop rotation area.

Firstly, conventional soil tillage drastically reduces the amount of newly formed organic matter compared with virgin soil; such newly formed (fresh) organic matter directly correlates with yields of crops [7]..

An increasing Ch:Cf ratio is evidence of soil organic matter ageing, and not necessarily of improvement in its quality. Lastly, the reduction of seasonal fluctuations in organic matter content shows that it is becoming less capable of self-regulation.

Thus, it is clear that cultivation reduces not only SOM but soil productivity in general. Table 2 shows a reduction in soil cation exchange capacity following cultivation in a field experiment on a typical clay loam chernozem in the Poltava region.Variants of fertilized and unfertilized plough land were compared with a long term (17 years) grassland. It can be seen that both fertilized and unfertilized variants of mouldboard plough land had a reduced CEC, especially in the 0-10 cm layer compared with grassland. On the unfertilized soil this reduction was 6.3 meq/100g, while on the soil with an annual application of 9 t/ha farm manure and N28P28K28 it was only 1.9meq/100g. The same tendency was observed in the 10-20 cm layer and in the deeper layers of the soil.

Our research shows how difficult it is to ensure a non-deficit balance of SOM using conventional tillage based on mouldboard ploughing. In crop rotations with 40% in inter-tilled crops during 17 years of experiment the most drastic reduction in SOM and total N content occurred on unfertilized plots, reaching 0.31% and 35 ppm respectively in the 0-40 cm layer. Farm manure (6t/ha) and mineral fertilizers ($N_{28}P_{28}K_{28}$) reduced the losses of SOM to 0.24% and total N to 15-28 ppm. But such fertilization did not ensure a non-deficit balance of SOM and N Our calculations show that it is necessary to apply 12-14 t/ha of farm manure to ensure a non-deficit balance of SOM in soils subject to mouldboard ploughing. Agricultural enterprises in Ukraine are unable to produce such an amount of farm manure, especially in the present day economic situation. Therefore, it is necessary to introduce more up-to-date technologies in crop production systems.

3. Conservation tillage and soil fertility.

To increase agricultural productivity, halt soil degradation and restore the full potential of Ukrainian chernozems, it is necessary to introduce a reform in land use and management, as well as in new crop production systems. For this purpose, a conservation system of farming is described whereby an increased reproduction of soil fertility is achieved. This system was developed at the National Agricultural University of Ukraine in 1974-1996. The system not only ensures higher and stable yields of crops but reproduces the potential fertility of chernozems and brings about a self-regulation function so that in this respect cultivated soils resemble their virgin counterparts [8,9].

Minimum non-plough tillage and plant residue management are the corner stones of the new conservation system of crop production. Crop residues of wheat, corn, sunflower, etc, will maintain a high level of energy for soil formation and become a means of soil protection against degradation. Our research shows that non-ploughing technologies increase nutrient element contents in the soil and instigate the self-regulating processes of soil productivity based on an increased intensity of seasonal and annual cycles of SOM and biogeochemical cycles of biogenic elements and energy. Higher seasonal changes in SOM content and a higher level of soil and agroecosystem self-regulation

increase soil fertility and crop yields, which become less dependent on weather conditions, rates of fertilizers and chemicals, and stabilizes the agrolandscape [1,9].
Conservation tillage incorporates crop residues in the upper (5-7 cm) of the soil, which is a well-aerated layer. This layer has the greatest humification coefficient and humus increases by more than one third when the residues are incorporated no deeper than 10 cm, as compared with their incorporation below 15 cm. A shallow placement of the residue drastically reduces organic carbon losses [1,9]. A non-deficit balance of soil humus is achieved and there is an increased reproduction of SOM. Using conventional systems of soil tillage, a non-deficit SOM balance in the central part of the forest-steppe zone of Ukraine can only be achieved by applying no less than 12 t/ha of farm manure but for conservation tillage this is reduced to 8 t/ha.
An understanding of the mechanism of SOM formation and the self-regulation processes involved has enabled great improvements to be made in crop production systems in Ukraine. Soil fertility can be controlled in the context of crop rotations and higher yields of crops can be obtained with lower inputs. The systems so developed protect the soil because they are based on the functioning principles of natural ecosystems where there is no tillage and plant cover is renewed naturally. Soil formation and production processes in virgin ecosystems allow 5 to 6 times more biomass to be obtained in crop rotations. Conservation technologies create soil conditions that are the most favourable for plant growth and development, including the most up-to-date high yielding varieties, hybrids and cultivars of crops.

4. Experimental results

The effectiveness of our approach has been proved in practical experiments at the "Obriy" co-operative farm in the Poltava region where mouldboard ploughing has not been used for 21 years and agricultural chemicals have been applied only in minimal quantities in the past 18 years. Minimum non-plough tillage (10-20 cm deep) was employed in all crop production systems. It was found that yields of crops became 2 times higher and fuel and chemical fertilizer requirements 2 times lower. Only seed material was treated with pesticides and overall the need for pesticides was reduced 5 to 8 times. In addition, soil tillage operations took up to 3 times less time than conventional tillage.
In the field experiments it was found that a discrete hierarchical order of changes in soil fertility occurred under the influence of soil protecting systems. Figure 1 shows the seasonal cycles in SOM content in a typical chernozem using conservation and conventional tillage crop production systems with the application of 12 t/ha farm manure and $N_{80}P_{75}K_{65}$. Even in the first three years, the conservation technology instigates a renewal of the seasonal cycles of humus in the soil. SOM contents changed only by 0.11% in the 0-20 cm layer in the plots under sugar beet with mouldboard ploughing in the period from the spring planting in May to the autumn harvesting in October, while on the non-plough tilled plots the amplitude of the changes reached 0.17-0.22%.
In the first years of conservation tillage the soil recovers its ability for cyclical changes of labile organic matter (LOM) and changes in the general quality of soil humus.

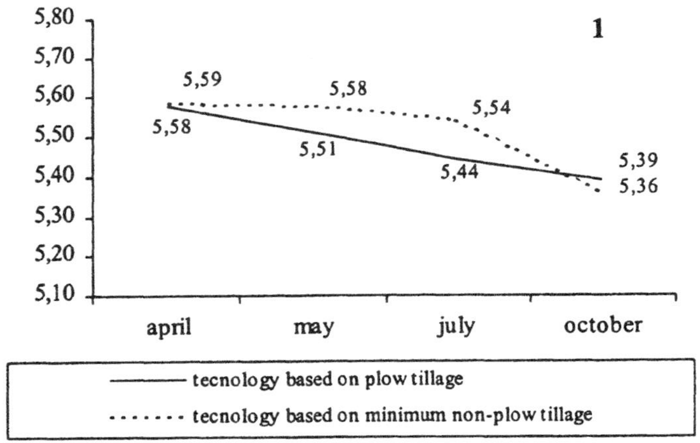

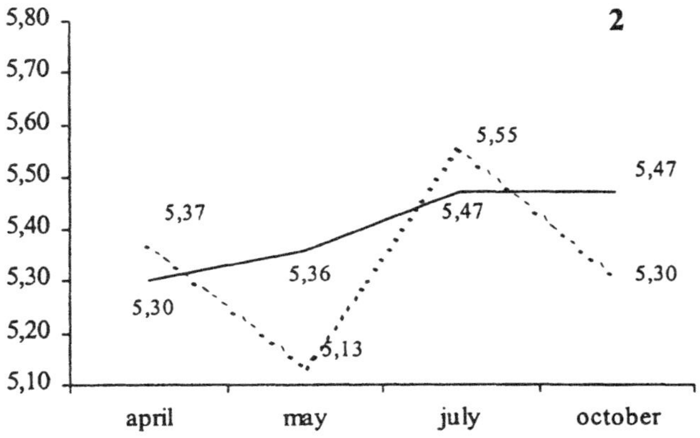

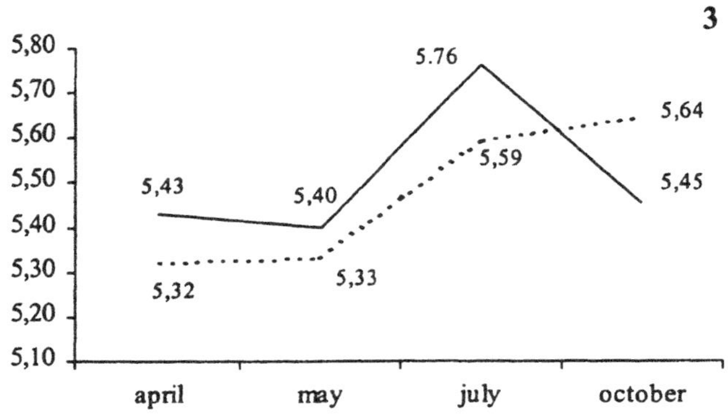

Figure 1. Seasonal cycles of humus in the 0-20 cm layer of a typical chernozem under the influence of different cropping systems in the first 3 years of experiment.

Field station of NAU in the Poltava region.

1. Sugar beet (1st year)
2. Sunflower (2nd year)
3. Annual grasses (3rd year)

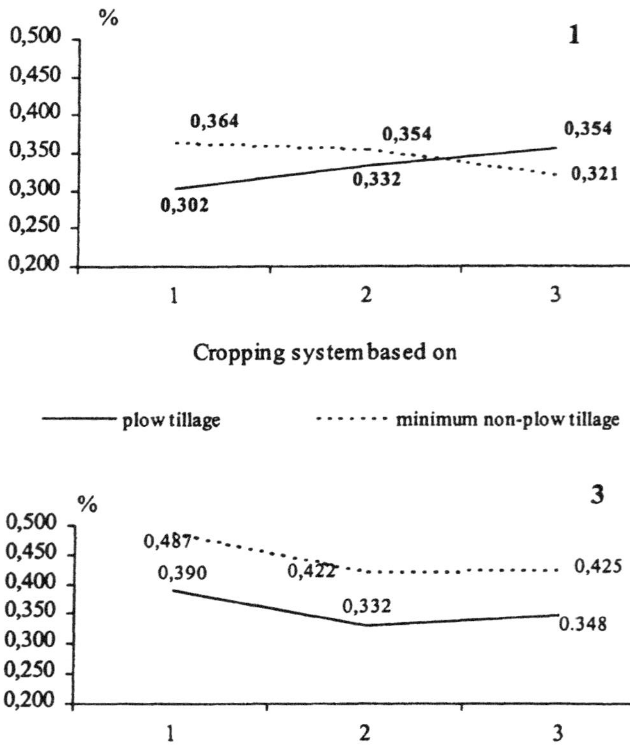

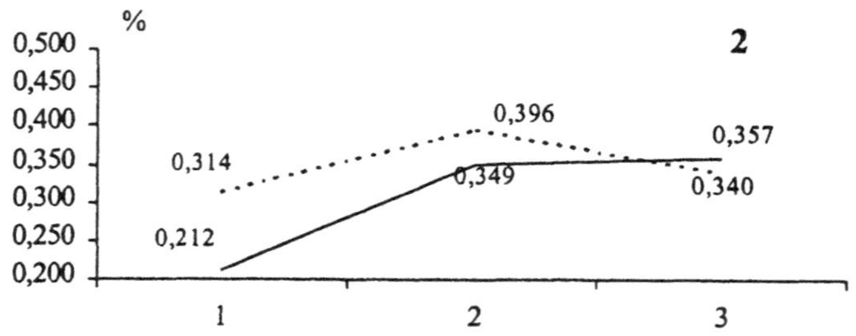

Figure 2. Seasonal cycles of labile soil organic matter in the 0-20 layer of a typical chernozem under different cropping systems in the first 3 years of experiment.

Note. 1. Beginning of vegetation.
2. Intensive growth.
3. Harvesting.

1. Sugar beet (1st year)
2. Sunflower (2nd year)
3. Annual grasses (3rd year)

Thus, under sugar beet with conventional tillage the content of LOM in the 0-10 cm layer has increased from spring to summer by 0.30 to 0.35 %, while with minimum tillage it decreased from 0.36 to 0.32 %(Figure 2). During the autumn-winter-spring period the SOM content and its labile form were completely renewed. It may be noted that even a short-term use of non-plough technologies increases fluctuations in the SOM content, especially soil humus. The instability of these fluctuations may be explained by the inability of the soil to come out of "shock" caused by previous mouldboard ploughing.

The first change in the quality of soil fertility occurs after 5 years of systematic non-plough tillage when the soil renews the internal connections that were destroyed by furrow inversion. Humus content becomes 0.10 to 0.12 % higher and the amplitude of the SOM fluctuations increases. Figure 3 shows that non-plough technologies decreased the content of humus in a crop rotation sugar beet ---> peas---> winter wheat from April to August by 0.24 %, while on ploughing it decreased only by 0.16 %. In the next period there was a gradual recompensation of humus. The maximum amplitude of seasonal change in humus content is characteristic of a typical chernozem under long-term grassland (0.31%).

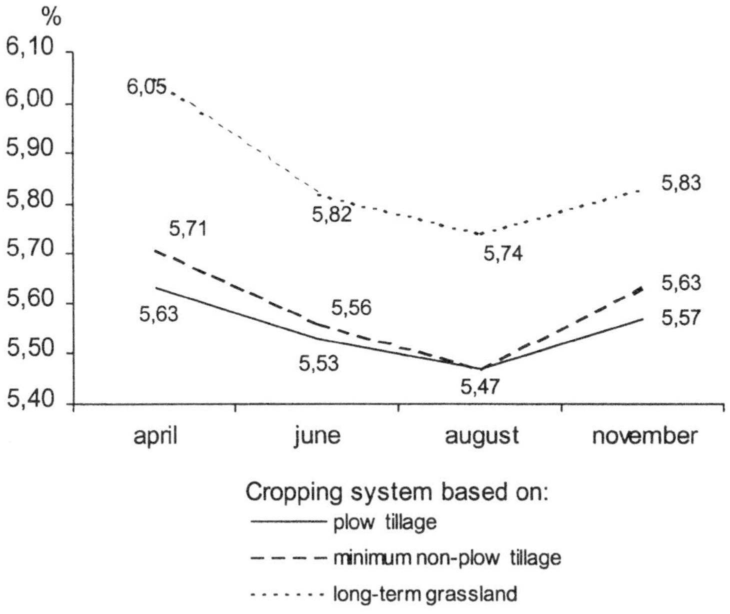

Figure 3. Seasonal cycles of humus in the 0-10cm layer of a typical chernozem influenced by various cropping systems for over 5 years.

During the next period of discrete change the amplitude of the seasonal fluctuations in the contents of humus and LOM continue to grow in the 0-10 cm layer (Figure 4). Under all three crops there was an advantage of non-plough technologies and long-term

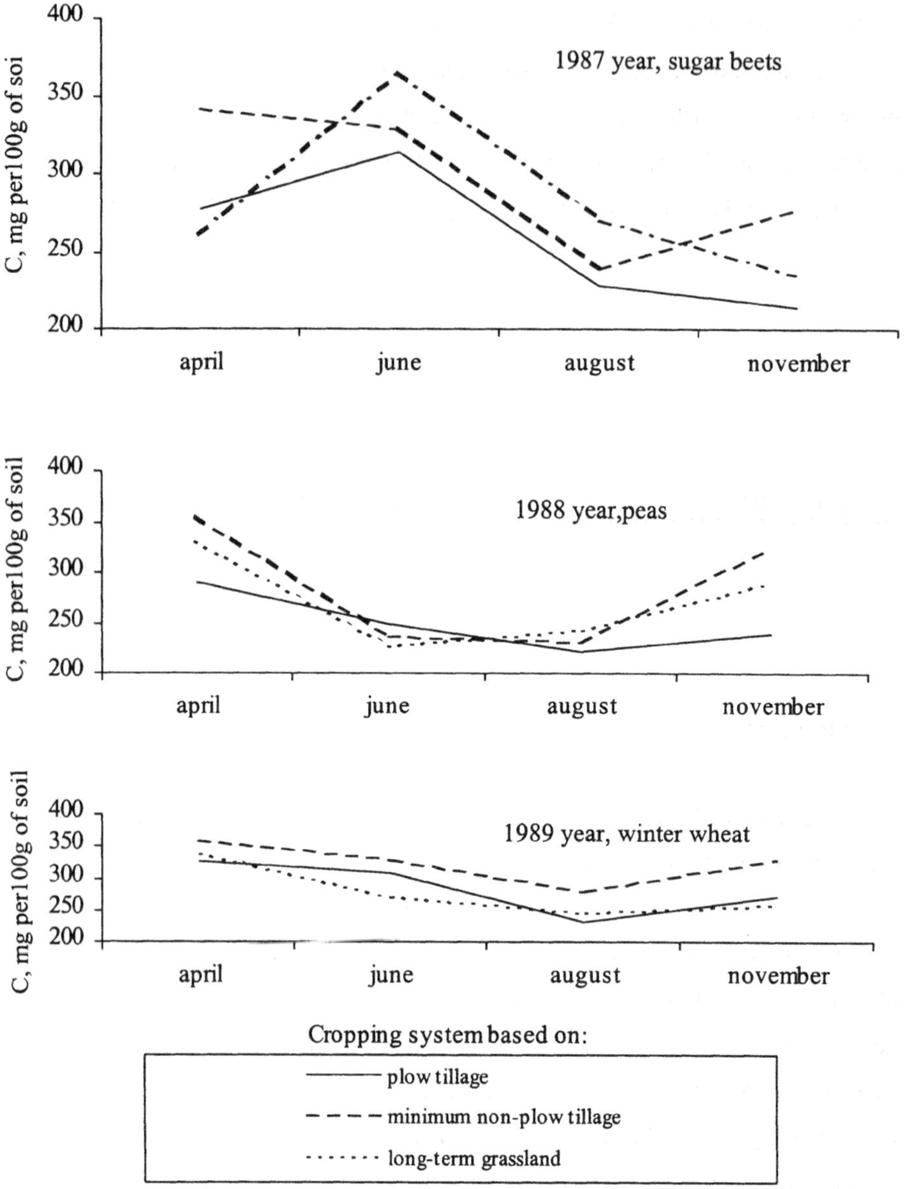

Figure 4. Seasonal cycles of labile soil organic matter in a typical chernozem (0-10 cm layer) influenced by various cropping systems for over 5 years

grassland compared to plough tillage. In terms of seasonal cyclicity of SOM, conservation tillage approaches that of long-term grassland so that the processes of soil formation may be considered as becoming more natural.
After 9 years of the non-plough conservation system the soil acquires the ability to self-regulate its fertility and these changes affect the bioproductivity of the soils. Within 5 years the crop yields have increased by 0.45-0.55 t/ha in grain units, compared to conventional tillage (Table 3). After a further 5 years, yield increments reach 1.2-2.0 t/ha compared with conventional technologies. Crop yields also become less dependent upon weather conditions and pesticides.

TABLE 3. Increase of crop yields in the "Obriy" farm of the Poltava region from 1966 to 1994 years under the influence of soil conservation crop production systems for various periods of time.

Years	Unit of Measure-ment	Crop yields and their increase							
		grain	for kinds of crops						
		crops a whole	winter wheat	spring barley	oats	peas	corn	Snflower	sugar beets
Crop yield 1970-1975	metric centner per ha	28,1	29,2	25,2	27,1	15,9	24,3	16,1	255
Increase of yield for :									
1975-1980	m.c/ha	4,1	9,0	3,4	0,8	14,0	2,4	-3,6	124
	%	16	31	13	3	88	10	-22	49
1981-1985	m.c/ha	7,2	5,3	5,4	10,3	2,0	25,8	-4,2	67
	%	28	18	21	38	13	106	-26	26
1986-1990	m.c/ha	28,2	34,3	28,4	38,3	5,5	34,8	12,2	140
	%	94	117	113	141	35	143	76	56
1991-1995	m.c/ha	25,2	33,9	27,6	31,1	12,5	-	11,9	159
	%	97	116	110	115	79	-	66	62

The third hierarchical change in the fertility of the soil occurs after 15 years of systematic use of non-plough technologies when the soil is restored completely to natural processes of soil formation. Yields of crops become 70-100 % higher compared with the initial yields and become ever less dependent upon weather conditions and agricultural chemicals. This is illustrated in the "Obriy" Cooperative Farm where in the last 5 years fertilizers were applied at rates of 10-30 kg/ha of NPK and pesticides not used at all, but where yields of grain crops were within 5.0-5.5 t/ha and sugar beet 40.0-45.0 t/ha.

5. Conclusions.

The hierarchical changes of soil fertility that occur in discrete steps as described above are the basis for the development of conservation systems of soil tillage and fertilization in Ukraine. Before the first hierarchical change occurs, the system of tillage should be of a non-plough type to varying depths especially on soil with less than 4.5% organic matter. In the next period, minimum non-plough tillage can be used for all crops and on soils containing more than 4.5% organic matter minimum tillage may be started even in the first years. The system of fertilization should include additional N dressings (N10) for each ton of straw. Thus the use of conservation systems of crop production is a prerequisite for an increased reproduction of soil fertility in chernozems, whereby the natural process of soil formation approaches that of its natural counterpart, at the same time ensuring the maintenance of soil quality, sustainable increased crop yields and a positive impact upon the overall environment.

6. References

1. Shykula M.K., Antonets S.S., Kapstyk M.V. et al. (1998) *Soil Fertility reproduction in Conservation Farming*, Oranta, Kyiv.
2. Myedvyedyev V.V. (1997). Reinnovation of ecological – reproductive and productive functions of soils as a most main stage for realization of Sustainable development of Ukraine. *Proc. of Agrarian science* **9**,16-20.
3. Gnatenco O.F., Petrenko L.R., Kapstyk M.V.et al(1998*). Soil degradation and their monitoring*, National Agricultural University, Kyiv.
4. Shykula M.K.(1993). The ways for increasing of Sustainability in Farming under conditions of economy reforming. *Changes in Land Use under conditions of economy reporming,* SOPS NASU, Kyiv.
5. Ponomaryova T.A., Plotnikova T.A. (1980). *Humus and soil formation*, Nauka, Moskow.
6. Lactionov M.I. (1974). *Law-governed natures of organic colloids transformation in Chernozems under agricultural management,* Kyiv. –Author's abstract for doctor's science thesis.
7. Chesnyak O.A., Chesnyak G.Yaand., Stepanenco A.Ya.(1970) The labile soil organic matter content in Typical Chernozems of Forest Steppe Zone of Ukraine as an indicator of their improvement under agricultural management. *Collection of scientific work of Kharkiv Agricultural University*, 87(124), 29-36
8. Shykula M.K. (1997). Mechanism of soil fertility self-regulation. *Proceedings of National Agraricultural University* **2**, 163-171.
9 Shykula M.K., Nazarenco G.V. (1997). *Minimum tillage of Chernozems and Reproduction of their fertility*, Agropromizdat, Moskow.

AN EROSION CONTROL SYSTEM FOR SUSTAINABLE LAND USE IN A LITHUANIAN CATCHMENT

B. Jankauskas and G. Jankauskiene

Kaltinenai Research Station of Lithuanian Institute of Agriculture
Varniu 17, 5926 Kaltinenai, Silale District, Lithuania

1. Introduction

About 52% of the terrain in Lithuania is on hilly, rolling relief, where the soil is erodible [10]. The last glaciation (about 12 thousand years ago) left a thin layer of erodible, glacial, clay loam moraine on the old basement rocks of the Zemaiciai uplands, and it is here that the soil erosion processes have been investigated. The Zemaiciai uplands are long moderately and strongly sloping hills in the central part, becoming short, gently sloping and densely grouped hills on the outskirts. Annual precipitation ranges from 800-858 mm in the highest parts of uplands to 750-800 mm in the lower parts.

Soil erosion is a hazard associated with ecological conditions and especially with agriculture where it is important for the long-term effects on soil productivity and sustainable land use. Erosion control is now a necessity in almost every country of the world under every type of land use [11]. The parent rock, soil texture, relief, velocity of wind, intensity and abundance of precipitation and plant cover are all natural factors that influence rates of erosion [9, 11, 20]. Tillage erosion and tillage translocation create favourable conditions for water and wind erosion in Lithuania and was assumed to be a primary cause [7, 12]. The loss of clay loam soil in Eastern Lithuania (Baltic upland) ranges from 1.3 t ha^{-1} under cereal grain crops to 56.6 t ha^{-1} under fallow [3]. For comparison, at Foulum (Denmark) soil loss ranged from zero under permanent grass to more than 25 t ha^{-1} under winter wheat drilled up- and down-slopes [14]. The mean annual soil loss in Switzerland ranged from 13 t ha^{-1} in the case of clay soils to 22 t ha^{-1} for a loess soil [15]. On the Swiss Jura Plateau the loss recorded from a loamy clay soil was 13.39 t ha^{-1} [16].

The high capability of perennial grasses and winter grains to resist erosion [7, 11] particularly in the context of reduced tillage systems [2, 17, 18] enabled field experiments forming an erosion control system for erodible soils in Lithuania to be designed.

M.J. Wilson and B. Maliszewska-Kordybach (eds.),
***Soil Quality, Sustainable Agriculture and Environmental Security in Central and Eastern Europe,* 277–283.**

2. Methods

An erosion control system for sustainable land use on the hilly, rolling relief of Lithuania was prepared on the basis of experiments carried out at the Vezaiciai branch and Kaltinenai Research Station of the Agricultural Institute of Lithuania [6]. Later, this system was supplemented by new research data of the author [7] and of other researchers at the Kaltinenai Research Station [1, 5, 8, 13]. Most experimental results were obtained by field experiments set up on the slopes, hill tops and footslopes of the hilly, rolling relief of the Zemaiciai uplands (approximately $55^0$34' N and $22^0$29' E). The prevailing soils were Dystric and Gleyic Podzoluvisols (FAO-UNESCO, 1994) slightly acid, loamy sand or clay loam (pH_{KCl} to 5.5), with low amounts of mobile P_2O_5 (50-100 mg kg^{-1}) and medium or slightly higher amounts of mobile K_2O (100-200 mg kg^{-1}). Four agrophytocenoses were investigated (six-course crop rotations) since 1982. The following crop rotations were compared:

I The field crop rotation: tillage crops 17%, grain crops 50%, mixture of clover timothy, referred to as CT.
II. The grain-grass crop rotation: grain crops 67%, CT 33%.
III. The grass-grain I crop rotation: grain crops 33%, CT - 67%.
IV. The grass-grain II crop rotation: grain crops 33%, mixture of orchard grass - fescue red (OF) 67% (Table 1).

TABLE 1. **Crop rotations and annual fertiliser regimes investigated at the Kaltinenai Research Station**

Crop rotation and average annual fertilising			
Field $N_{60}P_{65}K_{80}$	Grain-grass $N_{50}P_{60}K_{70}$	grass-grain I $N_{70}P_{60}K_{80}$	grass-grain II $N_{120}P_{60}K_{80}$
1. Rye,	1. Rye,	1. Rye,	1. Rye,
2. Potato,	2. Barley,	2. Barley,	2. Barley,
3. Barley,	3. Barley,	3. CT, I y.u.,	3. OF, I y.u.,
4. Barley,	4. Barley,	4. CT, II y.u.,	4. OF, II y.u.,
5. CT, I y.u.*,	5. CT, I y.u.,	5. CT, III y.u.,	5. OF, III y.u.,
6. CT, II y.u.	6. CT,II y.u.	6. CT, IV y.u.	6. OF, IV y.u.

* - years of use.

The field experiments were carried out on 2-5^0, 5-10^0 and 10-14^0 slopes. Instead of field crop rotation, perennial grasses of multiple composition for long-term use were grown on slopes of 10-14^0. The productivity level of the different agrophytocenoses was calculated by the content of metabolisable energy based on chemical analysis of each crop produced. The losses of soil due to water erosion were established by the method of rill volume measurement [19] and calculated as t ha^{-1} according to soil bulk density. Total precipitation varied from 635 to 1075 mm in the period of these experiments.

3. Results

The arable soils on this hilly, rolling relief were eroded by water and wind following tillage. The natural fertility of the soil on slightly, moderately and severely eroded slopes decreased by 21.7, 39.7 and 62.4 per cent, respectively (Table 2). A full set of 144 research data can be characterised by the equation:

$$y = 0.037x^2 - 0.1875x + 21.219; \;\; R = 0.7924 \qquad (1)$$

TABLE 2. **Dependence of barley yield upon gradient of slopes and degree of erosion**

Parts of relief	Degree of soil erosion	Yield* from 48 investigated plots		
		$t\ ha^{-1}$	decrease $t\ ha^{-1}$	in relative numbers
Plains	Non-eroded	18.9	-	100
Slopes of $2\text{-}5^0$	Slightly eroded	14.8	4.1	78.3
Slopes of $5\text{-}10^0$	Moderately eroded	11.4	7.5	60.3
Slopes of $10\text{-}14^0$	Severely eroded	7.1	11.8	37.6
Foot-slopes	Deposit soil	19.3	-	102.1
LSD_{05}		1.1		

* The average of 3 years grain and straw gross yield.

Perennial grasses provided full protection from soil erosion even on the slopes of $10\text{-}15^0$. Grass mixtures with a high percentage (90%) of lucerne were more suitable for hilly pastures, particularly on soils that were suitable for growing lucerne. The annual average yield of dry matter was 6.12 t ha^{-1} or 0.92 t ha^{-1} of digestible protein [7]. Later investigations showed that large areas of soils of the Zemaiciai uplands were not very suitable for growing lucerne because of excessive acidity and waterlogging in the subsoil. Therefore, it was necessary to establish more suitable long-term perennial grasses on these soils. The field experiments were successful and grass mixtures of high fertility for early, medium and late hay making or grassing were established. The annual average productivity of the most fertile hay meadow mixture during a 6 year period was 7.86-9.15 t ha^{-1} of dry matter. The productivity of pasture grassland was 5.56-7.07 t ha^{-1} [13]. The productivity of these grass mixtures did not decrease during a 6 year period, indicating that these grass mixtures can be sustained over a longer period.

The results of 12 year field experiments show, that the perennial grasses completely stopped water erosion even on slopes of $10\text{-}14^0$ gradient, while average annual rates of water erosion under winter rye, spring barley and potatoes were 5.4-29.6 t ha^{-1}, 18.0-59.7 t ha^{-1} and 44.4-196.2 t ha^{-1} respectively on the slopes of $2\text{-}5^0$, $5\text{-}10^0$ and $10\text{-}14^0$ (Fig. 1).

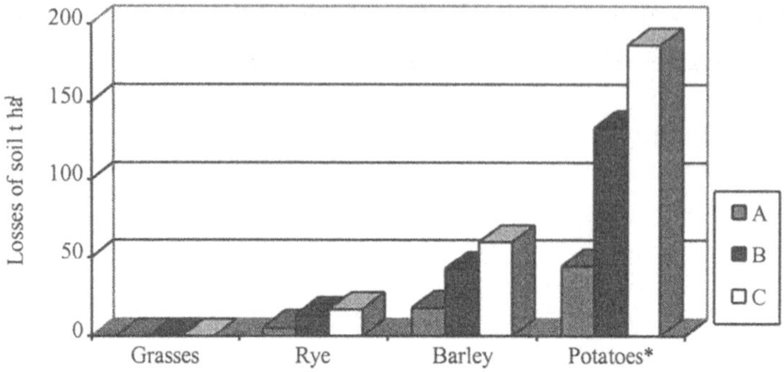

Figure 1. **Dependence of rates of water erosion on plant cover and slope gradient**
*The columns represent the average data of 1983-1994 on the slopes of: A - 2-5⁰, B - 5-10⁰ and C - 10-14⁰. * On the slope of 10-14⁰ potatoes were not grown. The data were calculated by the method of group comparison.*

The degree of cover of the different crops in the crop rotations determined the total erosion resisting capability of the agrophytocenoses investigated (Fig. 2). The average annual rates of water erosion under erosion-preventive grass-grain crop rotations decreased by 77.4-80.7% in comparison with the field crop rotation, while under grain-grass crop rotation it decreased by 21.5-24.6%. However even the grass-grain crop rotations could not completely stop soil erosion and annual soil losses by water erosion of 9.4-9.7 t ha^{-1} on slopes of 10-14^0 were recorded.

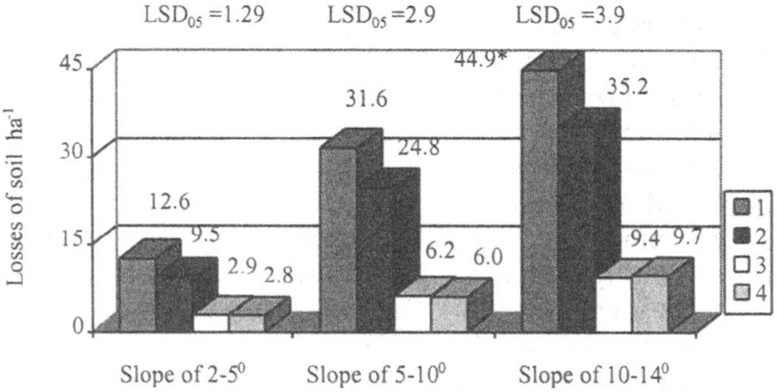

Figure 2. **Average annual (1983-1994) rates of soil erosion by water under different crop rotations**
*Crop rotations (agrophytocenoses): 1. Field; 2. Grain-grass; 3. Grass-grain I; 4. Grass-grain II. * Perennial grasses of multiple composition were grown instead of field crop rotation. The data were calculated by the method of group comparison.*

According to the mean data obtained over the twelve year period (1983-1994), the content of metabolisable energy accumulated by erosion-preventive, grass-grain crop rotations on the slopes of 2-5^0 and 5-10^0 were 11.8 and 27.7% higher than in the grain-grass crop rotation, or 14.1 and 38.7% higher than in the field crop rotation. The above mentioned research data enabled the design and recommendation of the different models of erosion-preventive crop rotations [7]. Therefore, a key recommendation is for the grassing of slopes over 10^0 and the use of erosion-preventive tillage and fertilising/liming on the slopes of $2\text{-}10^0$ in edition to erosion-preventive crop rotations.

Spraying with the herbicide Utal can be successfully used instead of mechanical stubble cultivation (particularly deep ploughing) during autumn soil tillage. Deep soil chiselling can be used instead of deep mouldboard ploughing. Rates of soil erosion were reduced by 1.6 to 8.9 times, while the productivity of cultivated plants remained at the same level [1]. Different rates of nitrogen fertiliser application on the various parts of the hilly, rolling relief [5], combined with liming of eroded acid soils [7] is also an important part of the erosion control system. An outline of our suggested erosion control system is presented in table 3.

TABLE 3. **Erosion-preventive grouping of erodible terrain and erosion-resisting measures**

Groups	Texture of soil* and gradients of slopes		Type of land use	Requirements for exclusions for the group	Recommended erosion-resisting measures
	S, LS, G	L, C			
I	Over 10^0	Over 15^0	Woods	To exclude slopes over 10^0 and 15^0. Slopes over 10^0 of heavy texture and over 5^0 of light texture are unsuitable for land reclamation.	To plant trees or shrubs, to carefully maintain perennial grasses.
II	$7\text{-}10^0$	$10\text{-}15^0$	Grass-land	In addition to indicated slopes to includ plots inconvenient for tillage from arable plots and to establish the pasture or grassland.	To plant perennial grasses for long-term use. To renovate the grassland by mixture of another composition. Cover crop must be annual grasses.
III	$5\text{-}7^0$	$7\text{-}10^0$	Arable land or grass-land	Similar to IInd group, but plots may be of suitable for tillage.	To put into practice the erosion-preventive grass-grain crop rotation. To apply erosion-preventive tillage.
IV	$2\text{-}5^0$	$3\text{-}7^0$	Arable land	Similar to IIIrd group, but 10% of light soil slopes up to 7^0 can be added.	To put into practice the erosion-preventive grain-grass crop rotation. To apply erosion-preventive tillage. To avoid growing of tillage crops and flax.
V	Up to 2^0	Up to 3^0	Arable land	Plains, suitable for tillage plots, and those remained after forming of I-IV groups.	To put into practice intensive field crop rotation. On the slopes of $2\text{-}3^0$ to apply erosion-preventive tillage.

- S - sand, LS - loamy sand, G - gravel, L - loam, C - clay.

The territory of a farm on hilly, rolling relief should be divided into 5 groups. The first (I) group covers highly erodible soil, i.e. slopes over 10^0 having sandy, loamy sand or gravel texture (light soil) and slopes over 15^0 having loamy or clay texture (heavy soils). We suggest that trees should be planted on such slopes. For Group II, the growth of long-term perennial grasses is recommended on light soil with slopes of $7\text{-}10^0$, on heavy soil with slopes of $10\text{-}15^0$ and on surrounding soil unsuitable for any other agricultural use. Erosion-preventive grass-grain crop rotation, including 50-80% of perennial grass is suggested for soils of the third (III) group, i.e. on the slopes of $5\text{-}7^0$ of light soils and on the slopes of $7\text{-}10^0$ of heavy soils. The above mentioned slopes should be arranged into a plot of land suitable for tillage. The fourth (IV) group covers slopes of $2\text{-}5^0$ of light soils and $3\text{-}7^0$ of heavy soils and represents utilisation of erosion-preventive grain-grass crop rotation including 33-50% of perennial grasses. When growing grain crops, it is important to use erosion-preventive tillage and optimum fertilisation on undulating and rolling relief. The fifth (V) group covers the remaining fields on the plains and on the gently undulating relief. A common field crop rotation can be used on these soils, but on slopes of $2\text{-}3^0$, erosion-preventive tillage is suggested.

4. Conclusions

1. Selection of optimum anti-erosion agrophytocenoses (long-term perennial grasses or erosion-preventive crop rotations) is the main framework of the erosion control system on hilly, rolling relief in the temperate zone.
2. Long-term perennial grasses stopped soil erosion even on the slopes of $10\text{-}15^0$. Erosion-preventive grass-grain crop rotations decreased soil losses due to water erosion on the slopes of $2\text{-}5^0$ and $5\text{-}10^0$ by 76.8-80.8% in comparison with field crop rotation, while grain-grass crop rotation decreased it by 21.5-24.6%.
3. Reduced tillage using effective herbicides can be applied on eroded slopes instead of stubble cultivation and deep mouldboard ploughing. It enabled additional decreases of soil losses under growing grain crops in erosion-preventive crop rotations.
4. Selective extra nitrogenous and organic fertilisation of eroded hill-tops and slopes encourages luxuriance of the vegetation and this increases the capability of the plant to preserve soil against raindrop impact and overland flow.

5. Acknowledgements

Authors heartily thank the organising Committee of NATO ARW for accepting this presentation, for covering the participation and travelling costs and Dr. M.J. Wilson for editing of this paper.

6. References

1. Arlauskas M., Feiza V. (1996) The problems of hilly agricultural land management and soil tillage, in E. Nugis (eds.), ***Sustainable agricultural development and rehabilitation***. Proceedings of International Symposium. Tallinn, pp. 77-83.
2. Arshad M. A., Frazluebbers A. J. and Azooz R. H. (1997) Long-term tillage effects on soil structure, hydraulic properties and organic matter in Northwestern Canada, in M. Fotyma at al. (eds), *Biblioteca Fragmenta Agronomica*, **Tom 2A**. Pulawy, pp. 43-46.
3. Bieliauskas P. (1985) Conservation farming on hilly relief, LZUM, Vilnius (In Lithuanian with summary in Russian).
4. FAO-UNESCO (1994) Soil map of the world. Revised legend with corrections. Wageningen.
5. Feiziene D. (1996) Differentiation of fertilisers application on the hills in Western Lithuania, in E. Nugis (eds.), ***Sustainable agricultural development and rehabilitation***. Proceedings of International Symposium. Tallinn, pp. 133-138.
6. Jankauskas B. (1994) Soil erosion and agrarian control measures. The work of doctor habilitatis of Agronomy, LZI, Dotnuva (In Lithuanian with exhaustive summary in English).
7. Jankauskas B. (1996) Soil erosion, Margi rastai, Vilnius (In Lithuanian with English summary).
8. Jankauskas B., Jankauskiene G. (1996) The erosion-preventive capability and productivity of different agrophytocenoses, in E. Nugis (eds.), ***Sustainable agricultural development and rehabilitation***. Proceedings of International Symposium. Tallin, pp. 119-126.
9. Jozefaciuk Cz., Jozefaciuk A. (1996) Erosion and antierosion melioration, Biblioteka monitoringu srodowiska, Warszaw.
10. Kudaba C. (1983) Uplands of Lithuania, Mokslas, Vilnius (In Lithuanian).
11. Morgan R.P.C. (1995) Soil erosion & conservation. Longman Group Limited, Harlow.
12. Loob D. A., Kachanoski R. G. and Miller M. H. (1995) Tillage translocation and tillage erosion on shoulder slope landscape positions measured using^{137}Cs as a tracer, *Canada Journal of Soil Science* **75**, 211-218.
13. Norgailiene Z. and Zableckiene D. (1994). Selection of different maturity of moving and grassing grass-stands for erodible clay loam soils, *Agricultural Sciences*, **3**. 72-76 (In Lithuanian with English and Russian summaries).
14. Olsen P., Hansen B., Hasholt B., Jensen N. H., Schjanning P. and Sibbesen E. (1995) Water erosion and surface transport of phosphorus on agricultural land in Denmark, in B. Linden (eds.), *The use of cach or cover crops to reduce leaching and erosion.* Proceedings of NJF seminar no. 245. Kniwsta, Sweden, pp. 132-141.
15. Prasuhn V. and Schaub D. (1991) The different erosion dynamics of loess and clay soil and the consequences for soil erosion control, *Z. Geomorph. N. F.*, Suppl. -**Bd. 83**. 127-134.
16. Schaub D. and Prasuhn V. (1993). The role of test plot measurements in a long-term soil erosion research project in Switzerland, in S. Wicherek (eds.), *Farm land erosion in temperate plains environment and hills.* Elsevier Science Publishers B. V., pp. 111-123.
17. Schumacher T. E., Lindstrom M. J. (1997) Agricultural management of environmentally sensitive lands a perspective from the Northern Great Plains, in S. Kukula (eds.), *Ochrana i vykorzystanie rolniezej przestrzeni produkcyjnej Polski*, Sesja I i II. Pulawy, pp. 109-117.
18. Skoien S. (1995) Soil erosion. Smaskrift 11/95, As, Norway.
19. Sobolev S. S. (1961). Soil conservation and increasing of its fertility, Nauka, Moscow (In Russian).
20. Zachar D. A. (1982) Soil erosion, Elsevier Scientific Publishing Company, Amsterdam-Oxford-New-York.

REHABILITATION OF ACID SOILS IN LITHUANIA BY LIMING AND LONG TERM MANURIAL TREATMENT

A. Pleseviciene and B. Jankauskas
Vezaiciai Branch and Kaltinenai Research Station
of the Agricultural Institute of Lithuania

1. Introduction

In the past world- wide concern about soil degradation has been focused mainly on wind and water erosion. But with growing populations and increasing industrialisation other forms of soil degradation have become more and more important. Soil degradation caused by air and water pollution with organic and inorganic pollutants is a matter of prime concern in industrialised as well as in developing countries. Also soil acidification as a natural process under humid climatic conditions is highly accelerated by acid deposition from anthropogenic sources. In many areas of the world, the activities of the mining industry, as well as those of modern agriculture have affected the environment and often require soil rehabilitation measures (4).

A large proportion of Lithuanian soils have an excessive acidity: 45.9% with pH (KCl) up to 5.5, only 29.4% from 5.6 to 6.0 pH. According to agrochemical soil investigations, carried out during 1994-1996, about 66.3% of soils in western Lithuania had an excessive acidity. By means of intensive liming, the proportion of soils with strong and moderate acidity had decreased to 29.7% in 1991 (7). Following this, however, the scale of liming has strongly diminished and the soil acidification is increasing again at the present time. The relative importance of each source of acidification (atmospheric deposition, crop growth and nutrient removal, addition of fertilisers) has changed as the soils have become more acid. The major cause of soil acidification at or near natural pH values is the natural input of H^+ from the dissolution of CO_2 , with subsequent dissociation of carbonic acid and mineralisation of organic matter (6). Analyses of soil and hay samples at Rothamsted Experimental Station in England indicate a slow but significant increase in aluminium in soil and a sudden and very large increase in the last 10-15 years (1).

Liming of acid soils is a basic means of soil management. Following liming of strongly acid soils in Lithuania, the average extra yield over 20 years was 3180 feed units ha^{-1} (5). The magnitude and duration of the effect of lime applications varies with soil type, initial pH, fertiliser nitrogen application and the crop grown, but generally such applications of lime are effective over the long term and maintain the buffer mechanisms that combat soil acidification (3).

M.J. Wilson and B. Maliszewska-Kordybach (eds.),
Soil Quality, Sustainable Agriculture and Environmental Security in Central and Eastern Europe, 285–291.

2. Methods

In our field experiments the soil was a strongly acid, loamy Dystric Podzoluvisol. There was 21.1% of clay (particle size <0.01mm) in the plough layer and 32.0% in the deeper soil horizons. The soil pH (KCl) was 4.0-4.1, hydrolytic activity was 5.8-7.1 cmol(+) kg-1, and mobile aluminium was 102-125 mg kg-1. It was concluded that the soil had been strongly degraded by acidification. The efficiency of different rates of cattle manure periodically applied has been investigated on acid (A) unlimed and limed (L) soil at the Vezaiciai branch of the Lithuanian Agricultural Institute since 1959.

The crop rotation up to 1980 was: 1.vetch and oats mixture, 2. winter wheat, 3. spring wheat, 4. perennial grasses (clover and timothy), 5. sugar beet, 6. field beans and 7. spring barley. The crop rotation since 1981 was: 1. winter wheat, 2. mixture of spring barley and peas, 3. oats, 4. fodder beet, 5. spring barley and 6 and 7. perennial grasses (clover and timothy).

The manured crops were winter wheat and beet (sugar and fodder) and the following rates of manure were used: 10, 20, 40 and 60 t ha-1 . The total rates of manure during full crop rotation 20, 40, 80 and 120 t ha-1 . However, a mixture of lupins and oats was grown for green manure and ploughed in (instead of 60 t ha-1 of manure) in 1959 when the experiment was first set up. Dressing of lime was done every 7 years under winter wheat using full rates of $CaCO_3$ according to hydrolytic activity. An optimal dressing of mineral fertilisers (ammonium nitrate, superphosphate and potassium chloride) was used according to the requirements of the crops grown.

3. Results and Discussion

The main purpose of our investigations was to rehabilitate the degraded properties of the strongly acid soil. A material improvement of all agrochemical properties of soil was shown following dressings of lime. The pH(KCl) level increased from 4.1-4.2 to 5.4-5.6 and remained at that level up to 1994. Exchangeable acidity decreased from 1.3-1.6 to 9.1-0.2 cmol(+) kg-1 , mobile aluminium decreased from 105-125 to 0-1 mg kg-1 and hydrolytic activity decreased from 2.8-3.2 to 2.1-2.6 cmol(+) kg-1. The base saturation increased from 2.8-3.2 to 10.2-12.5 cmol(+) kg-1 (Figure 1).

Manure had a lesser influence in changing these agrochemical properties. Nevertheless, there was a positive influence on many of the agrochemical properties of the acid (A) soil, but not on those of the limed (L) soil. Manure evidently decreased the exchangeable acidity and amount of mobile aluminium in the acid (A) soil, except for when, in the first crop rotation, green manure was ploughed in.
The amount of mobile P_2O_5 increased with increasing rates of manure in both (A) and (L) soils, while the influence of manure on the amount of K_2O was most evident in the (L) soil (Figure 2). The influence of manure was significant not only in the arable horizon but in the deeper layers of the soil too. From Table 1 it is evident that

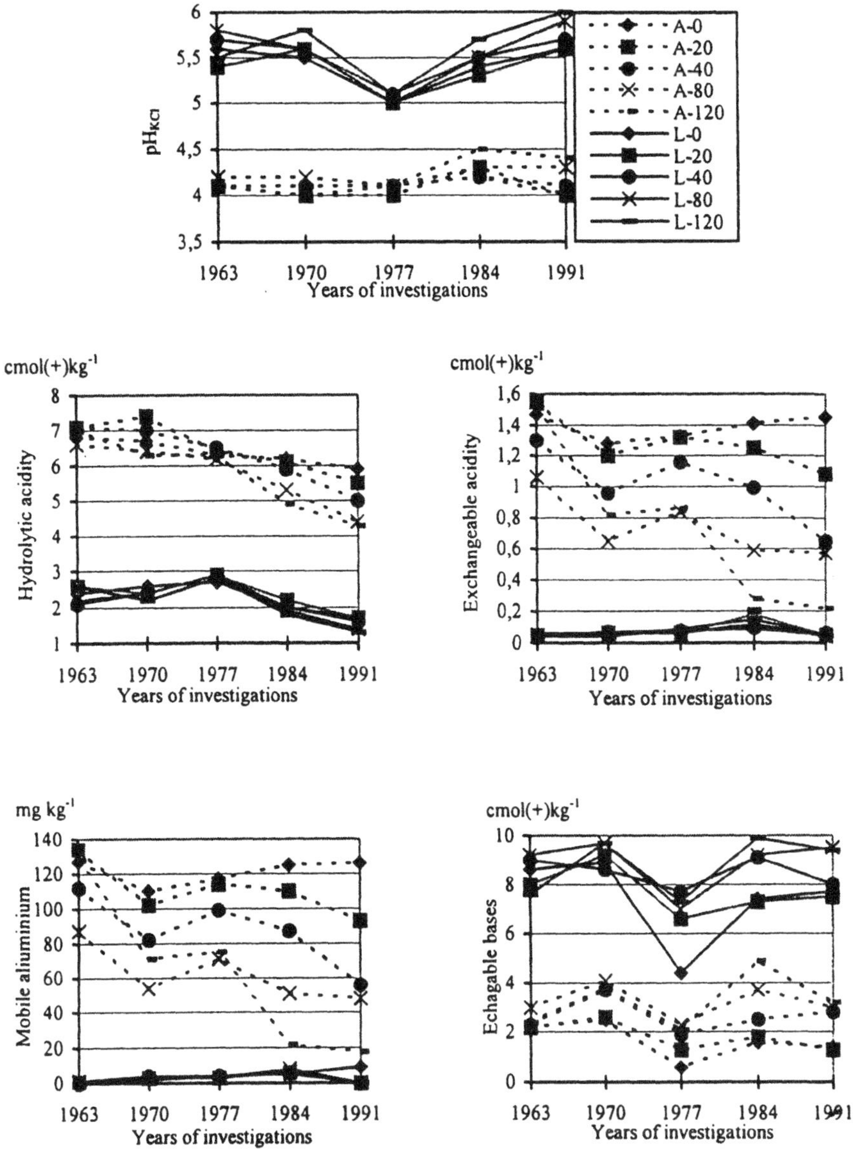

Figure 1. **Influence of lime and manure on agrochemical properties of strongly acid soil**
A - acid soil; L - limed soil. Numbers indicate rates of manure t ha^{-1}

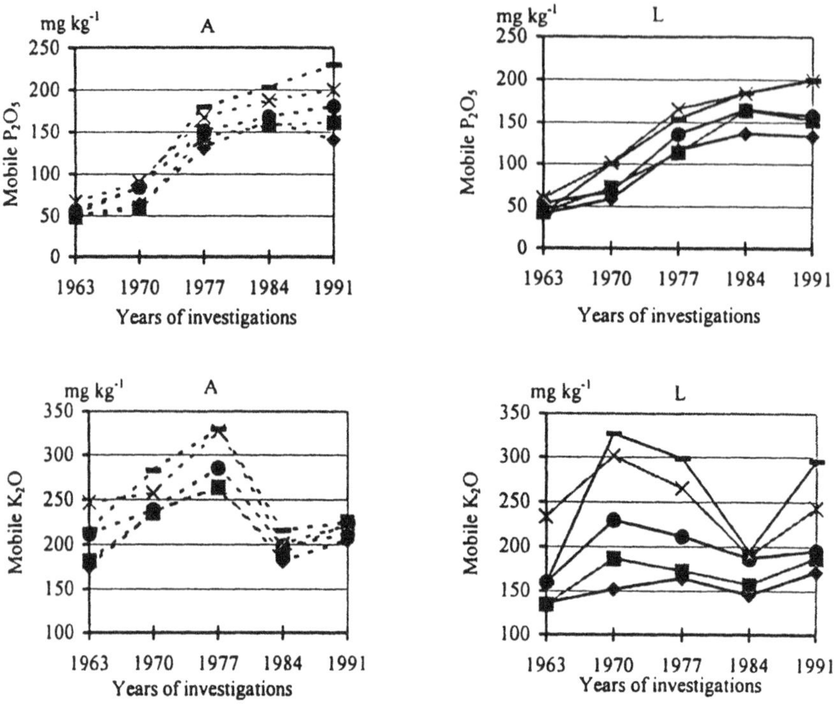

Figure 2. **Influence of lime and manure on amount of mobile P_2O_5 and K_2O in arable horizon of soil**
Meanings of legend the same as in Fig. 1.

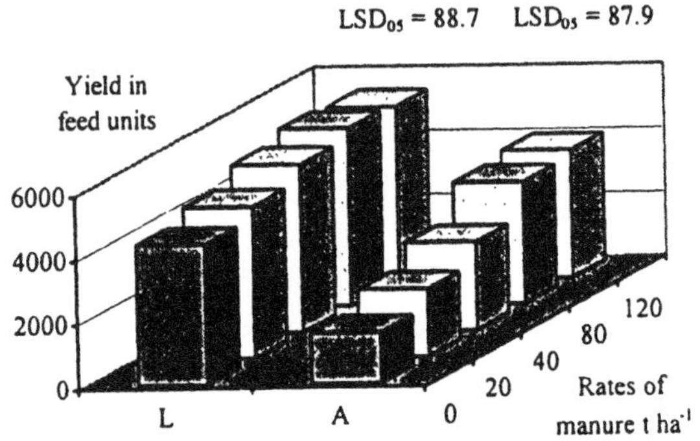

Figure 3. **Influence of lime and manure on average (35 year) productivity of crop rotation**
A - acid soil, L - limed soil

TABLE 1. **The influence of periodically used rates of cattle manure to amount of P_2O_5 and K_2O on acid and limed soil at the end of the fifth seven-course crop rotation.**

Vezaiciai, 1994

Rates of manure for full crop rotation, t ha^{-1}	Sampling depth in cm 0-25	25-45	45-60	60-75	75-90
			Amount of P_2O_5 mg kg^{-1}		
			Acid soil		
Unmanured	115.4	46.4	34.6	33.6	36.4
20	124.4	44.2	33.0	32.4	38.2
40	131.6	43.6	34.4	36.2	33.8
80	143.8	51.0	44.8	37.2	34.8
120	150.8	65.4	53.6	35.2	35.4
LSD_{05}	26.4	25.1	20.8	9.2	7.3
			Limed soil		
Unmanured	104.0	52.2	48.8	37.4	36.8
20	115.0	46.6	32.0	35.0	39.0
40	145.8	61.2	46.0	47.0	38.2
80	169.4	69.6	42.6	48.4	46.0
120	182.4	69.4	48.0	41.2	41.0
LSD_{05}	34.7	20.3	21.5	9.4	9.2
			Amount of K_2O mg kg^{-1}		
			Acid soil		
Unmanured	193.0	164.4	166.8	171.4	158.6
20	191.4	180.4	173.8	183.8	193.6
40	196.4	191.6	201.0	203.8	190.2
80	224.6	175.6	187.2	191.2	223.2
120	212.4	192.8	189.2	206.6	212.2
LSD_{05}	61.3	42.9	48.5	41.3	44.1
			Limed soil		
Unmanured	127.8	105.8	94.4	109.6	97.4
20	161.2	114.4	92.6	105.6	116.6
40	183.0	143.2	103.6	138.0	110.2
80	204.2	152.8	124.8	143.0	127.4
120	251.8	215.6	169.4	156.0	141.0
LSD_{05}	55.8	42.6	26.1	22.7	19.5

TABLE 2. **Influence of manure rates on the average amount of annual accu-mu-lated metabolizable energy during the first crop rotation.**

Vezaiciai, 1988-1995

Rates of manure for full crop rotation, t ha^{-1}	Amount off accumulated metabolizable energy, GJ ha^{-1} on acid soil: average annual	on acid soil: fall to 1 ton of manure	on limed soil: average annual	on limed soil: fall to 1 ton of manure
Unmanured	19.1	-	45.7	-
20	26.1	0.35	51.5	0.29
40	33.6	0.36	55.4	0.24
80	44.2	0.31	54.4	0.11
120	51.0	0.27	60.2	0.12
LSD_{05}	3.2		3.01	

the amounts of mobile P2O5 in the (L) soil increased even in the deeper layers at 75-90 cm, although at the same time manure tended to decrease the mobile K2O. Migration and accumulation of P2O5 and K2O was reduced at the end of the fifth crop rotation.

In general, cattle manure was more effective on the (A) soil and the yield of different crops and productivity of the crop rotations consistently increased with increasing rates of application. The greatest yield of feed units (Figure 3) or amounts of metabolizable energy accumulated (Table 2) was obtained on the limed (L) soil by manuring at rates of 40 to 80 t ha-1.

According to 1994 research data, the percentage of humus in the acid non-manured soil was 1.85%, whilst in the limed soil it was 1.88%. A significant increase in humus in the acid soil was observed only at a manure application rate of 80 t ha-1 (by 0.27%). Humus increased significantly on the limed soil following application rates of 80 and 120 t ha-1 by 0.26 and 0.32%.

An improvement of some soil physical properties was established when manure rates were larger than 40 t ha-1. Aeration porosity increased while soil density decreased.

4. Conclusions

1. Liming was essential for the rehabilitation of the strongly acid Dystric Podoluvisols investigated. Soil pH and base saturation increased whilst hydrolytic activity and exchangeable aluminium decreased to levels favourable to the growth of many agricultural crops.

2. The long term periodic use of cattle manure improved the agrochemical properties of the soils, especially under acid conditions. The improving influence of manure was less than that of lime.

3. The greatest productivity of crop rotation was obtained on soil that was limed and manured. The average annual productivity of crop rotation on limed and manured soils was 4724-5390 feed units/ha or 51.5-60.2 GJ/ha, while on manured acid soil the yield was only 2145-4240 feed units/ha or 26.1-51.0GJ/ha.

5. References

(1). ISCO 9th Conference. Conclusions and Recommendations. (1996), Bonn, Germany, 13pp

(2) Mazvila J., Eitmanavicius L and Vaisvila Z. (1993) Soil Acidity in Western Lithuania. Baltic Region. Agriculture in Acid Soils. Collection of articles. Vilnius, pp. 91-93

(3) Johnston A.E., Goulding K. W. T. and Poulson D. R. (1986) Soil acidification during more than 100 years under permanent grassland and woodland at Rothamsted. *Soil Use and Management*, **2,** 3-10

(4) Blake L., Johnston A.E. and Goulding K. W. T. (1994). Mobilization of aluminium in soil by acid deposition and its uptake by grass cut for hay; a chemical time bomb. *Soil Use and Management*, **10**, 51-55

.(5) Jankauskas B. (1996) Soil Erosion. 186pp (In Lithuanian with a summary in English)

(6) Goulding K. W. T., McGrath S. P. and Johnston A. E. (1989) Predicting the lime requirement of soils under permanent grassland and arable crops. *Soil Use and Management*. **2**, 54-58

(7) Goulding K. W. T and Blake L. (1997) Soil acidification and the mobilization of heavy metals caused by acid deposition and fertilizer application; natural and anthropogenic causes and effects of soil acidification. *Second International Scientific Symposium*, Lublin. 15-20

INTERACTING EFFECTS OF LIMING, FERTILIZATION AND PLANT BREEDING IN SOME LITHUANIAN SOILS IN RELATION TO ENVIRONMENTAL PROTECTION

J Staputis

Kaltinenai Research Station of the Lithuanian Institute of Agriculture

1. Introduction

Investigations in the field of soil liming are abundant but the interacting effects connected with the different chemical and physical properties of the soil, especially in relation to the field of environmental protection are insufficiently known. These effects were studied in three typical Lithuanian soils. The aim of the research was to (1) determine which soil factors have the greatest influence on the optimal reactions in relation to plant growth, (2) investigate the accumulation and distribution of nutrients in the plant biomass under different soil conditions and (3) determine the influence of the soil nutrients on the structure of the plant biomass grown under favourable conditions.

2. Materials and Methods

Investigations were carried out at the Vezaiciai Branch of the Lithuanian Institute of Agriculture. Liming and fertilisation trials were conducted using pot and laboratory methods on five soddy podzolic soils of differing texture and also on a peat soil. Thirty two pot trials were carried out involving 9540 chemical analyses of plants and 3160 analyses of soils. The crop rotation used was Spring Barley - Red Clover - Spring Wheat.

3. Results and Discussion

3.1 IMPACT OF LIMING ON THE YIELD STRUCTURE AND DYNAMICS OF NUTRIENTS IN THE SOIL

In general, the research data show that liming, fertilisation and plant breeding strongly interact and should be considered as a unity (1-5). The efficiency of soil liming depends upon the properties of the soil, particularly its texture, content and quality of humus, the amount of mobile Al, Fe and Mn and the content of available macronutrients (N, P, K, Ca, Mg) and micronutrients. Table 1 shows that the highest yield was obtained in the light loam soil at pH 5.5(KCl) although generally similar results were obtained over the pH range 4.6-5.5(KCl), both in terms of the relative distribution of the different parts of the biomass and in the partitioning of N, P and K.In the sandy loam soil at pH 5.0, there was little difference in the amount of P accumulated but the percentage of K (77.4), N (66.7) and Ca (19.4) taken up in the yield was significantly increased.

M.J. Wilson and B. Maliszewska-Kordybach (eds.),
Soil Quality, Sustainable Agriculture and Environmental Security in Central and Eastern Europe, 293–299.

Table 1. The influence of soil reaction on grain, straw and roots of spring barley biomass and on the proportion of accumulated nutrients (average over 3 years)

				Structure of biomass								
pH	Grain %	Straw %	Roots %	Portion of nutrients accumulated N% in grain	N% in straw	N% in root	P% in grain	P% in straw	P% in root	K% in grain	K% in straw	K% in root
4.1	12.0	64.0	24.0	22	56	22	30	45	25	5	85	10
4.2	24.4	58.5	17.1	31	54	15	31	56	13	12	78	10
4.4	36.4	54.5	9.1	53	41	6	66	30	4	31	65	4
4.6	41.7	51.0	7.3	69	28	3	60	37	3	36	62	2
5.1	42.1	51.3	6.6	69	27	4	65	31	4	35	63	2
5.5	43.3	50.0	6.7	66	30	4	65	31	4	34	64	2
6.2	35.2	57.5	7.3	59	36	5	62	30	8	26	70	4
LSD	1.61	1.40	0.44									

In the light loamy soil there was an increase in accumulated P (11.3-12.3) and further increases in K(88.9-96.9), N (83.3-96.6) and Ca (24.1-28.7) .Increases in nutrient uptake were observed in the medium loam soil corresponding to increasing amounts of clay and organic matter. In the forest soil, at an optimal pH of 4.7-5.1 the percentages of the various nutrients accumulated were P (16.6), K (82.1), N (139.9-149.4) and Ca (23.3-25.3). For the peat soil, where the optimum pH was most acidic (4.2-4.6), the proportion accumulated of all the nutrients, with the exception of Ca, was higher than that of the mineral soils. The soil reaction also had a major influence on the formation of the various structural elements making up the yield of spring barley (Table 2).

Table 2. The impact of soil reaction (pH) and fertilisation on the structural elements of the yield and accumulation of nutrients in spring barley in a light loamy soil.

		Influence of Different Factors (%)		
Plant parts		Soil reaction (pH)	Fertilisation (NPK)	Other factors
Number of shoots		53	14	33
Number of productive shoots		78	4	18
Length of ear		81	3	16
Weight of roots		66	10	24
Yield of grain		76	4	20
Yield of straw		75	11	14
Nutrients in yield	N	66	7	27
	P	78	2	20
	K	74	7	19

Thus, the length of the ear, number of productive shoots and yield of grain were about 80% higher in the limed soil compared with spring barley in the unlimed plots. The level of accumulation of nutrients in the yield also depended upon soil reaction, but the benefit of fertilisation on acid soils could be achieved only after liming. The investigation indicated that different plants or different varieties of the same plant did not react to soil liming in the same way. Furthermore, the efficiency of soil liming was limited not only by the initial reaction of the soil but by its chemical and physical properties as well. The physical and chemical properties of the soils before liming are presented in Table 3.

Table 3. The physical and chemical (mg/kg)properties of the soils before liming.

Soil type	Silt and Clay(%)	Clay (%)	Organic C (%)	Total N (%)	Mobile Elements Al	Mn	Fe	Charged Ions Ca	Mg
Loamy sand	7.85	5.27	1.4	0.110	1	18	86	561	97
Sandy loam	17.81	7.65	3.2	0.268	1	9	80	701	134
Light loam	27.10	9.25	2.5	0.225	55	14	250	321	109
Medium loam	34.72	14.4	1.1	0.140	53	178	280	200	97
Forest soil	25.52	11 6	2.7	0.210	133	18	1220	180	109
Peat	-	-	82.4	1.040	52	20	400	5731	1021

All the soils studied were acid (pH 3.8-4.4) and poor in mobile nutrients except for K. The Mn:Fe ratio ranged from 0.01 to 0.64 and was particularly inadequate for the development of spring barley, red clover and spring wheat, as the optimal ratio for these plants is 2.6, 3.1 and 1.9 respectively (3). Mobile aluminium was very low in the loamy sand and sandy loam soils (1 mg/kg soil) but it was higher in the other soils (52-133 mg/kg). The exchangeable Ca2+was adequate in the loamy sand and sandy loam soils but not in the other soils.

The effect of liming on the mobile nutrients in the soils is shown in Tables 4 to 6. Characteristic parabolic curves for mobile phosphorus accumulation were evident in the loamy sand and the sandy loam soils but not in the other soils. It was found that mobile potassium decreased after liming, especially in the Virgin Forest soil. It was demonstrated that maximum efficiency from liming was obtained when mobile Al ranged between 50 to 130 and mobile Mn ranged from 150 to 180 mg/kg soil. Sorbed calcium varied widely, from 700-800 in the loamy sand and sandy loam soils, 1000-1500 in the clay loam soil, 1200-2000 in the Virgin Forest soil and 5800-7400 mg/kg soil in the peat. Negative effects of liming occurred when high rates were used but these effects were eliminated by the addition of Mn and Cu.

Table 4. The impact of liming on mobile nutrients (mg/kg) in loamy sand and sandy loam soils after spring barley, red clover and spring wheat harvests.

Indices	Loamy sand					Sandy loam				
pH (KCl)	4.6	5.0	5.5	6.0	6.9	5.0	5.6	6.1	6.5	6.7
Mobile P	62/26**	80	80	87	95	77/15	64	56	71	71
Mobile K	197/137	202	165	190	154	194/78	159	116	159	157
Mobile N	14	9	10	11	11	21	21	20	26	25
Sorbed Ca	600	680	940	940	1480	820	1140	1340	1700	2080
Non-sorbed Ca (as % applied)	0	66.5	48.8	57.5	60.3	0	14.6	64.4	58.2	50.7

** before liming and fertilisation

Table 5. The impact of liming on mobile nutrients (mg/kg)in light loam and medium loam soils after spring barley, red clover and spring wheat harvests.

Indices	Light loam					Medium loam				
pH(KCl)	4.3	5.0	5.5	6.1	6.4	4.4	5.0	5.5	6.2	6.5
Mobile P	67/7	56	53	65	67	48/1	42	40	44	47
Mobile K	165/78	136	87	137	114	287/183	175	158	143	152
Mobile N	22	24	24	16	16	18	16	21	16	13
Sorbed Ca	440	920	1200	1560	1640	380	660	880	1220	1420
Non-sorbed Ca (as % applied)										

Table 6. The impact of liming on mobile nutrients (mg/kg) on a virgin forest soil and a peat soil after spring barley, red clover and spring wheat harvests.

Indices	Virgin forest soil					Peat soil				
pH(KCl)	3.9	4.7	5.1	5.4	6.3	3.8	4.2	4.6	4.9	5.3
Mobile P	53/1	52	50	56	74	48/18	59	62	58	66
Mobile K	205/78	58	70	70	86	174/161	111	110	84	166
Mobile N	33	23	22	24	19	38	29	34	36	28
Sorbed Ca	420	980	1220	1360	1740	5160	5660	6400	7400	9040
Non sorbed Ca (as % applied)	0	46.6	42.3	54.4	63.4	0	69.7	71.5	65.0	62.4

3.2 IMPACT OF LIMING ON THE ACCUMULATION OF NUTRIENTS IN THE ABOVE-GROUND PARTS OF PLANTS

A decision about the most favourable soil conditions can be made from the maximum yield level, the biomass structure, the amount of nutrients accumulated in the yield and the ability of the soil to sustain healthy plant growth. This can be considered as an optimal soil reaction with respect to cultivated plants. The proportion of nutrients accumulated in the above-ground parts barley, red clover and spring wheat is shown in Tables 7 to 9.

Table 7. The impact of liming on the accumulation of nutrients (%) in the above-ground parts of barley, red clover and spring wheat in the loamy sand and sandy loam soils.

Indices	Loamy sand					Sandy loam				
pH(KCl)	4.6	5.0	5.5	6.0	6.9	5.0	5.6	6.1	6.5	6.7
P assimilated	10.9	10.5	8.5	7.9	7.7	10.0	7.1	5.7	11.4	8.0
K assimilated	45.9	46.4	36.2	36.5	45.0	77.4	55.9	46.9	56.4	49.6
N assimilated	53.2	46.6	59.4	55.9	55.9	66.7	57.3	51.3	51.5	57.2
Ca assimilated	13.3	9.4	8.5	5.9	4.9	19.8	12.7	10.5	7.8	7.5

Table 8. The impact of liming on accumulation of nutrients (%) in the above-ground parts of barley, red clover and spring wheat in the light loamy and medium loamy soils.

Indices	Light loam					Medium loam				
pH(KCl)	4.3	5.0	5.5	6.1	6.4	4.4	5.0	5.5	6.2	6.5
P assimilated	5.0	12.3	11.3	8.6	8.4	3.9	18.0	19.4	21.6	21.3
K assimilated	43.7	88.9	96.9	79.2	68.9	86.6	152.5	190.0	156.4	145.8
N assimilated	29.8	83.3	96.6	81.3	92.1	28.6	95.3	112.1	123.4	130.7
Ca assimilated	11.1	24.1	21.7	15.6	12.2	21.1	29.8	31.9	24.3	21.5

Table 9. The impact of liming on accumulation of nutrients (%) in the above-ground parts of barley, red clover and spring wheat in the Virgin Forest and peat soils.

Indices	Virgin forest soil					Peat soil				
pH	3.9	4.7	5.1	5.4	6.3	3.8	4.2	4.6	4.9	5.3
P assimilated	7.2	16.7	16.6	15.4	15.7	35.7	39.1	38.8	36.6	24.0
K assimilated	65.5	83.0	82.1	78.3	79.0	107.4	101.2	105.6	89.6	85.1
N assimilated	72.1	139.9	149.4	149.7	146.7	128.4	186.9	212.9	216.7	195.4
Ca assimilated	14.5	25.3	23.3	14.6	13.8	4.1	7.8	7.1	6.6	4.4

It can be seen that the optimal reaction in the loamy sand occurs in the pH range 4.6-5.0. Under these relatively acidic conditions, it was found that about 11% P, 45% K 50% N and 12% Ca were accumulated in the yield (Table 7). For the sandy loam soil optimal conditions occurred at the somewhat higher soil pH of 5.0, where about 10%P, 77% K 66% N and 20% Ca were accumulated in the crops. In both the light loam soil and the medium loam soil (Table 8), most of the nutrients optimum uptake occurred in the pH range 5.0-5.5, although the highest proportion of N was assimilated at pH 6.5 in the latter soil. It seems that the amount of clay particles in the soil ,and perhaps the amount of soil organic carbon too, may play a significant role in the mechanics of nutrient uptake from these soils. In the forest and peat soils (Table 9), it was difficult to define an optimal pH range for nutrient uptake although the biggest effect seems to occur at a relatively low pH.

3.3 PLANT BREEDING AND ALUMINIUM TOLERANCE

At present there is no practical alternative to liming but in the future it may be that modern methods in molecular biology will be able to take advantage of genetic aluminium tolerance to reduce the need for this practice. Soil liming at the highest rates is a very expensive operation for improving soil fertility. Its object is to reduce or eliminate the impact of aluminium on the plant roots especially with respect to the growth of crops like wheat, barley, clover, sugar beet, lucerne, etc. Aniol (1) argues for the existence of aluminium tolerant genes in plants and if this is the case they could be utilised for improving the yield and quality of crops grown on acidic soils, thus removing the current requirement to apply lime.

4. Conclusions

It has been shown that:

- The properties of the soils examined strongly affect the plant nutrient supply and uptake from these soils.
- An optimal soil pH reaction can be reached by liming. Adjustment to this optimal reaction is a result of a complex interaction of physical, chemical and biological soil properties and relates directly to concepts of soil quality and sustainability.
- Liming of peat soil appears to be inadvisable and cannot be recommended from an environmental point of view.
- Modern methods of molecular biology aimed at utilising aluminium tolerant genes will impact directly on current agricultural practices based on soil amendments.

5. References

1. Aniol A. 1995. Aluminium tolerance in wheat; Plant breeding theories, achievements and problems.*Proceedings of an International Conference held on 14-16 July 1997.* Dotnuva Akademija, Lithuania. pp 14-22.

2. Èiuberkienë D. 1995. Dirvoþemio reakcijos ir træðimo átaka dirvoþemio agro- cheminëms savybëms ir augalø derliui; Lieutuvos Þemdirbystës instituto moskodarbai. Dotnuva Akademija, **48**, pp 23-29 (in Lithuanian)

3. Nebolsin, A. N and Nebolsina Z. P. 1997. Optimalnje dlia rastenje parametrykislotnosti dernovo-podzolistoj pochvy. *Agrochimija*. Nr 6, pp. 19-26 (in Russian).
4. Knaðys V. 1985. DirvoÞemiø kalkinimas. V. Mokslas, 262pp. (in Lithuanian).
5. Staputis J. 1996. The influence of interaction of liming and soil properties on yield and accumulation of N, P, K and Ca in plants and efficiency of nitrogen mineral fertilisers. *International Conference on Issues of Agronomy and Animal Husbandry Science*. Kaunas Akademija. Pp211-218.

PART V

SOCIO-ECONOMIC ASPECTS OF AGRICULTURAL SUSTAINABILITY AND ENVIRONMENTAL SECURITY IN CENTRAL AND EAST EUROPEAN COUNTRIES

Synopsis and Overview

Although the papers in this section refer mainly to Poland, it is believed that the issues discussed will be of general applicability to other CEECs because agriculture is operating within similar constraints in the context of a common background throughout the region. Thus, agriculture in all the CEECs is in a state of transition because of the shift towards the market economy and private ownership, in addition to declining levels of state subsidy. Consequently, agricultural output has fallen substantially in all the CEECs since 1989 and, perhaps more importantly for soil quality, there has been a sharp fall, in some cases amounting to a collapse, of inputs such as fertiliser and lime. Clearly, such changes may be expected to have major repercussions upon agricultural sustainability. This is part of the background to the paper by *Dalton* who also considers the likely impact of continuing trends towards trade globalisation. It is considered that, in the light of these changes and bearing in mind the severe budgetary constraints that all the CEECs face, there is a danger that the importance of sustaining the quality of the natural capital stock (such as soil) will be overlooked. Policies will need to be devised which give to issues such as maintenance and enhancement of soil quality, water conservation and biological diversity a proper weight when set against the more narrow goal of increased economic efficiency. *Dalton* describes the dilemmas and contradictions inherent in such policies when applied to the CEECs in the general climate of increasing resistance to any form of agricultural protection, the financial difficulties faced by the CEECs and complications ensuing from accession to the EU. There are no easy solutions to these problems but *Dalton* points out that for those CEECs seeking to join the EU there is an unprecedented policy opportunity to develop environmental or sustainable components for pre-accessionary modernisation programmes, which may lead to structural funding upon accession. Modernisation of agriculture in the CEECs need not necessarily only take the form of larger farms and fewer farm workers and there may be scope for pluriactive farming systems to become much more widespread. In the EU such systems tend to be associated with greater ecological diversity and a lower intensity of farming but a wider perspective on the part of both farmers and policy makers is required if this type of activity is to be fostered.

The challenges facing agriculture in Poland and other CEECs desiring to join the European Community are further analysed by *Duer and Krasowicz*. The development of sustainable agriculture should be environmentally sound, economically viable and socially acceptable. The choice of agricultural systems and the technology involved depends upon the socio-economic and ecological conditions in which the farmers are operating. Socio-economic indicators for the agricultural

M.J. Wilson and B. Maliszewska-Kordybach (eds.),
Soil Quality, Sustainable Agriculture and Environmental Security in Central and Eastern Europe, 301–302.

sector in Poland demonstrate that the country, along with other CEECs, lags well behind the EU countries and the USA in this respect. For example, the labour force engaged in agriculture is inversely related to the gross domestic product (GDP) per capita. In Poland, GDP per capita is only 13.5% that of the average value for the EU countries; again, 26% of the labour force are employed in agriculture in Poland compared with a mean value of <7% in the EU. It seems that the expected trend in agriculture in the CEECs must inevitably be towards increased effectiveness of labour and productivity of land, leading to an increased GDP per capita and with fewer people engaged in agriculture. However, *Duer and Krasowicz* argue that such a trend, which in the West has been associated with high inputs and financial support through guaranteed prices and subsidies, may not be compatible with the economic and ecological constraints under which agriculture operates in the CEECs. It is suggested that an improvement of alternative agricultural techniques based on traditional solutions and taking biodiversity into account should also be an important element in the future development of Polish agriculture.

These issues, with specific reference to Poland, are further developed by *Wos*. It is considered that in the conflict between liberal and environmental agricultural policies, the balance is always on the side of economic efficiency, irrespective of negative external effects. It is concluded that a new rationale is needed which attempts to reconcile the economic and environmental points of view and in this respect the CEECs may be in a special position. As these countries have only recently started to introduce market mechanisms and to implement an economic rationale, it may be possible to develop new criteria based upon long-term environmental considerations and the good of society. From this point of view, Polish agriculture has its strengths and weaknesses. Its strength is its "natural" as opposed to "industrial" character, making it more biologically diversified. On the other hand, economic constraints, combined with a generally low level of environmental education, mean that the ability of Polish farmers to implement environmentally friendly technologies is rather limited. *Wos* concludes by listing essential and specific measures that should be taken in order to protect agricultural ecosystems in Poland.

Finally, *Gedek* discusses a modelling approach for environmental assessment at the farm level. In cases where there is no previous experience, this approach may be the only method possible for the assessment of environmental policy measures. The proposed methodology consists of comparison between two models, with and without the policy measures included. Information is provided upon production structure of the farm and production means and profit levels caused by the introduction of the policy measures. Basic information is also obtained as to whether the proposed policy measures actually limit the environmentally negative effects of agriculture in the way intended. *Gedek* illustrates the use of the model by a hypothetical policy measure aimed at limiting the use of nitrogen mineral fertiliser. A general conclusion is that the effects of policy depend very much on farm type. The model promises to be a useful tool for the assessment of environmental policy measures but further refinements are necessary.

APPROPRIATE AGRICULTURAL POLICIES TO SUSTAIN AGRICULTURE IN THE CENTRAL AND EASTERN EUROPEAN COUNTRIES

G. E. Dalton

University of Aberdeen
(U. K. Know How Fund Project, on Integrating EU and Polish Agricultural Policy, Ministry of Agriculture, Poland)

1. Background

One of the main components of a sustainable agricultural system is the maintenance and care of the stock of natural capital. Recent experiences in CEE countries provide some important lessons as to the effect of large reductions in transfers from consumers and taxpayers to farmers, on the natural environment. The transition toward the market economy has been associated with a substantial fall in the level of agricultural output and a run down of all forms of capital stock including levels of soil fertility. These effects can be compared with those in the European Union where the high level of such transfers are often blamed for agricultural pollution and the destruction of habitats. This is not surprising given the objectives that most agricultural policies are designed to meet.

Agricultural policy is currently in a state of flux brought about by the anticipated costs of applying the common agricultural policy (CAP) to the applicant countries as well as the pressures for less trade distorting policies which are agreed and monitored within the World Trade Organisation (WTO). In addition, shifts in public opinion are occurring in the light of both environmental and food safety concerns which are modifying both the amount and form of public expenditure on agriculture. Moreover, doubts about the efficacy of achieving desired income support goals using existing instruments are being increasingly confirmed.

The prospects for policies which do support sustainable agricultural practices being adopted depends on existing conditions and policy priorities as well as the cogency with which the case for such policies is presented. Many regions within the CEEC's have endured extremely depressed conditions in both agriculture and the whole rural sector [1]. A common means of survival has been based on subsistence with low levels of input use, the abandonment of land, much less employed labour and a run down of livestock numbers [2]. There has been a great shortage of cash made worse by high rates of inflation, low levels of profitability and the need for further development of supporting institutions and infrastructure. There are thus urgent needs to restore the run-down state of the sector as well as to ameliorate some serious social conditions. An example, is the inadequate provision that has been

M.J. Wilson and B. Maliszewska-Kordybach (eds.),
Soil Quality, Sustainable Agriculture and Environmental Security in Central and Eastern Europe, 303–316.

made in the past for the pensions of retired farmers and other forms of social security for all farmers, which is now a burden on government finances.

One response to the above conditions has been a partial restoration in the levels of price support through import tariffs and domestic intervention buying. The OECD measure of this, the producer subsidy equivalent (PSE) shows that in Poland, for example, the PSE in 1996 was 28% as compared with zero in 1989 - 1991. This represents a surprisingly high level of transfers of resources from consumers to producers in a poor country where food expenditure is a high proportion of household expenditure. Only a small proportion of these transfers were financed from direct taxation [3]. Such a revealed level of public preference for supporting agricultural incomes through commodity price support is not an encouraging sign for additional transfers to enhance the environment. A case will have to be made which shows that such instruments will also improve farm family welfare.

Within the EU, agricultural support levels have been sustained at a high level through the use of direct payments on land and livestock in response to the international undertakings to reduce the level of price support and the amount and value of subsidised exports. More environmental policy instruments have been introduced especially for special features in identified regions. In addition, there is a tendency to renationalise the CAP through special national measures (with national contributions) and to maintain prices through restrictions on production through the use of quotas. Quotas are most important in the dairy and sugar sectors, but the full effect of higher prices and direct payments are at least moderated by the use of set aside for cereals and oilseeds and through limits on the amounts of subsidy payable, either at a national or farm level for both livestock and crops. There is also an overall ceiling on the total agricultural budget in the EU. The transparency of the amount of money involved in financing the so called compensatory payments for lower price support (introduced in the 1992 CAP reforms) which are paid for by taxation, is a threat to the persistence of these payments. External pressures on the system include EU enlargement, the financial conditions to be met in order to qualify for membership of the single currency and the next WTO round of discussions on further liberalisation of agricultural trade. Agricultural policy is undoubtedly about to change within the EU.

2. Trends in Agricultural policy

Accession to the EU will mean that CEE countries will adopt the CAP. This will involve implementing more complex policy instruments and depending on the evolution of the CAP prior to accession will probably mean higher levels of price support and a greater use of supply management measures. Some price convergence has already occurred for those products where recent world prices have been high (notably grain) and where protection has fallen in the EU and risen in CEEC's.

The agenda 2000 proposals which were announced in July 1997 [4] give the clearest clue as to the likely development of the CAP over the medium term. They are essentially an extension of the previous 1992 McSharry reforms of the CAP by the EU and are more in line with the WTO agreements on reducing support mechanisms which distort trade. The trend is for free market price levels for grain, oilseeds, beef and to a lesser extent milk where quotas will be retained, with support given through direct payments. These direct payments have so far been justified as compensation payments for lower support prices but the likelihood is that they will continue as the main means of support and will be made as independent as possible from the level of production. This independence from production, termed decoupling, is argued to have less distortionary effects on trade. Such a policy is in line with new policy developments in the USA and Canada and as such prepares the way for the forthcoming debate in the next WTO round. Given the importance of the agenda 2000, a short summary of its main contents follows covering market policies, rural development policies and structural funds, and environmental measures

2.1 MARKET POLICIES

The approach taken in the proposed changes to the market regimes is as noted above, for a further shift to direct income support especially for cereals and beef. This shift is supported by anticipated surpluses particularly of beef at the beginning of the next century. The cereal intervention price is recommended to be cut in one step to ECU 93.5/tonne from 119.9/tonne with a non specific crop area payment increased to 66 ECU/tonne (from 55 ECU /tonne) multiplied by the regional cereal reference yield/hectare of the 1992 reform. This payment will be lowered if market prices are sustained at a higher level than currently foreseen. In the case of beef, market support is to be reduced in steps, from ECU 2708/tonne to ECU 1950/tonne. Intervention will be replaced by support for a private storage scheme. Existing headage payments will be increased and a new headage payment for dairy cows will be introduced to compensate for lower cull cow and calf prices as well as lower milk prices. The main objective of these proposals is an improvement of price competitiveness in these sectors. No reductions over time are foreseen for the direct payments but the amounts suggested would not fully cover the gross revenue loss calculated on the basis of support price cuts. In addition agenda 2000 suggests allowing member States to modulate or impose an upper limit on the size of the direct payments per farm based on common criteria.

Thus, it would appear that the prospect of CEEC agriculture intensifying production in response to high EU support prices is unlikely in that the level of price support will not be sustained. External WTO pressures explain this approach. However, the other major proposal is that applicant countries would not be eligible for the direct forms of support in that it is argued they will not suffer any price reduction. This does ,however, contradict the concept of the single market and, it is also alleged, will result in unfair competition to the detriment of agriculture within the CEEC's. The expectations of farmers in applicant countries, of prosperity based on high levels of support upon EU accession, have changed to real worries about unequal levels of support within an enlarged EU with new member States losing out.

2.2 RURAL DEVELOPMENT

In spite of the fact that Agenda 2000 identifies rural development as a major challenge for the future, especially in applicant countries, the suggestions presented are mainly limited to a reorganisation of existing measures within the present Community policy framework, all within the present ceiling on structural expenditure of 46% of EU GDP for the period 2000 - 2006. The proposals are to reduce the present seven Objectives to three, two of which are regional. Objective 1 regions are to be redefined more rigorously in terms of the threshold of 75% of average per capita community GDP and in terms of their boundaries and financial allocations. A new Objective 2, for economic and social restructuring is to cover other regions with social problems such as declining rural areas. Specific Community initiatives will be reduced in number but will still include one for rural development, such as the LEADER programme.

All rural areas will from now on have access to rural development measures and there is likely to be a single integrated programme for each region. In Objective 1 regions such integrated programmes will be financed by regional and social funds as well as the horizontal guarantee section of the Agricultural budget. There is a great need for economic development in most of the rural areas of applicant countries because of low incomes, poor infrastructure, social deprivation and a high proportion of the total workforce dependent on agriculture. The immediate challenge is to design financial procedures and implement effective development policies. Whereas, the current EU procedures may appear to be complex and expensive to operate they do allow a great deal of national flexibility and a whole range of measures including modernisation of infrastructure and businesses, retirement, environmental and reforestation measures to be inserted into a single regional plan or framework.

2.3 ENVIRONMENTAL MEASURES

The agro-environmental measures proposed in Agenda 2000 are rather limited and include a proposal to make direct payments conditional on farmers respecting certain environmental conditions (cross compliance). This proposal is optional to member States and is yet another example of the renationalisation of the CAP. The concern amongst applicant countries and indeed within the WTO is that such agro-environmental measures proposed in Agenda 2000 are rather limited and include a proposal that such environmental payments become a cover for what are really income subsidies with richer nations being able to afford more. Moreover, in countries where subsidies are currently high, cross compliance types of payments will result in greater subsidies for the production of environmental goods than they will in countries where support is currently low.

According to Agenda 2000, targeted agro-environmental measures should be reinforced and encouraged through increased budgetary resources albeit from a low relative base. In addition, the document points to the considerable overlap between less favoured areas and locations of high natural value and envisages a gradual transfer of payments from compensation for disadvantage towards support to maintain and promote low input farming in areas with many natural resources. Such proposals are an opportunity for applicant countries to design environmental programmes according to their own circumstances and specific locational priorities. EU funding, as with structural measures, is dependent on some national finance, or own funding, being made available.

2.4 OVERALL OBJECTIVES OF AGENDA 2000

It is unlikely that all the proposals within the Agenda 2000 document will be adopted but precedent points to the fact that the direction of the proposals can be taken as being fairly certain. The underlying objectives inherent within the proposals can be stated as:-

improve the competitiveness of EU agriculture on both domestic and external markets;

ensuring the highest possible level of food safety and the promotion of food quality;

ensuring a fair standard of living for the agricultural community and a contribution to the stability of farm incomes;

integrating environmental goals into the CAP and to further develop the role that farmers can and should play in the management of natural resources and landscape conservation;

promoting the creation of complementary or alternative income and employment opportunities, both on and off the farm, for farmers and their families.

All these objectives fit with those revealed in the current policies of applicant countries. However, there are serious concerns about how well the proposals will in practice achieve these goals and especially whether they will exploit rather than sustain the agriculture of potential member States.

2.5 UNANSWERED QUESTIONS

Several basic questions remain unanswered. Firstly, is support in the sense of non market transfers necessary to sustain agriculture? The transition experience is not convincing for those in favour of a free market philosophy in that the whole sector in the CEEC's has been run down. However, the experience within the EU where support is at a high level is also not a convincing case for sustaining the natural environment or indeed other desirable features of the system.

Secondly, the spin-off effects on sustaining natural capital of policies designed to meet specific policy goals such as farm incomes are either not recognised or are not considered. This is especially true for issues concerning natural resources and is a major criticism of the Agenda 2000 proposals. Indeed there does not appear to be sufficient weight being given in the policy framing process to the large scale effects of the overall policy package on sustaining or conserving natural resources including the characteristics of the soil. Again the US system points the way where an index of some seven types of desirable outcomes is used to prioritise programmes and a bidding system by farmers is used to decide which land is eligible for such programmes [5].

Thirdly, a major part of the accession process is the adoption of the acquis communautaire or regulations which control the markets for all forms of food and agricultural inputs. It includes controls, especially at common EU borders, on plant and animal disease, hygiene measures and chemical residues in foods, some aspects of animal welfare and grading standards. All these measures assist in the operation of a large single market. Regulation by Governments of this nature does not usually involve public finance and the reality for many agricultural and food firms in the CEEC's is that the cost of adoption will drive them out of business. This inability to pay for the necessary changes also applies to the adoption of common environmental standards such as water quality norms.

CEE countries are in fact being required as a condition of membership, to adopt and pay for an expensive set of legislation which has been developed by a group of much richer societies.

Fourthly, institutional support for the sector is often poorly developed within CEEC's. For example the ownership of land and other forms of property rights are still being developed in many countries. These are most important in preventing over exploitation of natural assets. Much has been achieved in improving research, education, training and extension as well as the development of infrastructure and markets. But much remains to be done within tight budgetary constraints if the sector is to be able to take up the opportunities and withstand the threats of competition within a large single market. In fact a central argument running through a series of Commission documents is that investments to improve such measures should have priority within the CEEC's. However, the reality is that the amount of preaccession funding for such purposes, as proposed in the Agenda 2000 budget proposals, is small in relation to the need.

In the current debate over funding decisions surrounding agricultural support it is the case that despite much rhetoric there is too little emphasis on the sustainability issue. Large transfers to farmers are taking place even within CEE countries and it is not clear what the effect is on stocks of natural capital. The effects of different policy instruments such as quotas, price support, trade restrictions and direct payments on sustainability are not fully considered. In fact there is a logical argument that the number of policy instruments must equal the number of policy objectives if compromises are to be avoided. If this is accepted then what are appropriate sustainability policy instruments?

Pragmatically, it is neither possible nor necessary (fortunately) to set out policies to cover a range of objectives for a long time ahead. Instead we have the ability to observe systems and to learn and amend as time passes. The crucial quality in this process is to learn and adapt fast enough. There are enormous changes taking place in agricultural policies and also in the international nature of the market for food as globalisation gathers pace.

The main message of this paper is that enlargement should encourage soil scientists to reinforce their international perspective of sustainability in that the policy decision making process is becoming less parochial. The issues of soil nutrient status, water quality, conservation and biological diversity, as well as the wholesomeness of food, are on the agricultural policy agenda but without appropriate facts and analysis they are likely to be inadequately funded.

3. The Agricultural Situation and Outlook

Monitoring and anticipation are important functions of successful system management. Some of the influences on the sustainability of agriculture are well outside the control of the sector such as movements in exchange rates, the level of employment and real interest rates. It is also clear that as the economies of the CEEC's develop the availability of relatively skilled and cheap labour within the agricultural sector will decline and it would not be wise to base any long term strategy on this current competitive advantage. However, in the meantime the fact that relative prices for inputs are different to those within the EU, especially the ratios between labour, land and capital means that to simply copy EU agricultural systems would almost certainly be a non optimal solution. This conclusion would be reinforced by the generally lower level of output prices in central Europe. We would thus expect land and labour to substitute for fertiliser and other chemical inputs in the search for profit which incidentally would perhaps produce more sustainable systems. These price relationships will certainly change.

Another important influence on sustainability is trade in agricultural products. Most CEEC's have important trade in food products to and from the EU and latterly there has been a growth in trade with states of the former Soviet Union. Many of the exports to more eastern countries are for highly processed foodstuffs [6]. The loss of trade to competitors would have significant effects on the prosperity of many sectors of the rural economies of Central Europe. Enlargement will intensify this competition as will developments within the WTO agreements and it is thus clear that to survive the agricultural and food sectors must be able to compete on both price and quality. Enlargement and trade liberalisation is also an opportunity for growth within the sector and given the resources available this could well happen. This issue raises another dilemma. Does growth in the sector conflict with the objectives of sustainability? The transition experience is that decline does not maintain the land, the people or man made capital, whereas so called high levels of intensity are generally thought to be harmful. So on the one hand agriculture must become competitive to survive and prosper but if the basis of this competitive position is the erosion of natural capital, then it is not sustainable. It would also seem unwise to follow an extensive system of agriculture when most pundits are prophesying that the demand for food world-wide will continue to be strong. No-one can be certain of this since it is usual for such views to gain credence on the back of a commodity boom[1]. But many environmental policies do restrict output if only because such policies are attractive to conservationists(usually rich ones). There is also a common interest with protectionists, who are strongly represented within all agricultural communities, to curtail supplies in order to drive up prices (with all the negative effects such price levels have on input use). It is thus quite conceivable in the political bargaining in the enlargement process that environmental arguments may be used to meet the objectives of the managed market lobby. And yet change is a crucial aspect of sustainability as the whole transition process so strongly emphasised.

Changes in the system of agricultural support also provide a warning that the amounts of agricultural subsidies could well decline during the next generation. Direct payments are a real and visible call on taxpayers (unlike import levies) and involve substantial deadweight costs (taxes reduce incentives for economic activity) as well as being expensive to administer. This a serious issue in the enlargement process in that the expense of setting up new administrative systems to support agriculture in a country like Poland with millions of small farmers is not cheap and the prospect of such monies being found from an EU budget (a common pool) has receded with the agenda 2000 package. This will influence the ability to argue for funds for environmental measures.

Enlargement does however have two positive effects on the case for sustainability although they will require a great deal of energy to fulfil. Firstly, it can be hypothesised that the relatively poor voters of the CEEC's are not likely to give much weight to the sustainability argument when it comes to spending money. There is the possibility that common EU funding will be driven by a greater interest and ability to support sustainability measures. There are still the difficulties of countries finding their own national contributions in such programmes. Secondly, the emergence of large political groups of countries with a common agricultural policy, such as the EU (failing which we could see CEFTA develop), improves the prospect for internationally sensible agreements. To illustrate this point, nutrient loss from agriculture into the various river and sea systems (Baltic, Black Sea, Mediterranean and North Sea) could be modified by a relocation and changes to agricultural systems. Is it possible to reduce nutrient loss by a relocation or a spreading out of livestock populations and using more land and less fertilisers to produce crops within an enlarged EU? These issues are rarely mentioned in the important policy making committees and lobbies. The same arguments apply to wildlife and biological diversity. However if these issues are not addressed, is agriculture sustainable if the life of Europe's seas and rivers is at risk?

4. More Specific Policy Dilemmas

The need for change is often overlooked in the search for short term solutions in agricultural policy. A cynical view is that policies of substantial transfers to agriculture are maintained, come what may, but the rationale changes. This is best illustrated in the continual debate over economic efficiency and equity. Price support is justified for all farmers on the basis of the needs of small farmers even though they benefit least. Socially driven policies such as retirement schemes also benefit those who contribute nothing namely those who can use the freed resources. Special assistance for the disadvantaged actually discourages change and worse may attract new entrants to the sector. Environmental schemes often have to compensate farmers for the loss of support they would have received had they continued a fully commercial policy.

In all of this anticipated discordant debate, it will be helpful to both understand and quantify the full environmental effects (costs and benefits) of policy driven decisions. Someone has to pay for these costs either now or in the future.

The application of decoupled ideas of support is also not clear cut. In theory, a direct payment will not influence the intensity of input use since it is independent of yield and for most products, price. However, in that crops and livestock have to be raised to receive the payments, they do sustain production levels at historically based quantities. The maintenance of this system within the EU while not introducing it within the CEEC's will distort markets in that additional production will be maintained. A transitional period simply puts off the adjustment needed by the EU's farmers at the expense of cutting off opportunities for those in the CEEC's.

Another most controversial topic is scale of enterprise. History has shown that large state farms for the most part do not work and that they lack the necessary resilience to survive large external shocks. It is important to be clear whether they did not work because they were state run or because they were large. Currently, there is much emphasis on the need to restructure agriculture based on simple measures such as apparent labour productivity as in the small farm sector of Poland. Increasing farm size may not produce efficiency automatically as the evidence for economies of scale in agriculture is weak. Most farm populations in the world have large numbers of small farms operating alongside larger ones. What is clear is that the opportunities to enhance productivity through better knowledge and more effective markets will probably lead to a concentration and possibly some specialisation of production. Moreover, the landscape of small and fragmented farms with village settlements as opposed to that of industrialised farm factories may be a better basis for some types of tourism.

The implementation of policies concerned with land use will tend to be focused on fewer large farms in that the distribution of control, even in regions with many small farms, is usually highly skewed. The costs of implementing land management agreements with individual farmers can nevertheless be so high that it may be necessary administratively to incur the distortion (the costs of producing environmental goods vary between farms) of blanket schemes in which standard payments are made. Social policies will concentrate on a larger number of smaller farmers. The arguments for different policy instruments according to farm size begin to take shape. In the agenda 2000 proposals, income support measures are being looked at according to modulation principles whereby support is limited to a given level of output along with a debate about limits to the total amount of support each farm can receive.

Milk production can be based on both large and small scale farms. The proposal by the Commission to introduce quotas to the CEEC's which will be kept in place until 2006 is especially problematic in that there will be pressure to base this quota on current levels of production.It will probably be more difficult for individual farms to develop their dairy enterprises and for the sector as a whole to respond to market opportunities as the economy grows and trade opportunities arise. Dairy cattle are the main source of beef animals so that quotas will hamper adjustments in the beef market too. The basis of mixed farming using rotational grass and ruminant livestock is thus artificially constrained. The idea of lower milk prices will assist to some extent in that high milk prices encourage high yielding systems of production which are based on concentrates. The transition experience does underline the responsiveness of CEEC farmers to market forces. If they can be designed, it would thus seem sensible to harness the contribution of markets to the development of sustainable systems. There is still much to do in getting existing markets to work better and indeed creating them. The prospect of EU membership as set out in the single market legislation will mean the adoption and implementation of a whole raft of legislation on product standards. The ideas connected with food quality, traceability and assurance are developing quickly and these will need to be applied within the CEEC's too. The atomistic nature of firm structure in most parts of the food chain may well require the development of industry wide food quality schemes in order to both secure outlets and obtain price premia. Such schemes will need to be verified according to EU standards. At a practical level this will require the development of systems of farm and livestock registration which coincidentally is also required for modern income support measures. Registration is only one example of a real role for government in the operation of markets and policies. A strong case can also be made, at least initially for the provision and dissemination of information, and also for education and training including awareness of a fairly suspicious agricultural community.

5. Future Measures

The development of specific environmental schemes for particular localities needs to be accompanied by more general measures within CEE countries to urgently restore and modernise much of the capital stock of the agricultural sector. This can only be fully financed if the sector is prosperous which means that the agricultural industry of CEE countries has to be capable of becoming internationally competitive.

5.1 RESTORATION AND MODERNISATION

Agricultural systems in the CEE countries have been run down as a result of the traumas of transition. Restoration is required to sustain the potential contribution of the sector. This potential has to be judged in the light of effective demand and the ability of the systems to compete in a large single market. Restoration has to be accompanied by an improvement in efficiency (modernisation) if the external changes facing agriculture are to be met. Restoration will result from a multiplicity of investment decisions, in which many farmers and other agribusinesses forgo present consumption in the hope of future gain. Investment is driven by both expectations about the future and constraints on funding. The latter are heavily influenced by high real interest rates and inflation which exacerbates cash shortages within all parts of the economy and especially in agriculture where production cycles are long. Catch up technology which has become available, as a result of market liberalisation may however not require large amounts of fixed long term capital. Opportunities to specialise which may be considered as a result of reduced market and political risks may also be feasible and attractive changes. The response to cash shortages and capital scarcity by many Governments is to subsidise credit for agriculture. This popular measure can only assist those who take out loans, normally the larger farmers. Such loans are often highly attractive financial transactions and so have to be rationed in some way or other. They can crowd out the development of commercial types of lending and thus significantly distort the market for capital including the incentives for saving. Moreover, subsidised capital will in theory substitute for plentiful labour. New capital goods are often, on the other hand, a carrier for new technology.

5.2 PROSPERITY

Investment is mostly financed from their own resources by farms and firms. Prosperity is thus a facilitator as well as encouragement for taking risks about the future. Prosperity can be defined as the surplus which is generated between the value of output sold or in stock and the value of inputs purchased and used over a single production cycle.

The usual way agricultural policies are used to increase this surplus is to raise the value of output by increasing product prices. Less commonly, the costs of inputs are lowered both by direct subsidies (credit subsidies) and by various tax concessions such as for fuel and VAT. Politically such measures are the usual focus of attention. The reality is, as shown in a recent comprehensive study [7], that productivity growth and favourable exchange rate movements are quantitatively at least of equal importance to future prosperity than even EU levels of subsidies. In the CEEC's yields are generally half of those in Western Europe, labour productivity is possibly at about a quarter of EU levels, price risks are also high and there are substantial opportunities to exploit size economies, to raise quality with the incentive of substantial price premia and to reduce transaction costs. As the economies grow and trade is promoted, there will be opportunities to meet increased demand and more closely fit production to market requirements. In addition the cost of resources such as land and labour are still low by international standards. Sensitive and holistic policies under the sustainability banner should not block off these opportunities.

Most CEE countries have a large proportion of their working population engaged in agriculture. Expectations by farm family members of at least income parity with their urban relatives will grow as economic growth occurs and the terms of trade (buying power) of agricultural products for purchased inputs deteriorates. An essential part of enhancing agricultural incomes is thus for rapid rates (in EU terms) of transfer of labour out of agriculture to other sectors. This will depend on income earning opportunities in other areas of economic activity and the degree to which obstacles to the mobility of farm people can be overcome.

5.3 REGULATION, INTERVENTION AND MARKET SOLUTIONS

There is probably a predisposition towards regulation, the use of the law and sanctions to impose standards. The effectiveness of such measures will depend on the probability of being detected, convicted and the severity of the sanction. Within a community of nations it is difficult to derive regulations which are consistent for all countries and which do not distort trade. Regulation imposes the costs, of say pollution, on the transgressors according to the polluter pays principle.

Such measures are attractive to government in that they do not involve budgetary expenditure but too wide an application of such measures can impose such a level of costs as to reduce competitiveness, especially if imposed unilaterally. Another popular measure is to provide capital subsidies for the provision of facilities to contain silage effluent and manure so as prevent runoff into streams and groundwater. It can be argued that this is a subsidy which increases the amount of silage or pollutant produced and hence is a subsidy for potentially polluting systems. It should be possible to design support schemes which encourage the use of less polluting practices.

The approach can be criticised in that inherently, rational actors may simply follow the rules as closely as possible without adopting the spirit of the law. Moreover, for all to fulfil the law is often a suboptimal cost solution. The cost of all farms meeting environmental regulations will vary and this fact has led to the idea of tradeable permits. Permits, to pollute which in aggregate sum to less than the absorptive capacity of the environment, through tradeability, allow those farms with high compliance costs to pollute more than those with lower costs.

There is much debate concerning cross compliance measures where a farmer in order to receive income support has to fulfil certain environmental conditions. The most common one is stocking rate limits but as applied they are so crude as to be almost meaningless. Moreover stocking rate limits may not be always consistent with sustainability goals. Such restrictions do constrain the amounts of subsidy that can be claimed.

A more general application of the cross compliance idea is where, for example, planning permission or capital subsidy is dependent on some matching investment in the environment, not necessarily at the same location as the productive investment. In agriculture, state support for an investment project would be conditional on implementing measures to counter any negative environmental effects. A spin-off of such an approach is the potential growth in awareness of environmental issues amongst the agricultural population.

The use of market incentives to promote sustainable systems results in each actor having a stake in and being rewarded for their individual contribution to sustainability goals. To set such schemes up is not easy. For example, products produced according to some sustainability criteria will need to be produced in sufficient bulk, which will likely require collective action, in order to justify the costs of differentiation within the whole distribution chain. Such products must be identifiable and the characteristics claimed be recognisable. Production of a sustainable line must produce a profitable premium at each stage in the food supply chain and which ultimately depends on the willingness of the consumer to pay for it.

Common support arrangements throughout an enlarged community including conversion premia, rules for grades and designation and registration of origin will all assist in the setting up of market incentive schemes. However, unless the costs of externalities imposed by agriculture are internalised (unless the polluter pays) there will be little incentive to relocate production or to farm in a more sustainable way.

Estimating and allocating these costs is a difficult task, but without at least some approximate information it will be impossible to influence the policy framing process and to design practical schemes which actually do work and support sustainable agriculture.

6. Conclusions

A sustainable agriculture will not occur on its own. It is dependent on too many uncontrollable events. To be sustained, agriculture must continually change in response to markets and political preferences. To be sustained, agriculture must continually change in response to markets and political preferences. As shown by the transitional experience it is possible for agriculture to survive major change by running down capital. However, resilience based on this strategy has produced an urgent need for restoration for all forms of capital including the state of soils.

It is argued that while prosperity is necessary for investment and modernisation, such prosperity need not be based on price support. There would appear to be good prospects for the profitable development of agriculture within the CEEC's and opportunities for labour to move to other faster growing areas of economic activity.

Development in the CEEC's depends on open markets and liberal policies which do not freeze output growth potential with supply and trade control measures. It is anticipated that policies to support equity or fairness in agriculture will not be based to the same degree as in the past on measures which distort markets. Also enlargement of the EU will potentially open up great opportunities to improve the natural environment affected by agriculture in a much more global way.

A wide range of policy measures to enhance the sustainability of agriculture are possible. No single approach is recommended since criticisms can be made of all of them. A multiple approach may well be necessary. It is also clear that some collective if not government action in the CEEC's will be required to sustain agriculture in that completely liberal policies produce unacceptable situations as has occurred during transition.

The claim is made that sustainability issues are hardly considered in the current EU policy formulation process. This is a serious threat to sustainability and an opportunity for soil scientists to objectively describe sustainability issues and to develop options for their solution.

7. References

1. Davies, A.S.[1], Slee, W.[2] and Tisenhopfs, T.[3]; (1997), Social and economic conditions in rural Latvia - A report for the government of Latvia and the World Bank. [1]Scottish Agricultural College Economics Division,[2]the Department of Agriculture, University of Aberdeen and [3]the Institute of Philosophy and Sociology, Latvian Academy of Sciences.

2. E U Commission, Directorate General for Agriculture (1995). Agricultural situation and prospects in the central and eastern European countries. Summary Report Working Documents

3.OECD (1997). Agricultural Policies in OECD Countries. 10[th]edition.Monitoring and evaluation 1997. OECD. Paris.

4. EU Commission. (1997). Agenda 2000 proposals. Vol.1 - Communication for a stronger and wider union Strasbourg, 15th July 1997

5. Osborne, T. 1997. New Conservation Reserve Program Criteria Enhance Environmental Gains. Agricultural Outlook. Economics Research Service, USDA. October 1997 15-18

6. Piskorz, W and Leszko, D, (1997), An analysis of trade in agricultural and food products with CEFTA countries. An Attempt to Assess the Liberalisation of Trade with CEFTA Countries. Foundation of Assistance Programmes for Agriculture, Agricultural Policy Analysis Unit. Warsaw, April 1997.

7. Muench, W and Guba, W, (1997). Poland's Accession to the EU. Market Effects and Budgetary Costs Revisited. Report for Agricultural Policy Unit, Warsaw, 1997 from the Institute of Agricultural Economics University of Gottingen, Germany.

FUTURE OF POLISH AGRICULTURE: POSSIBILITIES AND PROBLEMS

I. Duer, S. Krasowicz
Institute of Soil Science and Plant Cultivation
Poland, 24-100 Pu awy

1. Introduction

The strategy for the development of Polish agriculture should depend on its adjustment to the rules acting on the world economy. Poland as a country associated with the EU, a member of the OECD and a signatory of various international agreements is under an obligation to develop sustainable agriculture. But first of all, the main goal of agriculture is to secure food-self sufficiency. The choice of agriculture system and quantity of food production, as a result of technology, depends on socio-economic conditions and the ecosystem, in which farmers are operating.

2. World background of agriculture

A specific character of world agriculture, at least over two generations, is the socio-economic reduction of value of agricultural production. This is distinctly expressed in decreasing employment in the agriculture sector (USA - 2,9, EU - 6%), or reduced expenditure on food (USA and EU - 15-20%). Food expenditure below 30% of total consumption means that the relation between agriculture and food consumption breaks down. Food is then treated as a normal good, not a special one [9]. Some characteristic numbers about Polish agriculture on EU countries background are presented in tab.1.

TABLE 1. Polish agriculture on the EU background.

Specification	Poland	EU countries
1. Agricultural land area - mln ha	18,6	144,8
2. Farm number – mln	2,1	6,9
3. Average farm area – ha	7,0	17-18
4. Agricultural land area per capita – ha	0,48	0,38
5. Agriculture contribution in GNP - %	6,9	3,0
6. Employment in agriculture sector - % of working population	25,0	6,0
7. Number of consumers per one food producer	10,5	≈ 40
8. Food expenditure - % of total consumption	40	15-20

M.J. Wilson and B. Maliszewska-Kordybach (eds.),
Soil Quality, Sustainable Agriculture and Environmental Security in Central and Eastern Europe, 317–324.

Many outstanding differences between Polish agriculture and EU countries are easily seen (employment, number and farm area, food expenditure). The effectiveness of Polish agriculture is a result of work load in this sector and the level of technological development. Because the workshop refers to sustainable development of agriculture in Central and Eastern Europe, we will compare the agriculture situation in countries that are our partners in the process of joining the European Community, namely the Czech Republic and Hungary, and with EU and USA.

TABLE 2. Land area and population of selected countries on EU background -1995

Country	Land area		Population		Inhabitants
	Km^2	relation to EU %	mln	relation to EU%	$/km^2$
EU (15)	3233130	100,0	370,4	100,0	115
Poland	312685	9,7	38,6	10,4	123
Hungary	93032	2,9	10,3	2,8	111
Czech Republic	78864	2,4	10,3	2,8	131
USA	9372614	290,0	257,9	70,0	28

EU countries represent a third of the European continent and more than half of the total population (tab.2). Overall, the EU is one of the most important elements of the world economy and is effective in competition with USA.

Average yield of grain production *per capita* is regarded as a measure of agricultural productivity. Arable land area and grain production *per capita* in Poland, Hungary and the Czech Republic is greater than the average of EU countries, but grain production per employee in the agriculture sector is much lower (tab.3). There is a wide margin of work load and extensive technologies used in agriculture, mainly in Poland.

TABLE 3. Average yield and grain production per capita in 1995 [8]

Country	Arable land ha *per capita*	Average yield $t \cdot ha^{-1}$	Grain production per one farmer/ t	Grain production t *per capita*
EU (12)	0,24	5,0	24,0	0,5
Poland	0,38	3,0	6,0	0,7
Hungary	0,48	3,9	23,8	1,1
Czech Republic	0,30	4,4	-	0,7
USA	0,73	4,7	140,3	1,1

If we take gross national product (GNP) *per capita*, as a criterion of welfare, the value in Poland is only 13,5% of that in EU (tab.4).

In the creation of GNP in the EU, the leading role is played by service, trade and industry, whereas the contribution from agriculture is marginal (3%). On the contrary agriculture represents about 7% of GNP value in Poland, and even more in Hungary (10%). These differences are reflected in employment in different sectors of the economy: 6% of labour is engaged in EU agriculture, less than 3% in USA and 25% in Poland (tab.4). Overall, the socio-economic conditions in Poland and the slow transformation in agriculture do not indicate that there will be a significant reduction of total employment in this sector, but probably full time employment will decrease, in favour of part-time farmers.

TABLE 4. Gross national product (GNP) and percentage of labour in agriculture in 1995 [7]

Country	% of labour in agriculture	Agriculture contribution in GNP (%)	Gross national product *per capita* (ECU)	%
EU (15)	6,0	3,0	17342=100	
Poland	25,0	6,9	2337	13,5
Hungary	19,3	10,0	3373	19,4
Czech Republic	9,8	5,6	3393	19,6
USA	2,9	2,2	20295	117

Available data show that technical progress in world agriculture has lowered the efficiency in the use of energy inputs per unit of crop output. The decreasing energy output/input ratio in agriculture means, that the food security of mankind is increasingly dependent on rapidly shrinking oil reserves, and results in increasing environmental impact per unit of crop output. In other words, current technological developments in food production are not sustainable. On the other hand, it is well-known that traditional farming techniques achieve high output/input energy ratios, but these techniques have almost disappeared in developed countries, and are rapidly disappearing in developing countries.

In order to verify whether certain agricultural production techniques are appropriate or not for a given society, the influence of that society's population and technological development on its agricultural food production should first be analyzed.

According to Giampietro [3] two important factors shaping the development of agricultural production should be examined:

1. Population density, or demographic pressure (AP_{DP}) which influences the amount of arable land *per capita*;
2. The level of technological development of society or socio-economic pressure (AP_{SEP}), which can be quantified by:

- average income *per capita*,
- the fraction of total human time, that is allocated in the primary sectors of the economy.

The indicators AP_{DP} and AP_{SEP} represent the minimum density of biophysical inputs, per hectare and per hour, that would be needed (on average) in the agricultural sector, to achieve a food self-sufficiency in a society.

From the comparative analysis of Giampietro [3] selected indices for the countries compared here are shown in tab. 5.

Countries with high demographic pressure (AP_{DP}), like EU countries, are likely to have a large production of food energy per hectare in the agricultural sector. In other words, technology used in agriculture, expressed in terms of yield/ha is affected by existing demographic pressure. Similarly, the country with high value of socio-economic pressure (AP_{SEP}), like USA is also likely to have a large production of food energy per hour of work. This means that technology used in agriculture, expressed in terms of labour productivity, is affected by socio-economic pressure. For Poland, the value of demographic pressure (AP_{DP}) is smaller than actual land productivity (AP_{ha}), indicating that Poland is self-sufficient in food production.

TABLE 5. Productivity of land and labour expected and actual values according to demographic and socio-economic pressure, referring to 1991 [3]

Country	Demographic pressure (10^3kcal ha^{-}year^{-1})	Socio-economic pressure (kcal h^{-1} year^{-1})	Actual labour productivity* (kcal h^{-1}year^{-1})	Actual land productivity (10^3 kcal ha^{-1}year^{-1})
EU (12)	12413	26900	22100	10211
Poland	6548	6600	8500	8499
Hungary	5307	12900	24000	9924
Czech Rep	8168	14600	16000	8904
USA	3550	64800	111700	6120

*/ output of food energy includes only crop production

Giampietro [3] maintains, that countries with net food exports have a AP_{DP} value smaller than AP_{ha}, and/or a AP_{SEP} value smaller than AP_{hour}. Whenever AP_{DP} exceeds AP_{ha}, or AP_{SEP} exceeds AP_{hour} at a national level, then this society usually imports food.

Countries with high demographic pressure (AP_{DP}), like EU countries, are likely to have a large production of food energy per hectare in the agricultural sector. In other words, technology used in agriculture, expressed in terms of yield/ha is affected by existing demographic pressure. Similarly, the country with high value of socio-economic pressure (AP_{SEP}), like USA is also likely to have a large production of food energy per hour of work. This means that technology used in agriculture, expressed in terms of labour productivity, is affected by socio-economic pressure. For Poland, the value of demographic pressure (AP_{DP}) is smaller than actual land productivity (AP_{ha}), indicating that Poland is self-sufficient in food production.

Giampietro [3] maintains, that countries with net food exports have a AP_{DP} value smaller than AP_{ha}, and/or a AP_{SEP} value smaller than AP_{hour}.

3. Trends in world agriculture development

General trends in the evolution of food production techniques for different types of socio-economic systems can be represented by land productivity (kg of grain/ha) and labour productivity (kg of grain/hour), as illustrated on fig.1.

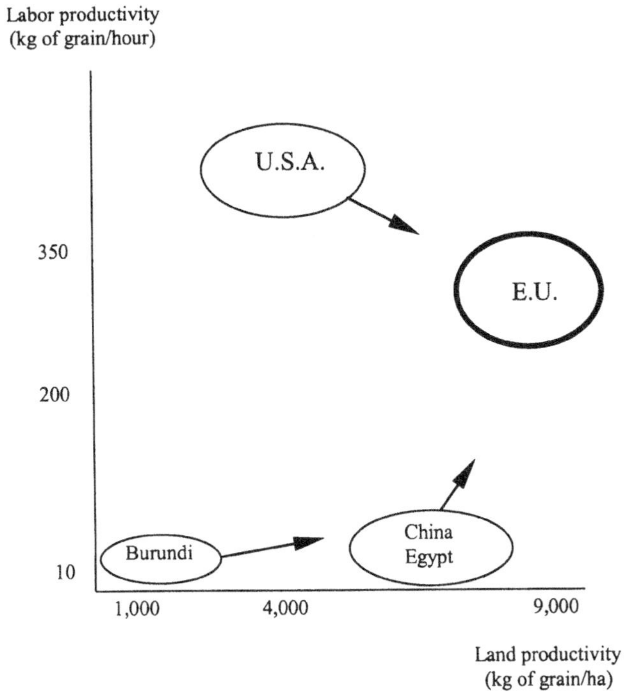

Figure 1. Expected evolution of agricultural production systems in terms of labour productivity (kg of grain throughput per hour of labour) and productivity (kg of grain throughput per ha of land) [3].

A descriptive character of four types of socio-economic systems is shown in tab.6.

TABLE 6. Main types of socio-economic systems [3]

Combination of pressure	Arable land ha *per capita*	Gross national product *per capita* $ USA/year	Characteristic for countries
Low AP_{DP} and low AP_{SEP}	>0,5	< 1 000	Africa (case Burundi)
Low AP_{DP} and high AP_{SEP}	>0,5	>10 000	USA, Canada, Australia
High AP_{DP} and low AP_{SEP}	<0,1	< 1 000	China, Egypt
High AP_{DP} and high AP_{SEP}	0,2	>10 000	EU countries

According to existing trends in population growth and economic development for these types of socio-economic system, it can be anticipated that:

- In African societies like Burundi, the population is growing faster than GNP/*per capita*. It might be expected that, they will move towards the situation typical of China.
- In societies like USA, GNP *per capita* will remain high and population growth will be relatively slow. Consequently, only a slow decrease of surplus production for the international market might be expected;
- In China, a tendency for increasing GNP *per capita* is expected. The huge population size will be maintained with absorption of labour from the agriculture sector to other sectors of economy;
- In the EU countries, high material standards will be maintained with minimum population growth, and it is expected that the impact of food production on the environment will be reduced.

4. Scenario of agriculture development for Poland

According to statistical data [1,8] and the indices developed by Giampietro [3], Poland has medium demographic and socio-economic pressures and is considered as a self-sufficient country (tab.5). The Polish economy shows a tendency for increasing GNP *per capita* (on average 5% yearly, from 1994) and decreasing unemployment (tab.7), although the absorption of labour engaged in the agricultural sector (25%) is still very slow, because of the poor development of the service sector.

TABLE 7. Dynamic of changes in GNP and % of unemployment

Country	Changes of GNP (%)				Unemployment % of working population			
	1994	1995[a)]	1996[b)]	1997[b)]	1994	1995[a)]	1996[b)]	1997[b)]
EU (15)	2,8	2,5	1,4	2,5	11,3	10,9	10,9	10,8
Poland	5,2	7,0	5,5	5,0	16,0	14,9	14,0	13,0
Hungary	2,9	1,5	2,0	3,0	10,4	10,4	10,0	10,0
Czech Rep	2,6	4,8	5,6	5,8	3,2	3,0	3,1	3,2
USA	3,5	2,0	2,3	2,0	6,1	5,6	5,6	5,4

a) estimate; b) prognosis

Whenever socio-economic development calls for a reduction of labour in agriculture, society must confront the need for an increase in farm labour productivity, to maintain food security. As shown in tab. 3, average grain production per farmer in Poland is 6 t, in EU countries it is 24 t and in USA it is 140 t. Such differentiation of agricultural labour productivity is the consequences of dramatic simplification of natural ecosystems (on average) in EU countries, use of monoculture in the USA and traditional techniques in Poland. Continuing changes in the socio-economic context of farming are driving technological progress in agriculture, toward solutions incompatible with biological processes in the environment [10,11]. It is realised, that current strategies of technological development in agriculture are unable to solve problems of food security in the long term, both in developed and in many developing countries. Discussion of sustainability of agricultural production should carefully consider ecological and socio-economic constraints to food production.

In developed countries, high-input agriculture is widely practised and is accompanied by strong socio-economic development and a "free environment" for farmers as well for investors in agro-industry. Exploitation of a "free environment" was and still is financially supported by the rest of society, through guaranteed prices and different subsidies. Modern high-input agriculture depends economically on the rest of society. But it should be kept in mind, that only developed countries are able to support the agricultural sector [2,3]. .

Developing countries, to which Poland belongs, have never had sufficient economic resources at their disposal to help support individual farmer families. On the contrary, the agricultural sector for many years was taxed to support other sectors of the economy. Nevertheless, the changes of economic system in Poland and the continuous although slow increase in socio-economic pressure, as well as accessibility to western technologies will lead to the adoption of high - input techniques. This can be observed already on large specialised farm, but unfortunately it can be destructive of the environment in many cases. If socio-economic pressure forces the adoption of high-input technologies, they should be used only on ecosystems that are less sensitive to cropping pattern simplification [5,6].

The creation of long term sustainability of agricultural production implies more emphasis will be placed on ecological constraints. Agriculture can have a positive influence on the environment and is profoundly rooted in the culture of Poland. Here there is still remarkable evidence of the shaping of the agricultural landscape, based on stabilising a symbiotic relationship between man and the land. Solutions carried out with great care and consideration of soil conservation, rotation and intercropping, have created agricultural activity that is compatible with environmental protection.

Alternative agriculture techniques based on traditional solutions and taking biodiversity into account should be undoubtedly improved and managed so as to significantly increase food production per hour of labour and per hectare of land, always bearing in mind socio-economic sustainability. In addition to the economic and ecological context, cultural identity, number of people living from agriculture, climatic conditions favourable to biodiversity and desirability of preserving natural ecosystems are decisive factors in determining Polish uniqueness. Technological development in agriculture must be based on the farm, which is the management unit with a biological base and represents human interest and natural processes. Agriculture can not be defined only in terms of economic efficiency, as is treated in the liberal vision of the Common Good, which conflicts fundamentally with sustainable development. The small space and short time scale economic analysis should be balanced by long term and world wide perspectives when discussing the future of Polish agriculture from the view point of sustainability.

5. Conclusions

Summarising consideration of the future of Polish agriculture it should be borne in mind that agriculture has a special role in the socio-economic system. Agriculture should not only produce food, but should also preserve a healthy environment, which involves among other things, a higher level of nutrient cycling, control of pollution of fresh water and management of the agrolandscape in general. The terms of reference for Polish agriculture, as well as for some other CEECS, in sustainable development are as follows:

- maintain national food - self sufficiency;
- production of safe and healthy food;
- preservation of environment;
- incorporation of regenerative technologies, which make the best use of locally available and human resources;
- the simultaneous growth of crops and livestock;
- more reliance on local management skills and knowledge;
- acceptance that the farm is the management unit with a biological base and represents a broad spectrum human interest in our society.

6. References

1. Basic Statistics of the Community (1995), Eurostat.
2. Duer I. (1996) Mo liwo ci rozwoju niskonak adowych technologii w produkcji rolnej. Ekonomia i rodowisko, 1-2 (8-9), 179-185.
3. Giampietro M. (1997 Socioeconomic constraints to farming with biodiversity. Agriculture, Ecosystems and Environment. 62, 2-3, 144-167.
4. Kawecka-Wyrzykowska E., Synowiec E. (ed). (1997) ABC Unii Europejskiej, Warszawa.
5. Krasowicz S., Harasim A. (1997) Post p technologiczny w produkcji ro linnej a integracja Polski z Uni Europejsk . Mat. IV Kongres SERiA, Szczecin, 816-826.
6. Hunek T. (1996) Strategiczne programowanie rozwoju rolnictwa. IERiG Warszawa, 3-45.
7. Mroczek W. (1997) Struktura gospodarki i handel Unii Europejskiej z krajami trzecimi w latach 90. Instytut Koniunktur i Cen Handlu Zagranicznego. Unia Europejska: Integracja Polski z Uni Europejsk , Warszawa, 26-41.
8. Rocznik Statystyczny GUS. (1995), Warszawa.
9. Schaller N. (1993) The concept of agricultural sustainability. Agriculture, Ecosystems and Environment, 46, 89-97.
10. Szemberg A. (1996) Prognoza przemian w strukturze obszarowej i zatrudnienie w rolnictwie polskim do 2010 r. IERiG Warszawa, 3-32.
11. Wo A. (1996)Prognoza przekszta cestrukturalnych polskiego rolnictwa do r. 2020. IERiG Warszawa, 3-122.

OPPORTUNITIES FOR SUSTAINABLE AGRICULTURE IN THE CEECs WITH PARTICULAR EMPHASIS ON POLAND

AUGUSTYN WOS, Prof. Ph.D.

Institute of Agricultural and Food Economics
Warsaw

1. Introduction. Conflict between liberal and environmental agricultural policies.

The last decade of the 20th century has witnessed a thorough re-orientation of agricultural policy towards economic liberalism. This is how the major actors on the world agricultural markets have reacted to costly intervention policies persistently pursued from the 30s on. Among the advocates of liberal agricultural policy there are huge exporters (including mainly the USA), GATT/WTO, and the European Union. The group is complemented, though not without resistance, by the countries of Central and Eastern Europe (CEECs).

Liberal agricultural policies encourage capital consuming intensification, concentration on production, increased engagement of resources, etc. Such a system of agricultural development threatens the environment because of the increased use of environmental resources per unit of agricultural product. This process is reinforced by the fact that domestic competition must move to an international level. This is a new factor because the process may farmers to behave in an anti-environmental way, leading to a decrease in the quality of foodstuffs they produce. On the other hand, consumers require the agriculture and food industry to produce higher quality products and, for example, the demand for organic food is increasing and becoming more and more pervasive. In fact, organic food is setting a standard and this trend seems to be irreversible.

All these factors create a conflict between the logic of the economic system - which encourages capital consuming intensification, increasing production as well as the use of environmental resources - and the demand structure for food which is gradually moving towards so called organic food. The EU has tried to mitigate the conflict created by production and market policies by launching budget-consuming environmental schemes. These issues are governed by regulation no 2078/92 which provides the EU Member States with the possibility of supporting organic farming.

M.J. Wilson and B. Maliszewska-Kordybach (eds.),
Soil Quality, Sustainable Agriculture and Environmental Security in Central and Eastern Europe, 325–333.

All this demonstrates a dichotomy of the environmental policies pursued by the EU, WTO and other countries. On the one hand, severe pro-efficiency mechanisms are promoted, with all their negative effects on the environment and on the production of organic food. On the other hand, various environmental schemes help counter the accumulation of these negative effects. The balance of those two forces is easy to foresee. Measures designed to encourage environmentally friendly food production are always limited. They have never been able to successfully counteract the phenomena created by a liberal economic and market policy. Apart from a single segment of pro-environmental activities, the whole economic mechanism is set to raise economic efficiency (maximising of profits), irrespective of the negative external effects. The advantages and disadvantages of such policies do not balance each other, particularly with regard to the environmental. Such a situation will lead to conflicts and promote regional inequalities and may well affect Poland and other CEECs.

An analysis of economic systems and the experience of recent years leads to the conclusion that a new rationale is needed, different from that existing in the West. We need a new rationale, with criteria for assessment encompassing not only economic factors, but also taking into account the environmental point of view. The merger of both these views will be rather difficult. The narrowly understood criteria of current economic rationality are so deeply rooted in the minds of decision makers at all levels that it will be very difficult to substitute them with new criteria of complex assessments, which in the long term also respect the views of the whole of society.

In this respect the CEECs are in a special position. They have only started to introduce market mechanisms and implement the criteria of economic rationality on a micro scale, but they have already realised that this approach is too narrow and biased. Perhaps they should stop half way in the transition from an extensive to an intensive economy.

In the light of these fundamental issues, the situation of Polish agriculture, and to some degree also other CEECs will be assessed from the point of view of sustainable development criteria. In relation to the EU countries, Polish agriculture has both strengths and weaknesses.

2. Strengths and weaknesses of Polish agriculture from the point of view of sustainable development.

2.1 STRENGTHS

(1) From the point of view of environmental protection and sustainable development, Polish agriculture has a great advantage over the majority of EU countries. It is more “natural”, less intensive and less prone to environmental destruction. Results of research show that only about 3% of arable land in Poland should be excluded from food production because of pollution, and the rest of the region fulfils all the necessary conditions for the production of organic food.

(2) Because of farm fragmentation and non-contiguous plots, Polish agriculture has more advantages in terms of biological diversity than the mono-cultural agriculture of highly developed states. Paradoxically, if the problem of food production in Poland is viewed from an environmental standpoint, the features accounting for its structural and technological underdevelopment turn out to be advantages. If the situation in agriculture is assessed on the basis of the criteria of sustainable agriculture, instead of on the basis of a very one-sided and imperfect criterion, namely the level of labour efficiency, it turns out that our developmental gap in relation to the EU agriculture is smaller than commonly perceived. Everything depends on taking advantage of these “underdevelopment factors” in the future.

(3) In the 90s, despite a decrease by 1/3 of the use of mineral fertilisers and chemical pesticides in Poland, there has been no decrease in crop production. This suggests that, in the past, mineral fertilisers were misused, with consequent disastrous effects on the environment. These effects include not only the pollution of soil and water. The use of energy and raw materials for the production of fertilisers (especially nitrogen fertilisers) has dropped, and this has affected the emission of toxic gases to the atmosphere and the production of waste. The effects of a partial de-intensification of agriculture, which was forced by an economic crisis, has proved to be advantageous to the national economy and to the natural environment. It can be maintained that the social cost of producing a unit of food has diminished. The costs include not only direct farm resources, but also deterioration of the environment in which production takes place. If we consider all these elements including externalities, the balance of change from a social point of view, may be positive.

(4) The issue of food quality should be understood in a wide context, including not only 'technological quality', i.e. alimentary and trade characteristics given to the products during industrial processing, but also biological quality, given to the products in the process of agricultural production, called 'organic production'. Considering both elements it can be concluded that the competitive position of Polish agriculture may even improve in the future. It should be borne in mind that the market will evolve and that demand will change towards organic products, which are regarded as healthier than 'conventional' ones.

(5) Environmental technologies are by their nature labour-intensive. From this point of view Polish agriculture has an advantage over the agriculture systems of the EU Member States because there is a structural surplus of relatively cheap labour which has little significance from the point of view of agricultural production costs. But in this respect the situation is also changing. The times of a cheap labour force in agriculture are coming to an end. General economic development and progress - which promotes alternative income sources - will almost certainly increase the price of agricultural labour so that this advantage which Polish agriculture currently enjoys on the international markets will become less and less evident.

2.2 WEAKNESSES

As pointed out above, natural and environmentally sustainable Polish agriculture has certain advantages over the industrial agriculture of the EU countries. But such natural agriculture, by its nature, is subject to destructive outside factors resulting from the overall economic situation as well as the use of old technologies.

Environmental forces are of a global character and as a result it is impossible to protect agricultural environmental resources except in a global context. With respect to agriculture, it is important effectively to carry out national protection schemes concerning resources and the environment. It is particularly important to carry out activities aimed at reducing the emission of toxic gases and dust to the atmosphere, the pollution of waters with heavy metal, toxic compounds and salts leading to salinisation. Non-agricultural factors are estimated to account for about 4/5 of all pollution accounting for the poor quality of agricultural products. These external (including foreign) sources of pollution must be first monitored and eliminated.

All this demonstrates a low level of environmental protection on a global scale, adding to the weaknesses of Polish agriculture from the point of view of this discussion. This was vividly demonstrated by the floods during the summer of 1997. The greatest weaknesses, though, include energy-consuming and resource-consuming industrial technologies. They all result in a high level of pollution affecting the quality of soil, water and air and while such pollution only rarely reaches a critical point, nevertheless a state of potential threat continually exists. In recent years the state of the environment in Poland has significantly improved, as a result of implementation of new technologies outside agriculture. Statistical data from 1990-1995 show that the emission of industrial and urban sewage to ground water decreased by 12,2% and the emission of industrial air pollution decreased too (dust, 62.9%,gases, [excluding SO2] 32.3% and SO2 25.7%. Industrial environmentally unfriendly waste decreased by 14.7% and the area of devastated and degraded land decreased by 22.9%.

Moreover, the quality of food has improved, which is demonstrated by the smaller percentage of rejected samples in the overall number of samples recently examined by public health authorities.

A reduction in international tension has increased overall prosperity and the growth of GNP, and a perception of improvements in food safety and quality is also an element of prosperity. Polish progress in this field is evident.

The second weakness of Polish agriculture is the inability of farms to accumulate capital, which significantly limits investment possibilities. This refers particularly to investment in infrastructure aimed, for example, at improvement of the water balance of soils, improvement of water quality, and of preventing water and wind erosion of land. Research shows that only one in ten farms is able to produce an economic surplus which could be used for its own investment. This structural disadvantage of Polish agriculture, correlated with a generally low level of environmental education, diminishes farmers' abilities to implement environmentally friendly technologies. Despite the fact that organic production is not resource- consuming (in the balance of direct costs), it calls for prior infrastructure inputs, which are impossible to put into effect by economically weak farmers. Thus, environmentally friendly activities on a micro scale call for outside support, especially in the form of cheap credits and know-how.

Summing up this discussion on the strengths and weaknesses of Polish agriculture it seems evident that in its current state, Polish agriculture approximates to the standards of sustainable agriculture to a higher degree than that of the EU Member States and in this

particular aspect is superior. Following entry into the world market and participation in globalisation, it is evident that Polish agriculture will come under the pressure of criteria related to micro-economic effectiveness, which will inevitably result in the loss of its environmental features. We are at a cross-roads in terms of choosing a development strategy for agriculture. The strategic question is whether to follow a universal path, pursuing a model of industrial development and paying a huge price through the loss of environmental resources, or whether to stop half way by protecting the existing environment and counting on the fact that in the future the pay off for environmentally friendly agriculture will be higher. The logic of strategic thinking calls for choosing the second strategy. However, knowledge of the reality - especially the power of the economic mechanism of rationality and its current effectiveness on a micro scale - makes it doubtful that an environmental strategy will succeed. People easily communicate when they discuss events in the distant future, but lose this perspective when they must take decisions on an every-day basis. In such situations, short term profit maximisation tends to take precedence over the long term criteria of global thinking. This is why, in terms of choosing a strategy for Polish agricultural development, it is difficult to be optimistic. It is more probable that we will follow the path of the highly developed EU countries rather than stop half way, with a conscious determination that the environment which is not yet degraded should be protected. Everything possible should be done to slow down this process of environmental degradation to the maximum extent.
The third part of this paper concerns a programme minimum, i.e. the undertakings which absolutely have to be carried out in order to stop the processes of degradation of the agricultural environment.

3. Programme minimum

What can and should be done in Polish agriculture as quickly as possible in order to stop the destruction of agricultural ecosystems is summarised below:

1. Special environmental protection should be offered to environmentally valuable areas and arable lands throughout the country. These lands constitute the basis for the development of Polish agriculture. In view of economic development and changes in the demand structure (towards organic food) the importance of these areas will continue to increase and care should be taken that this opportunity is not lost. Such areas should in particular be excluded from any schemes promoting a narrow specialisation of farms, which leads to mono-culture and eliminates biological diversity. Instead, environmentally valuable areas should be subject to half-extensive production (eg. cattle farming) and moderate fertilisation with

mineral fertilisers. For these areas, special recommended schemes for environmentally friendly farming should be developed, enabling farmers to get decent incomes despite lowered production intensity.

(2) Professional advisory services should develop a catalogue of good agricultural practices and help farmers implement them. At the farm level these practices should include: crop rotation, the balance between organic and mineral land fertilisation, choice of an adequate system of land cultivation and integrated plant protection. In addition, advice should be given on modern agricultural technology, the structure of plantation and crop rotation, fertilisation, regulation of water balance in soil, the use of heavy machinery and equipment and optimal relations between animal and plant production. Such advice should take account of environmental protection including the protection of the most valuable renewable resources.

(3) A programme of farm sanitation should be carried out systematically. The programme should include the supply of drinking water and farm water, farm or village sewage systems, treatment plants, waste dumps, neutralisation of toxic chemical substances and their metabolites.

(4) In order to protect arable areas against different types of degradation, it is desirable from economic and environmental points of view that a minimum of 640 thousand ha and a maximum of 1.2 million ha of poor soils should be forested. Extensive research at IUNG on the negative effects of soil acidification (in addition to the importance of maintaining a desirable balance of nutrients and soil organic matter) indicate that it is essential to develop and implement soil improvement schemes in order to raise soil pH, so as to combat acidification and mineralisation.

(5) Within the framework of the programme of land protection against water and wind erosion there is a need for the implementation of pilot schemes in areas particularly prone to erosion, as well as programmes for re-cultivation and development of gullies.

(6)Improvement of arable land should be adjusted to the needs of agro-ecosystems. The following should be done in particular: (a) determine the need for further land improvement as well as check the current state of equipment, machinery and systems in use, (b determine the future of derelict or abandoned land, (c) develop regional programmes for re-naturalisation of land and adjust them for the purposes of environment, recreation and conservation, (d) develop realistic programmes for improving the quality of permanent pastures and develop appropriate irrigation and drainage schemes.

(7) With regard to rationalisation of soil fertilisation the following is recommended: (a) in order to maintain ecological balance, it is essential that the amount of fertiliser elements supplied to the soil corresponds to the amount taken up by the plants; an excess of absorption over supply guarantees that negative environmental effects are minimised, (b) the variety of mineral fertilisers should be enlarged, (c) professional advice with regard to fertilisation should be developed, (d) legal norms as to mineral fertilisation are essential.

(8) The programme of adjusting plant protection to the requirements of the environment and the requirement of agricultural production includes the following: (a) liquidating storage bunkers for unwanted pesticides, (b) increasing the use of plant protection measures to 1,2 - 1,4 kg of active substance per 1 ha, (c) introducing new, safer products to the market, (d) introducing new biological plant protection agents to the market, (e) increasing the area of crops protected in accordance with EU norms, including integrated crop production technologies.

(9) Technologies of animal production should be adjusted to the regimes of environmental protection. This requires the following: (a) the creation of sufficient animal genetic resources (particularly for bovines, pigs and poultry), (b) taking care of animal welfare, (c) the promotion of moderately intensive animal production technologies.

(10) A programme of agricultural production with integrated and environmentally friendly methods should be carried out. The essential conditions for environmentally friendly agriculture include: (a) adaptation of organic farm registration to IFOAM (International Federation of Organic Agriculture Movements) and EU rules (regulation no 2092/91 of June 24, 1991), (b) financial support to organic farmers, (c) promotion of research and development units concerning organic farms and dissemination of the results of such research.

(11) Food processing should be adjusted to EU standards by 2005. The following recommendations should be taken into account within environmentally friendly activities: (a) production schemes of manufacturing plants must respect the rules of environmental protection, (b) EU legislation concerning these issues should be quickly disseminated and implemented; in this matter Poland is lagging behind.

(12) The protection of the environment in agriculture should be integrated (at regional levels) with the creation of a specifically Polish farm countryside, which includes: (a) rational division of arable land,

pastures, forests and environmentally valuable land, (b) protection and upkeep of land possessing special environmental and natural features, (c) maintenance of a typical landscape architecture, which is an element of Polish cultural heritage.

ASSESSMENTS OF ENVIRONMENTAL POLICY MEASURES: A FARM APPROACH

Stanislaw Gedek
Agricultural Academy
ul. Akademicka 13, 20-934 Lublin

1. Introduction

Environmental problems have recently come to be seen as very important to society. Mrs Thatcher described protection of the environment even as *one of the great challenges of the late twentieth century* (The Speech to the Royal Society, September 1988). This results in environmental issues influencing voting behaviour and, as an outcome, the presence of environmental regulations in political programs. It is reflected in policy proposals and regulations, changing also business conditions. The last effects should be also taken into account when environmental policy measures are assessed.

Despite the impression given in some of the environmental literature, heavy industry is not the only source of environmental degradation. Also agriculture is, at least potentially, a source of serious environmental problems, although less spectacular ones.

Intensive agriculture can affect the quality of the environment in many ways. The most visible and probably the less harmful effect is landscape transformation. Other effects include contamination of ground water with pesticides, nitrogen, phosphorus and potash, diffusion of harmful substances and transformation of ecosystems. Even acid rain is partially caused by modern agricultural activity. Legislation is one way to control these negative external effects and assessment of its results is a decisive factor for developing efficient policy.

The ultimate object of any policy directed to agriculture is the farm. All the changes in agricultural sector, resulting from policy, will take place in farms. The farm is a very complex structure however. If policies are to be efficient, the implications of any particular regulation, as well as the system of regulations, have to be determined, analysed and evaluated for all types of farm. Despite the importance of analysis at the micro level, until quite recently, researchers had concentrated their analysis of institutional changes at the sector level only[1]. Some models for farm level modelling of institutional changes have been developed but their use for environmental policy assessments is very limited.

---

[1] For a typical macroeconomic approach of the environmental policy modelling see Pearce and Turner [26]. A general comparison of both approaches in agricultural policy analysis can be found in Van Dijk *et al.* [9].

M.J. Wilson and B. Maliszewska-Kordybach (eds.),
***Soil Quality, Sustainable Agriculture and Environmental Security in Central and Eastern Europe*, 335–346.**

The aim of this paper is to present a methodology which can be used to analyse and evaluate environmental policy effects at a farm level.

2. The state of art

One of the first attempts to analyse the effects of institutional changes at a farm level was that of Kempinski [16]. He applied a simple budgeting approach to evaluate the impact of the UK entry to the European Common Market for an average British farm, obtaining quite interesting results. Later approaches employed more sophisticated techniques - mostly linear programming.

In the early 70's several universities independently developed linear programming models for evaluating and analysing institutional changes in agriculture at a farm level. The best known results of these efforts are APMAA Model (*Aggregative **P**rogramming Model of Australian Agriculture*) developed at the University of New England in Australia [21, 25, 34] and the model developed at the University of Newcastle upon Tyne in the United Kingdom, known as the Newcastle Model [30]. Models developed at other universities have been less widely discussed, including the Polish SGGW model[40], (the name comes from Polish abbreviation of the Agricultural University of Warsaw, where the model was developed), and models developed in Denmark [1], USA [20, 29] and Germany [14], where work started even in the middle of the 60's. Quite recently a model has been developed at Wageningen Agricultural University in Holland called The MIMOSA system (***MI**cro **MO**delling to **S**imulate changes in **A**griculture*)[41].

The models mentioned above have had a numerous applications in agricultural policy appraisal. The Newcastle Model was used mostly for evaluation and analysis of various aspects relating to the United Kingdom entry into the European Community. An excellent illustration of this approach is that of Davey and Weightman [8]. The model has also been extensively used elsewhere, and a list of applications has been given by Thomson and Buckwell [30]. The main goal of the APMAA models was to evaluate institutional changes in Australian agriculture. A good example of its use is described by Wicks, Parton and Beesley [39]. The SGGW model, developed mostly for applications in extension services, has also been used for policy and its applications [28]. There is no report on the use of the use of any of the above models for environmental policy modelling with the exception of the MIMOSA system [41].

A general assumption (applied rather implicitly), underlying the farm level approach of policy modelling is the concept of the farmer as an „adaptive man", who tries to adapt his farm organisation to the changes in the whole business environment, including the institutional ones [41]. This means that the above approach assumes that a farmer's reaction to the institutional changes consists of two parallel actions: the adoption of the institutional changes themselves and the adaption of the whole farm to the new situation created by the changes as the farmer tries to meet his objectives, usually profit maximisation or a family wealth maximisation. The methodology of policy measures assessment consists then of comparing the results of the two models : the basic one (pre-institutional change) and the one where the impact of the assessed policy measure (or set of measures) is included.

Most of the models described above are so called "supply" models. The compared results are then the quantity of various commodities produced. The more complex models include also a comparison of the farmer's profit. Generally, models check to what extent anticipated policy goals can be met and what are the "by-products" (*e. g* farm profit reduction) of the evaluated policy measures. In the case of environmental policy measures assessment a comparison of pesticide use [41,42] or nitrogen losses has been made [2, 41, 42].

The core of each of described model is a linear programming problem:

$$\text{maximise: } \mathbf{c}^{\mathrm{T}}\mathbf{x}, \tag{1}$$

such that:

$$\mathbf{A}\mathbf{x} \leq \mathbf{b}, \tag{2}$$

$$\mathbf{x} \geq 0, \tag{3}$$

where: **c** is the vector of objective function coefficients, **x** - the vector of decision variables, **A** - the matrix of input-output coefficients (technology set) and **b** is the vector of constraint or right hand side (RHS) coefficients. In the case of the farm models, decision variables form the set of activities and objective function coefficients are gross margins or cost per unit of respective activity. In the case of policy modelling, vector **b** contains also coefficients reflecting institutional restrictions, not only capacities and technical restrictions as is in the case in DSS models. Matrix **A** contains usually a number of logical coefficients, forming for example rotational constraints restricting the ratios of some crops in the sown area.

In the linear programming models all the relations are linear, as the name says. This assumption is sometime criticised as inconsistent with the neo-classical microeconomic production theory, which assumes curvilinear relationships with diminishing marginal returns, expressed by the functions, at least twice differentiable. As proved long ago by Weinschenck [36, 37, 38], the application of linear programming models to the farm profit maximisation problem is not as great a simplification of the neo-classical microeconomic production theory as implied in the depth of difference in the meaning of words „curvilinear" and „linear". On the other hand there are some indications that at least some production functions may be of a linear response and plateau (LPR) form [4, 7, 27, 32] and the lack of sufficient empirical data usually makes impossible the estimation of production functions parameters which can be „operationally meaningful" on a farm level [31]. The other weakness of the neo-classical microeconomic production theory consists of the fact that there is no procedure which can be used to test the hypothesis of relevance of any production function to other relationships which take place in the farm. The hypothesis *must be taken on faith* [33]. This fact stimulated the nonparametric approach to production analysis which does not need any assumption of a function differentiability [13, 33] and sometime leads to linear relationships [6]. For a more general critique of the neo-classical microeconomic production theory see Georgescu-Roegen [12] and Upton [32].

The model (1)-(3) gives solutions in real numbers. There are some variables which can have only integer values (number of tractors, number of cows, *etc.*) however and rounding the results to the nearest integer may lead to serious mistakes. A remedy for this inconvenience is so called integer programming. In integer programming either all the elements of the **x** vector are given in integers in the optimal solution or only selected ones. Despite the fact that there are well developed algorithms solving pure and mixed integer programming problems, this technique was not used in the policy appraisal a farm level. The reason was a considerable increase of computer time in compared with regular linear programming. The problem was even more serious due to large number of representative „farms". Computer capacity has increased quite substantially, however, in the last few years and the trend has continued. The problem of computer time does not seem to be a serious constraint for the use of integer programming now and will be less so in the future. Computer packages containing integer programming modules are also available. Even the Microsoft Excel spreadsheet (4.0 and higher) has the SOLVER function which enables the solving of integer linear programming problems.

Agricultural activity is highly exposed to risk. This fact has to be taken into account if farm level policy modelling is to be a useful approach. The risk module is a very important part of APMAA model [10, 18, 35] and this contributed greatly to the development of risk problems in linear programming models in general. Also the SGGW model has its risk module although it was not used in its policy modelling application [3, 19]. There is no report of risk incorporation in the Newcastle Model or in the MIMOSA system [41]

Incorporation of risk (or uncertainty) into linear programming models has always been a very important task in the programming development of the models. Quite a wide range of risk incorporation methods based on different approaches, different rationales and concerning all elements of the model have been used(1)-(3)[2]. Only some are considered as possibly suitable for policy modelling at the farm level, although some have never been reviewed. This element of farm level policy modelling is the area where a consolidation of impressive results is really needed.

Application of linear programming models to agricultural (including environmental) policy modelling at a farm level leads to a problem which does not exist in DSS applications of this technique. It consists of the fact that the technology matrix **A** in the case of a real farm can be filled up with coefficients obtained empirically on a farm level. Because in policy modelling a „farm" should be representative for a certain type of farms, *a good deal of judgement is necessary to produce consistent coefficient values typical for farms in each group*[3]. This is probably the most difficult part of the policy

[2] For a revue see Jablonowski and Wrzosek [15] and Gedek [11].
[3] Thomson and Buckwell [30] p. 2.

modelling at the farm level and the area where mistakes are most likely. No precise method as to how to ensure proper *judgement to produce consistent coefficients* has been developed and is quite unlikely that it will ever be.

The main assumption underlying the concept of the policy modelling on a farm level is that each type of farm may react differently to the applied policy measure. It is impossible however to construct a model for each and every farm. It means that some kind of representative „farms" or farm models has to be developed to draw any conclusion. The methodology applied to obtain this objective consists of grouping all the farms into homogeneous classes. The basis for constructing these classes are physical resources (land, labour, etc.). The most popular technique of obtaining such homogenous groups of farms was cluster analysis [5, 30, 41], as well as regression analysis [17]. A more sophisticated method, using a formalised procedure of aggregation, was proposed by Önal and McCarl [22, 23] and generally a good deal of work has been done in this area[4]. However, as in the case of risk incorporation in policy modelling, there is a need for consolidation of the results obtained and a precise comparison of methods.

The value of farm approach policy modelling consists first of all in depicting the direction of changes rather than absolute numbers [24]. There is no possibility to conduct any experiments to check the results of a given policy, however, and usually there are no data from previous applications. Simulation models are, therefore, the only source of information. Even such limited information is of great value, particularly as analysis at a sector level will provide no information more precise than that obtained from a farm level approach.

The information obtained from a farm level policy modelling approach is of various kinds. The most important in the case of the supply model are the level of farm income and the structure of production. This information is only a *by product* in the case of environmental policy modelling. More important are reduction of pollutant emission [41], particularly when the evaluated environmental policy measures belong to the so called *environmental tax* category. In the case of *fixed standards* environmental policy measures, the information about changes in the production structure and in farm income level are very important as indicators of environmental policy costs.

The problem (1)-(3) solution given by most computer programs provides a wider range of information than just the $\mathbf{x}$ vector elements (the production structure) and the $\mathbf{cx}$ value (the income of a farm). Amongst these are so called *shadow prices*, which can be consider a marginal productivity of resources. If the simulations show an increase of the land shadow prices for instance as a result of the evaluated policy application in the model, a real increase of land prices can be expected if the policy is applied. There is no report of the use of shadow prices use in any policy evaluation. Another type of information which can be obtained from the models and which has not been currently utilised, consists of input use. The only input information which has been utilised in farm level policy modelling relates to the level pesticide use [41,42] and the level of

[4] For a revue see Önal and McCarl [23]

N use [2, 41,42]. Figures concerning other inputs may be also of some value as changes in use means changes in demand and then changes in input price. It seems than that farm level policy modelling has greater potential than is currently realised.

The farm level policy modelling approach does of course have some limitations. These result from the fact that this approach can not consider, or can consider only to the certain degree, the feedback from the sector to the farm [41]. This means that the farm level and sector level approaches are not competitive but rather complementary. A more complete picture of any policy, impact including environmental policy should be attained when both approaches are adopted.

3. The example

The aim of this section is to illustrate how a farm economic approach can be used to assess environmental policy measures. A rather simple, hypothetical example has been chosen for this reason. The advantage of such an option is that it enables a general description of the methodology.

The hypothetical policy measure to be evaluated as an example is the limit on N use in mineral fertiliser. This measure belongs to the fixed standards approach of environmental policy. As it is assumed that the farmer will obey the regulation, the only things to be evaluated are changes in income level and production structure. These are the by products or costs of the applied policy.

In the example the results of proposed policy application for two farm models - with 20 and 35 hectares of a good quality arable land are checked. Labour resource was the same in both models. Financial resources were not included for simplicity. Linear programming was used to model the structure of production with the objective function of the model maximised for farm gross margin. Table 1 shows the model matrix.

It was assumed that the farmers in both cases could make a choice among such alternatives as wheat, barley, rape, potatoes, sugar beets and forage crops: maize and clover. In the case of wheat and barley two levels of mineral nitrogen fertilisation were utilised (see constraint No 15 of Table 1), so that *wheat I* and *wheat II* and *barley I* and *barley II* represent different objective function parameters too (see the objective function row of Table 1).

The only animal production activity considered in both models was milking cows. It was assumed that one cow requires 30 tons of rough feed (on a green matter basis) yearly. Rough feed was supplemented by concentrates at a fixed rate in the case when maize or beet tops are fed to the cows (see constr. 14). The integrity requirement was imposed on the variable *number of cows* so mixed integer linear programming model was applied.

Labour was balanced in three periods - first crop of hay (constraint No 9), grain harvest (constr. No 10) and potato and sugar beet harvest (constr. No 11) as it is likely that the

TABLE 1. The model matrix

Objective function [PLN]		1500	1600	1300	1400	1100	2100	2700	2500	-400	0	-16			
variables / constraints	Const raint. No.	Cows	Wheat I [ha]	Wheat II [ha]	Barley I [ha]	Barley II [ha]	Pota toes [ha]	Sugar beets [ha]	Rape [ha]	Clover [ha]	Maize [ha]	Rough feed [T]	Concen trates [100 kg]	sign	R H S
Variable. number		*1*	*2*	*3*	*4*	*5*	*6*	*7*	*8*	*9*	*10*	*11*	*12*		
Land	*1*		1	1	1	1	1	1	1	1	1			≤	20
Max. of cereals	*2*		1	1	1	1								≤	12
Max. of wheat	*3*		1	1			-1		-1	-1				≤	0
Max. rape	*4*		-1	-1	-1	-1			1	1				≤	0
Max. of sugar beets	*5*							1						≤	4
Max. of potatoes	*6*						1							≤	5
Max. of clover	*7*				-1	-1				1				≤	0
Max. of clover	*8*									1				≤	2
Labour - period III	*9*	5					8	165		15				≤	225
Labour - period IV	*10*	8	10	10	10	10			10	15				≤	345
Labour - period V	*11*	11	8	8			110	50	7		70			≤	480
Rough feed	*12*	30						-50		-40	-70			≤	0
Rough feed II	*13*							50			70	-1		≤	0
Concentrates	*14*											0,2	-1	≤	0
Nitrogen	*15*		120	70	100	60	150	250	180		250			≤	N

lack of labour occurs only during those three periods. Labour requirements coefficients reflect technologies currently applied in Polish family farms. Rotational constraints consist of constraints 2-8 (see Table 1). Objective function coefficients were hypothetical, reflecting however Polish reality.

The policy evaluation consists of comparison of optimised results attained for a basic situation, without any mineral N use limits and those with mineral N use limitation. The results were obtained for two farm models - 20 and 35 hectares. All the right hand side parameters in table 1 refer to the 20 hectares model. In the case of the second model respective RHS parameters were changed. There were 5 levels of limits applied - 150, 120, 100, 90 and 70 kg N/ha. The results were obtained by the variation of right hand side parameter N of constraint 15 which was equal to limit applied times number of hectares.

The results are shown in Table 2. One general conclusion can be stated from these results. At least some consequences of applied policy depends on the type of farm. For example in the 20 hectares model the lower the N use limit the smaller is the number of cows; in the 35 hectare model there is an opposite trend. In the 20 hectares model lowering of the N use limit has practically no influence on sugar beet area, while in the 35 hectares model there is quite a substantial decrease of sugar beet area. The other conclusion is that there can be some surprising effects which cannot be predicted without model approach. For example it could hardly be expected that the limit imposed on mineral N use will in the case of N applied at 150kg/ha increase the area of sugar beet, which is the most heavily N fertilised crop in the example. The same applies for rape and potatoes which are also quite heavily N fertilised. Both conclusions, drawn from this simple example show the need of farm level environmental policy evaluation.

The figures from Table 2 can also be used as an example of how the model results can be utilised in policy assessment. Analysis of the results shows clearly that a mineral N use limit reduces considerably farm profit in both cases, more or less in proportion to the N use reduction.

It also shows that such a policy will result in reduction of wheat production, but not necessarily of grain. The decrease of wheat production is overlapped in some cases by barley production. In some farms applying the evaluated policy may decrease milk production. It also shows that such a policy will result in reduction in wheat in the 20 hectares model as compared to the basic situation. In other cases the same policy may increase milk production. In the 35 hectares model one of the results of applied policy is an increase in the number of cows. The result for whole milk production will depend on the proportion of both types of farms in the whole sector. It is also obvious that N reduction will decrease the rape production in both cases, particularly with lower N limits. As the input requirement is considered, .the reduction of feed concentrates use can be expected, mostly due to the changes in the type of roughage fed to cows (clover replaces maize and sugar beets tops) as a result of applied policy.

TABLE 2. A comparison of two models optimisation results

Variables	20 hectares						35 hectares					
	Limit of N [kg/ha]						Limit of N [kg/ha]					
	∞	150	120	100	90	70	∞	150	120	100	90	70
Cows [No]	8	3	4	5	5	5	2	2	3	4	6	6
Wheat I [ha]	8.13	5.88					16.18	13.80				
Wheat II [ha]		3.30	7.84	7.03	5.88	4.30		2.40	15.94	12.98	12.02	8.93
Barley I [ha]			1.60									
Barley II [ha]				4.64	6.12	7.70			0.85	6.44	3.50	12.17
Potatoes [ha]							1.40	1.41	1.50	1.98	1.96	1.88
Sugar beets [ha]	1.12	1.27	1.10	1.03	1.03	1.03	1.24	1.20	1.12	0.99	0.77	0.77
Rape [ha]	8.13	9.18	7.84	5.03	3.87	2.30	6.88	16.20	14.04	9.23	8.93	3.45
Clover [ha]			1.60	2.00	2.00	2.00			0.85	1.77	3.50	3.50
Maize [ha]	2.62	0.37	0.02	0.27	0.27	0.26					0.02	0.02
Concentrates [100 kg]	52	23	16	18	18	18	12	12	11	10	8	8
Fallow land [ha]					0.83	2.41			0.70	1.61	4.30	4.37
Object. function [PLN]	**45400**	**43895**	**40143**	**35791**	**33045**	**28782**	**75425**	**74688**	**66924**	**58990**	**55439**	**46990**

There is one effect of applied policy that has to be mentioned. With lower N use limits, in both models, there was fallow land (see Table 2). This means that the main goal of N use reduction policy - a reduction of N emission to ground water may be only partly achieved. The reduction of N use for the whole farm does not necessarily mean the reduction of N use of every hectare if fallow land is always located far from the farm yard.

The information provided by the model shows the costs of evaluated policy. It is a political decision to make the final assessment. The above conclusions are however no better than the assumptions of the model: prices, coefficient of the matrix **A**, the list of alternatives etc. Nevertheless, even this simple, hypothetical example shows that a farm level policy environmental modelling approach can be useful and sometime it is the only way to assess environmental policy measures. It should also be mentioned that farm level analysis must be supplemented by sector analysis to draw a full picture of the policy results.

4. Summary

1. A farm level policy modelling approach seems to be a very interesting proposition. In some cases, where there is no previous experience, this approach is the only method for the assessment of policy measures. There are numerous examples of successful use of this technique. A farm level policy examination should be supplemented, however, by a sector level investigation.

2. A methodology for farm level policy modelling has been developed since the middle of 60's at several universities. Achievements in this field are very impressive. A sophisticated model has been developed and the progress of computer technique makes possible wider use of this methodology. Several aspects of the models used for this purpose has been developed on a different basis. This is related mostly to the problem of considering risk and aggregation problems. The undoubtedly distinguished achievements in these two areas should be reviewed and consolidated.

3. Unlike typical agricultural policy problems, environmental policy has seldom been modelled using a farm level approach. The few examples available are very promising nevertheless. Especially interesting is work done by Ada Wossink at Wageningen Agricultural University [41]. The models should be refined, however, and a priority is the addition of a risk module.

4. The utilisation results of the models, particularly in environmental policy modelling, has been very limited. In the example, developed for this paper, it has been shown that models can produce a wider range of information than usually utilised. A methodology for a broader application is yet to be developed.

5. For various reasons the environmental aspects of agriculture will undoubtedly become more important in the future. It is very important to have the tools for the assessment of policy measures and a farm approach seems to be necessary for the development of the full picture of environmental policy results. Much has been achieved but there is still a lot to be done.

References

1. Anderson, F. and Stryg, P.E. (1976) Inter-regional Recursive LP Model Used in Forecasting Danish Agricultural Development up to 1985, *Eur. Rev. Agric. Econ.* **3**, 7-21.
2. Berentsen, P.B.R. and Giesen, G.W. J. (1993) A Comparison at Farm Level of Different Governmental Policies to Reduce N-Loses in Diary Farming, Paper presented to the VIIth EAAE Congress, Stresa Italy.
3. Borkowski, B. and Wrzosek, P. (1993) Modyfikacja i zastosowanie metody Charnesa-Coopera do planowania struktury produkcji gospodarstw rolniczych, *Roczniki Nauk Rolniczych, Seria G* **86,** 87-98.
4. Boyd, D.A., Tinker, P.B., Draycott A.P. and Last, P.J. (1970) Nitrogen requirements of Sugar Beets on Mineral Soils, *J. Agric. Sci.* **74**, 37-46.
5. Buckwell, A.E. and Hazell, P.B.R. (1972) Implications of Aggregation Bias for the Construction of Static and Dynamic Linear Programming Supply Models, *J. Agric. Econ.* **23**, 119-134.
6. Chavas, J.P. and Cox, T.L. (1988) A Nonparametric Analysis of Agricultural Technology, *Am. J. Agric. Econ*, **70**, 304-310.
7. Curnow, R.N. (1973) A Smooth Population Response Curve Based on an Abrupt Threshold and Plateau Model for Individuals. *Biometrics* **29**, 1-10.
8. Davey, P.B. and Weightman, P.W.H. (1971) A Microeconomic Approach to the Analysis of Supply Response in British Agriculture, *J. Agric. Econ.* **22**, 297-319.
9. Dijk, G. Van, Hoogervorst, H. Bauwens, N. and Tijdink, T. (1986) A Framework of Agricultural Policy Analysis, *Sociologia Ruralis* **26**, 109-129
10. Francisco, E.M. and Kennedy, J.O.S. (1975) A Note on the Use of Tshebysheff's Inequality in Linear Programming, *J. Agric. Econ.* **26**, 265-267.
11. Gedek, S. (1984) Methods of Considering Risk in Programming Models Used in Agriculture, *Annales UMCS Sect. H* **XVIII**, 365-379.
12. Georgescu-Roegen, N. (1972) Process Analysis and the Neoclassical Theory of Production, *Am. J. Agric. Econ.* **54**, 279-294.
13. Hanoch, G. and Rothschild, M. (1972) Testing the Assumptions of Production Theory: A Nonparametric Approach, *J. Polit. Econ.* **80**, 256-275.
14. Heidhues, T. (1966) A Recursive Programming Model of Farm Growth in Northern Germany, *J. Farm Econ.* **48**, 668-684.
15. Jablonowski, S. Wrzosek, P. (1991) Some Methods of Stochastic Programming and their Application in the Problem Considering Risk in Farm Production Planning, Paper presented to the conference: Organisation eines freien Markets in der Landwirtshft, Warsaw, 12-13 September.
16. Kempinski, T. (1966) *Entry into the European Common Market and British Agricultural Income*, Bulletin 114, University of Manchester, Department of Agricultural Economics.
17. Kennedy, J.O.S. (1975) Using Regression Analysis to Reduce Aggregation Bias in Linear Programming Supply Models, *Aust. J. Agric. Econ.* **19**, 1-11.
18. Kennedy, J.O.S. and Francisco, E.M. (1974) On the Formulation of Risk Constraints for Linear Programming, *J. Agric. Econ.* **25**, 129-145.
19. Marszalkowicz, T. (1983) Uwzglednianie ryzyka w optymalnym planie przedsie-biorstwa, *Zagadnienia Ekonomiki Rolnej* **1/83**, 48-58
20. Miller, T.A. (1974) Economic Adjustment Research for Policy Guidance: An Example from Agriculture, [in:] G. Judge and T. Takayama (eds.) *Studies in Economic Planning over Time and Space*, North Holland, Amsterdam.

21. Monypenny, J.R. and Walker, N. (1976) APMAA '74: a Minimum Level Aggregative Programming Model of New South Wales Agriculture, *Rev. Mktg. Agric. Econ.* **44**, 3-18.
22. Önal, H. and McCarl. B.A. (1989) Aggregation of Heterogeneous Firms in Mathematical Programming Models, *Europ. Rev. Agric. Econ.* **16**, 499-513.
23. Önal, H. and McCarl. B.A. (1991) Exact Aggregation in Mathematical Programming Models, *Cdn. J. Agric. Econ* **39**, 319-334.
24. Orcutt, G., Mertz, J. and Quincke H. (1986) *Microanalytical Simulation Models to Support Social and Financial Policy*, North Holland, Amsterdam.
25. Parton, K.A.(1979) The Use of a Decision Maker's Utility Function in a Linear Programming Analysis of Agricultural Policy, *Aust. J. Agric. Econ.* **23**, 207-219.
26. Pearce, D.W. and Turner, R.K. (1990) *Economics of Natural Resources and the Environment*, Harvest Wheatsheaf, New York etc.
27. Perrin, R.K. (1976) The Value of Information and the Value of Theoretical Models in Crop Response Research, *Am. J. Agric. Econ.* **58**, 55-61.
28. Saganowska, E. (1986) Symulowanie reakcji producentów rolnych na planowane zmiany cen i zaopatrzenia gospodarstw w srodki produkcji, *Zagadnienia Ekonomiki Rolnej* **6/86**, 48-58.
29. Sharpless, J.A. (1969) The Representative Farm Approach to Estimation of Supply Model, *Am. J. Agric. Econ.* **51**, 353-361.
30. Thomson, K.J. and Buckwell, E.A. (1979) A Microeconomic Agricultural Supply Model, *J. Agric. Econ.* **30**, 1-11.
31. Upton, M. (1979) The Unproductive Production Function. *J .Agric. Econ.* **30**, 179-194.
32. Upton, M. and Dalton, G. (1976) Linear Production Response, *J. Agric. Econ.* **27**, 253-256.
33. Varian, H.R. (1984) The Nonparametric Approach to Production Analysis, *Econometrica* **52**, 579-597.
34. Walker, N. and Dillon, J.L. (1976) Development of Aggregative Programming Model of Australian Agriculture, *J. Agric. Econ.* **27**, 243-248.
35. Webster, J.P.G. and Kennedy, J.O.S. (1975) Measuring Farmers Trade-offs between Expected Income and Focus-Loss Income, *Am. J. Agric. Econ.* **57**, 97-105.
36. Weinschenck, G. (1957) Grenzwerttheorie und Kalkulationverfahren in landwirtschaftliechen Betrieb, *Berichte über Landwirschaft* **35**, 801-830.
37. Weinschenck, G. (1961) Die Ermiettlung des Betriebsoptimums mit Hilfe der verbesserten Differenzrechnung, *Agrarwirtschaft* **10**, 65-83.
38. Weinschenck, G. (1964) *Die optimale Organisation des landwirtschaftlichen Betribes*, Verlag Paul Parey, Hamburg.
39. Wicks, J.A., Parton, K.A. and Beesley, P.A. (1977) An Analysis of the Impact on the Australian Agricultural Sector of Changes in Availability and Productivity of Labour, Paper presented to the 6th Conference of Economists, Hobart.
40. Wieckowski, W. (1979) *Optymalizacja planu produkcji przedsiebiorstwa rolniczego przy uzyciu rozwiazan standardowych*, PWN, Warszawa.
41. Wossink, G.A.A. (1993) *Analysis of Future Agricultural Change: a Farm Economics Approach Applied to Dutch Arable Farming*, Agricultural University Press, Wageningen.
42. Wossink, G.A.A., de Koeijer, T.J. and Renkema J.A. (1992) Environmental-Economic Policy Assessment: A Farm Economic Approach, *Agricultural Systems*

PART VI

CONCLUSIONS AND RECOMMENDATIONS

The Workshop was concluded by a plenary session in which participants discussed and agreed the conclusions that had emerged from the information presented in the formal sessions and, in addition, formulated recommendations for future action. These are listed as follows:

(1) Soil quality encompasses the capability of the soil to produce safe and nutritious crops in a sustained way over the long term without adverse environmental effects. It is best understood in the context of soil function and may be viewed as the sum of productivity and resilience where the latter is defined as the ability of the soil, through the operation of natural soil processes, to recover after disturbance caused by external and/or internal stresses.

(2) Sustainable agricultural development may be viewed as a long-term goal encompassing satisfactory production of food and fibre in an environmentally sound, economically viable and socially acceptable way.

(3) Environmental security incorporates the notion that the components of the environment, including agroecosystems and the soil as part of these ecosystems, are not threatened with degradation or loss of functionality in the short-term. Healthy ecosystems tend to be predictable, flexible and with a high degree of internal control.

(4) In terms of the proportion of agricultural soil affected, the most serious threat to soil quality in the CEECs that was identified in the Workshop, concerned degradation by water/wind erosion and acidification. Both threats are related to declines in soil organic matter and nutrient status, trends that have been exacerbated since 1989 by the dramatic decrease in inputs of fertilisers, lime and other agrochemicals to agricultural soils.

(5) The Workshop concluded that the contamination of agricultural soils in the CEECs by heavy metals and organic pollutants generally affects much smaller areas than are affected by erosion and acidification. The extent of this contamination varies from country to country and is usually confined to so-called "hot spots". Thus, serious heavy metal contamination (meaning that such soils should be excluded from crop production) accounts for <1% of Polish agricultural soils. In several CEECs (Poland, Czech Republic, Slovakia) the level of organic contamination of soils in rural areas is comparable to that of the background levels found in Western European countries, but hot spots may occur near emission sources. Radionuclide contamination of agricultural soil is more extensive in Ukraine and in Russia affects ~4% of agricultural soils.

M.J. Wilson and B. Maliszewska-Kordybach (eds.),
Soil Quality, Sustainable Agriculture and Environmental Security in Central and Eastern Europe, 347–349.

(6) Maps showing the sensitivity of soils to various threats have been compiled in many CEECs but it is recommended that the information should be presented in a more refined way and where necessary on a risk assessment basis.

(7) Many CEECs hold abundant data relating to soil quality derived from extensive networks of monitoring sites. However, in order to clarify ongoing changes in soil quality that may be affecting the region as a whole, it is recommended that such information should be brought together in a more consistent way. The possibility of creating a harmonized monitoring network in line with those existing in Western Europe should be considered.

(8) Methods of managing agricultural soils in the CEECs, as well as helping to prevent soil quality decline, should be applicable over wide areas. For this reason there appears little alternative but to employ traditional agricultural practices such as fertilisation, liming, crop rotation, use of cover crops, etc, in order to build up soil nutrient status and carbon content, as well as to improve soil physical characteristics.

(9) Other more novel practices such as the application of organic wastes and sludges to agricultural soils may be useful in relatively restricted areas as a way of improving soil quality or even of remediating contaminated soils.

(10) Conservation tillage of agricultural soils is not widely used in some of the CEECs because of yield reductions and possible adverse effects on soil physical properties relating to increases in soil density. In areas that are very susceptible to soil erosion, however, reduced tillage is useful or even essential and, for example, is widely practiced in the Ukraine.

(11) There may be scope for increased use of organic farming systems in the CEECs, particularly as farmers in these countries are now accustomed to working in the context of minimum inputs. However, the economic viability of organic farming systems in the CEECs is doubtful under current economic circumstances.

(12) It is not thought likely that precision farming will be widely taken up in the CEECs at the present time although may change in the future,

(13) It was concluded that there was an urgent need for wider education and training of agricultural advisers, specialists, researchers and farmers in the CEECs so as to help in the development of appropriate sustainable agricultural systems. One way of doing this is through the establishment of an International Centre for Agricultural Training and Education aimed specifically at the CEECs and it is recommended that detailed proposals should be developed and funding possibilities sought for such a project.

(14) Since 1989, inputs of fertilisers, lime and other agrochemicals to agricultural soils have been reduced to very low levels in all CEECs, as a consequence of the withdrawal of state subsidies and the shift to private ownership of land. Many farmers have had to "mine" the natural fertility of the soil, thus leading to a decline in its quality. It is recommended that policies should be developed in the CEECs that would help to arrest this decline and to promote sustainable agricultural development.

(15) For those CEECs that have applied to join the EU, the measures proposed for the reform of the CAP in Agenda 2000 are of particular importance. Although there are some proposals in Agenda 2000 that reinforce the importance of agri-environmental measures, in general the proposals do not take into account the development of sustainable agriculture in the CEE applicant countries. It is recommended as a matter of urgency that these countries develop and promote environmental or sustainable components for pre-accessionary modernisation programmes.

(16) It is concluded that in order for such proposals to be convincing and acceptable to policy makers, the issues need to be clearly identified and argued from a European perspective. In addition, the options involved in addressing these issues need to be analysed in a quantitative way, emphasising the costs and benefits in sustaining natural capital stock. It is recommended that this approach should be developed through collaboration between economists and agricultural/soil scientists.

(17) It was concluded that the protection, maintenance and enhancement of soil quality and the development of sustainable agriculture in the CEECs was unlikely to be promoted in the context of liberal agricultural policy and market incentives alone. Ultimately, environmental/agricultural sustainability may require some degree of regulation and subsidisation.

LIST OF WORKSHOP PARTICIPANTS

ALCOCK Ruth
Institute of Environmental Sciences,
Lancaster University,
Lancaster, LA1 4YQ,
UK.

ANDRONIKOV Sergei
V. V. Dokuchaev Soil Institute,
Pijevsky per. 7,
109017 Moscow,
RUSSIA

BACHINGER Johann
Centre for Agricultural Landscape and Land Use Research (ZALF) e. V.
Institute for Land Use Systems and Landscape Ecology,
Ebelswalder Str. 84,
D-15374 Müncheberg,
GERMANY.

BALL Bruce C.
Scottish Agricultural College (SAC),
Soils Department,
West Mains Road,
Edinburgh, EH9 3JG,
UK.

BONDAREV A.G.
V.V.Dokuchaev Soil Institute,
Pijevsky per. 7,
109017 Moscow,
RUSSIA

BURCZYK Piotr
Institute for Land Reclamation and Grassland Farming,
ul. Czeslawa 9,
71-504 Szczecin,
POLAND

CHANEY Rufus
US Department of Agriculture,
Agricultural Research Service,
Environmental Quality Laboratory,
BARC-WEST bldg b007,
Beltsville,
MD 20705,
USA

CLAY Sharon
South Dakota State University,
Plant Science Department,
Brookings,
SD 57007,
USA.

CURLÍK Jan
Soil Fertility Research Institute,
Gagarinova 10,
827 487 Bratislava,
SLOVAKIA

DALTON Graham
ADF Project "Integrating Agricultural Policies of Poland and EU",
Ministry of Agriculture and Food Economy,
ul. Wspólna 30,
00-930 Warszawa,
POLAND

DAVIS Craig
Ohio State University,
School of Natural Resources,
210 Kottman Hall, 2021 Coffey Road,
Columbus,
Ohio 43 210
USA

DEXTER Anthony
Institute of Soil Science and Plant Cultivation (IUNG),
ul. Czartoryskich 8,
24-100 Pulawy
POLAND

DUER Irena
Institute of Soil Science and Plant Cultivation (IUNG),
ul. Czartoryskich 8,
24-100 Pulawy,
POLAND

FOTYMA Mariusz
Institute of Soil Science and Plant Cultivation (IUNG),
ul. Czartoryskich 8,
Pulawy 24- 100,
POLAND

GEDEK Stanislaw
University of Agriculture, Lublin,
Institute of Economy and Business Organization,
ul. Akademicka, 13,
20-934, Lublin,
POLAND

HANEKLAUS Silvia
Institute of Plant Nutrition and Soil Science,
Federal Agricultural Research Centre,
Bundesallee 50,
D-38 116 Braunschweig,
GERMANY

HELIOS-RYBICKA Edeltrauda
University of Mining and Metallurgy,
Faculty of Geology, Geophysics and Environmental Protection,
al. Mickiewicza 30,
30 059 Kraków
POLAND

HOLOUBEK Ivan
RECETOX-TOCOEN & Assoc.
Masaryk University.
Kotlarska 2,
611 37 Brno
CZECH REPUBLIC

JANKAUSKAS Benediktas
Kaltinenai Research Station of the Agricultural Institute of Lithuania,
Silale District,
LT-5926,
LITHUANIA

KAPSHTYK Mikhailo V.
National Agricultural University,
Department of Soil Science and Soil Protection,
Geroyiv Oborony 15,
252 041 Kiev,
UKRAINE

KICINSKA-SWIDERSKA Alicja
University of Mining and Metallurgy,
Faculty of Geology, Geophysica and Environmental Protection,
Al. Mickiewicza 30,
30-059 Kraków,
POLAND

KRASOWICZ Stanislaw
Institue of Soil Science and Plant Cultivation (IUNG),
ul. Czartoryskich 8,
24 100 Pulawy,
POLAND

LAZARZ Maja
Institute of Soil Science and Environment Management,
University of Agriculture at Lublin,
ul. Leszczynskiego 7,
20-069 Lublin,
POLAND

MALISZEWSKA-KORDYBACH Barbara
Institute of Soil Science and Plant Cultivation (IUNG),
ul. Czartoryskich 8,
24-100 Pulawy,
POLAND

MURÁNYI Attila
Research Institute of Soil Sciences and Agricultural Chemistry
of the Hungarian Academy of Sciences,
Herman Ottó 15,
Budapest H-1022,
HUNGARY

NÉMETH Tamás
Research Institute of Soil Sciences and Agricultural Chemistry
of the Hungarian Academy of Sciences,
Herman Ottó 15,
Budapest H-1022,
HUNGARY

PATERSON Edward
Macaulay Land Use Research Institute,
Craigiebuckler,
Aberdeen AB15 8QH,
UK

PEDERSEN Carl Aage
The Danish Agricultural Advisory Centre,
National Department of Plant Production,
Udkaersvej 15,
DK-8200 Aarhus,
DENMARK

PODLESAKOVÁ Eliska
Research Institute for Soil and Water Conservation,
Zabovreska 250,
156 27 Prague 5,
CZECH REPUBLIC

PRANAGAL Jacek
University of Agriculture at Lublin,
Institute of Soil Science and Environment Management,
ul. Leszczynskiego 7,
20-069 Lublin,
POLAND

SHIKULA M. K.
National Agricultural University,
Geroyiv Oborony St. 15,
252041 Kiev,
UKRAINE

STAPUTIS Juozas
Kaltinenai Research Station,
Lithuanian Institute of Agriculture,
Silale District,
LT-5926,
LITHUANIA

STUBBLES Russell
College of Agriculture and Biological Sciences,
Horticulture, Forestry, Landscape and Parks Department,
South Dakota State Unuversity,
Box 2140A NPB 254C,
Brookings,
SD 57007-4730,
USA

STUCZYNSKI Tomasz
Institute of Soil Science and Plant Cultivation (IUNG),
ul. Czartoryskich 8,
24-100 Pulawy,
POLAND

TERELAK Henryk
Institute of Soil Science and Plant Cultivation (IUNG),
ul. Czartoryskich 8,
24-100 Pulawy,
POLAND

VÁRALLYAY George
Research Institute of Soil Sciences and Agricultural Chemistry
of the Hungarian Academy of Sciences,
Herman Ottó 15,
Budapest H-1022,
HUNGARY

WILSON M. Jeffrey
Macaulay Land Use Research Institute,
Craigiebuckler,
Aberdeen AB15 8QH
UK

WOS Augustyn
Institute of Agricultural and Food Economy,
ul. Swietokrzyska 20,
00-0002 Warszawa
POLAND

ZDANOWICZ Agata
ADF Project "Integrating Agricultural Policies of Poland and EU",
Ministry of Agriculturs and Food Economy,
ul. Wspólna 30,
00-930 Warszawa,
POLAND

SUBJECT INDEX

COLOURED PLATES

Fig. 1. STATE OF SOIL POLLUTION WITH HEAVY METALS

Figure 1 from Terelak and Motowicka-Terelak, p. 43.

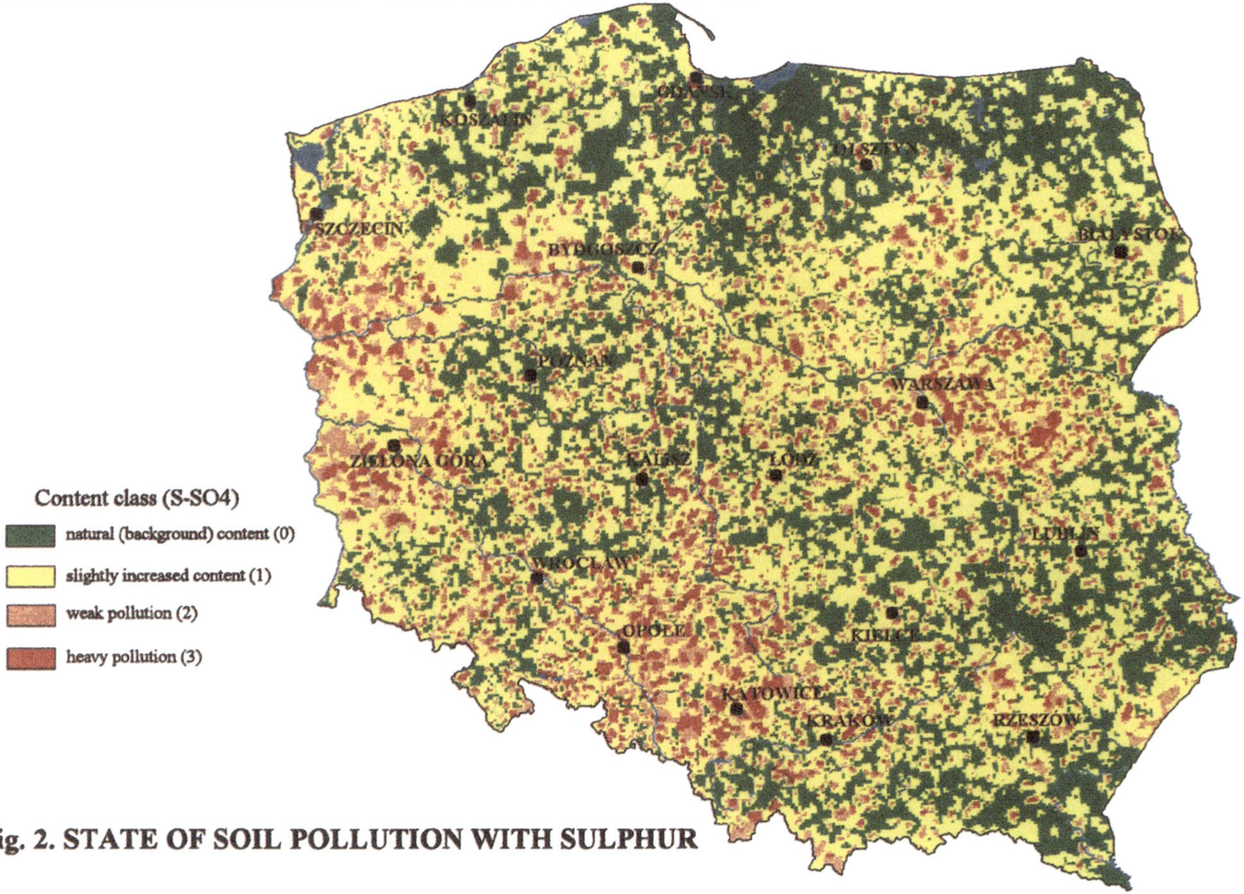

Fig. 2. STATE OF SOIL POLLUTION WITH SULPHUR

Figure 2 from Terelak and Motowicka-Terelak, p. 45.

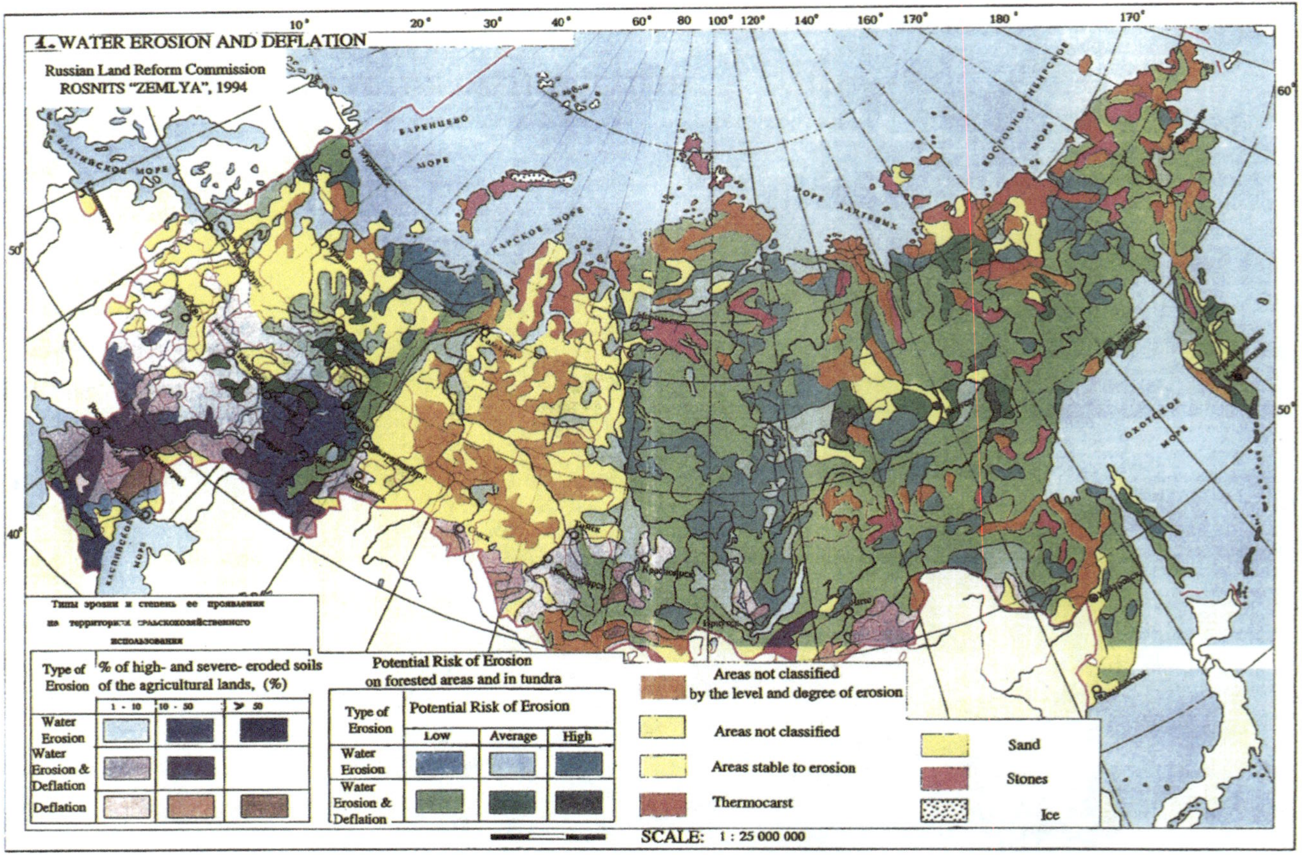

Figure 1 from Andronikov, p. 89.

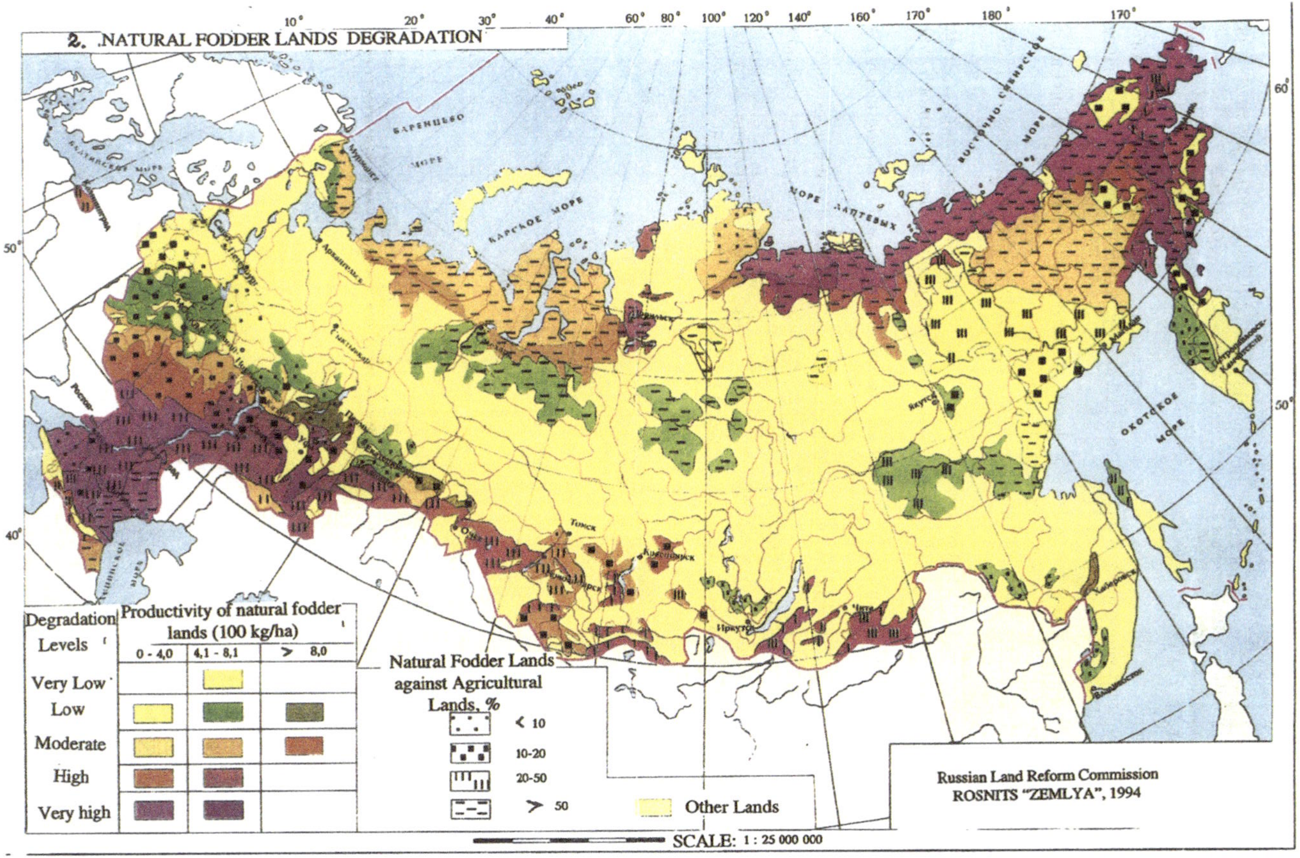

Figure 2 from Andronikov, p. 91.

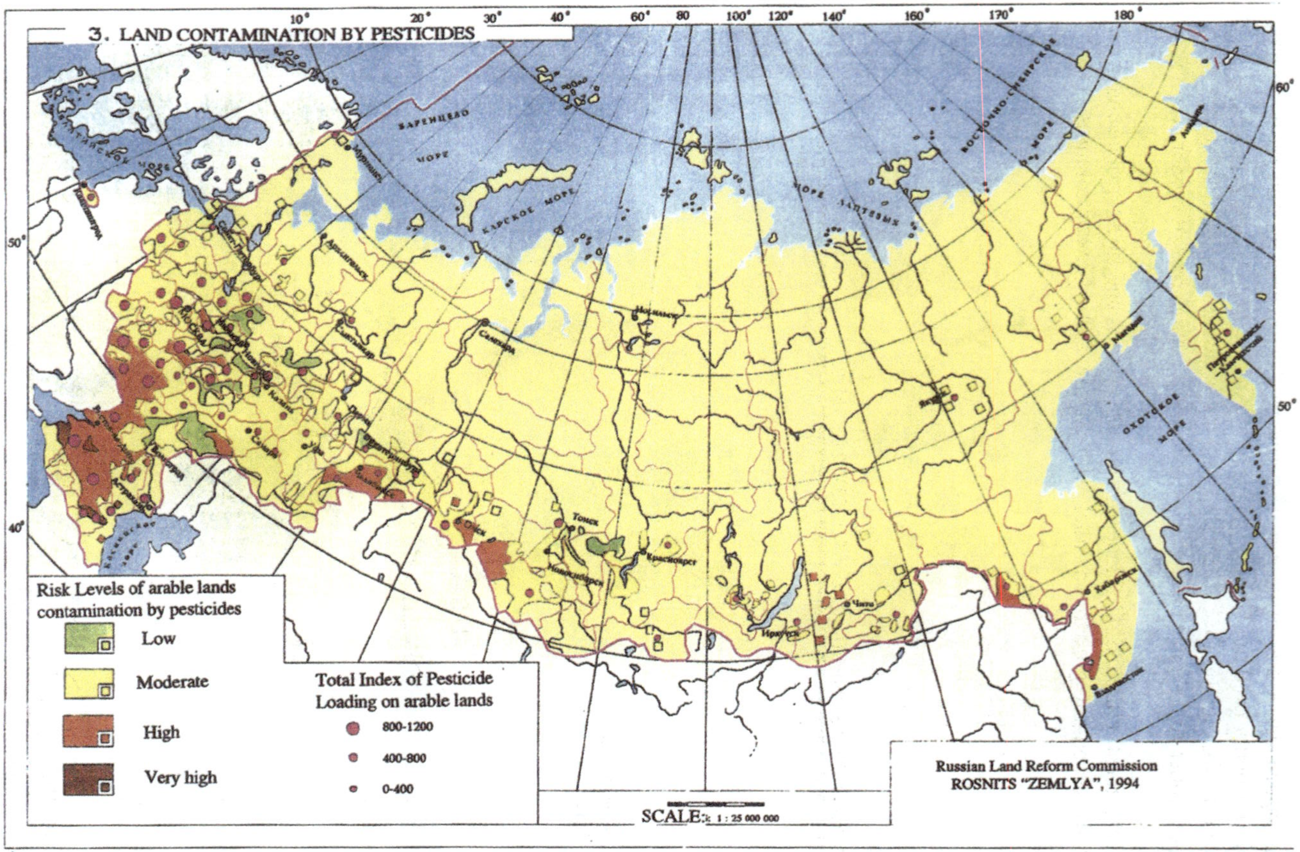

Figure 3 from Andronikov, p. 92.

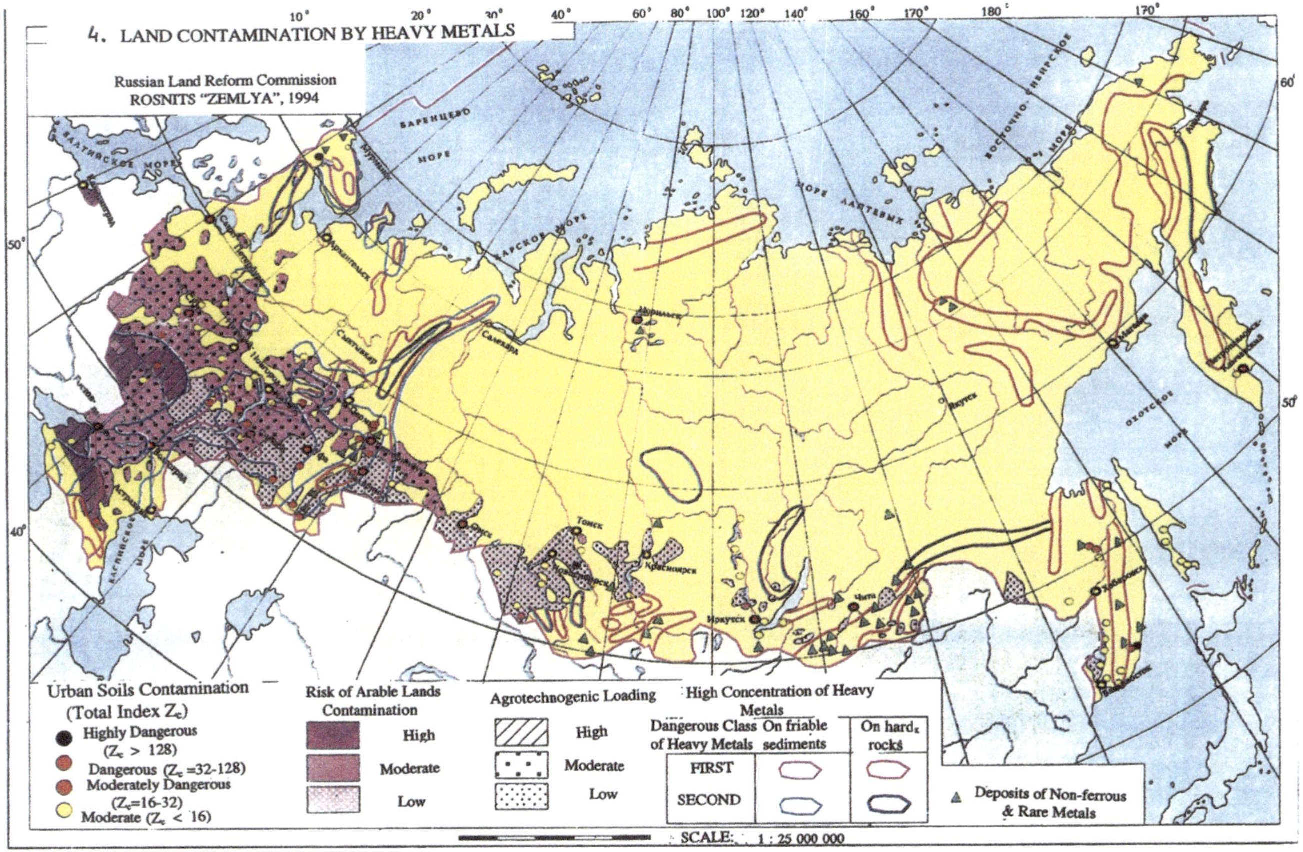

Figure 4 from Andronikov, p. 93.

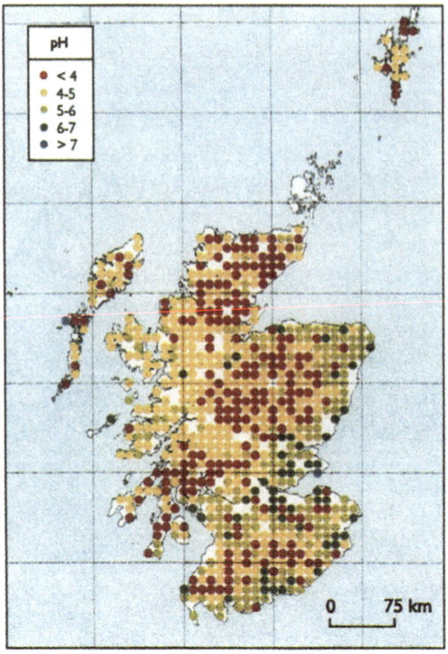

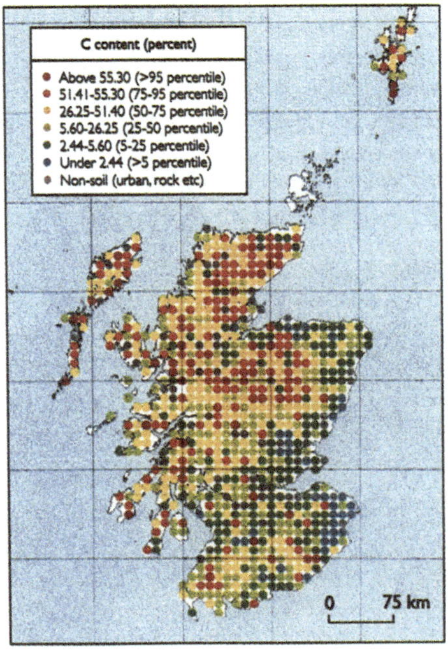

Figure 1 from Paterson et al., p. 194.

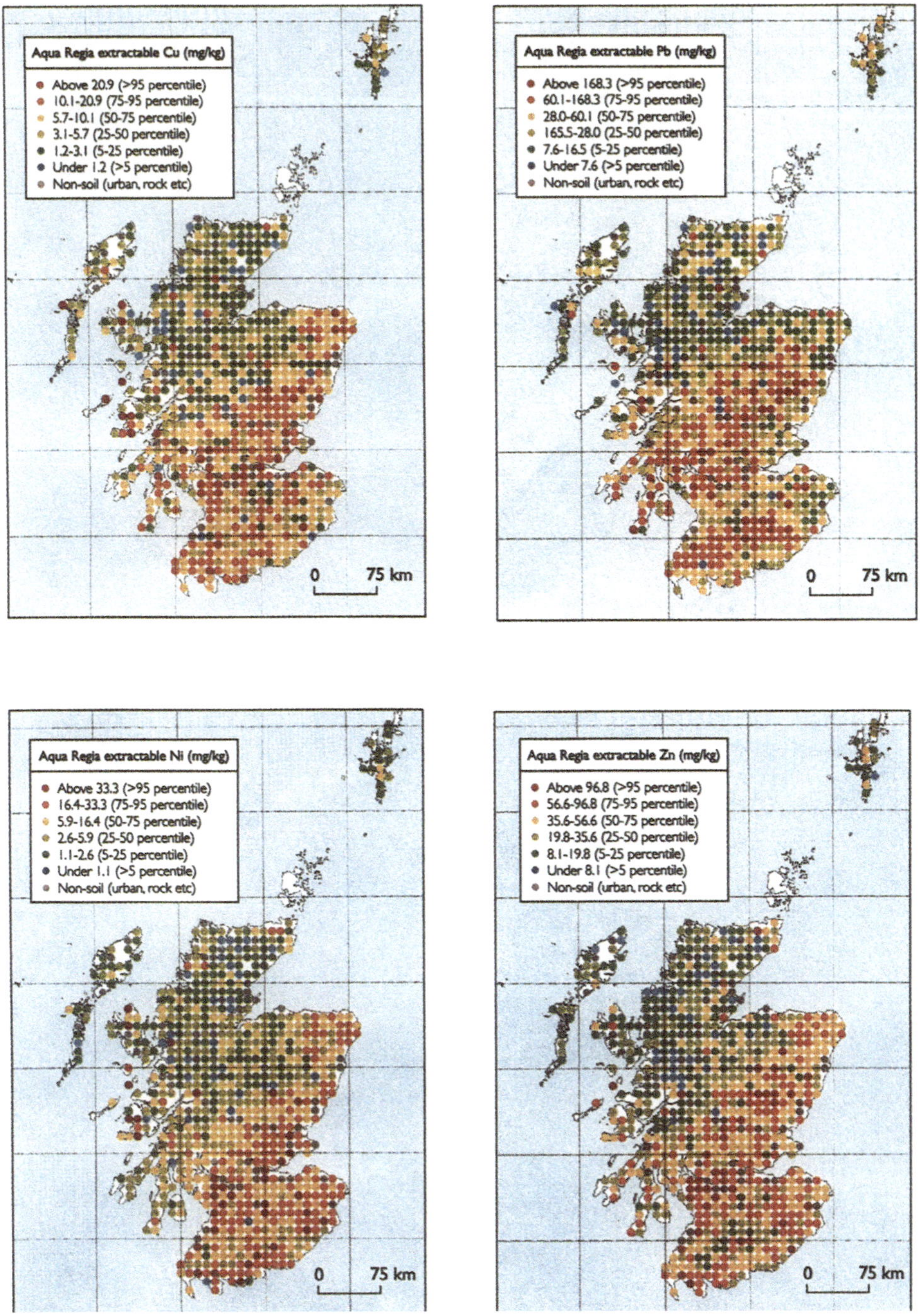

Figure 3 from Paterson et al., p. 199.

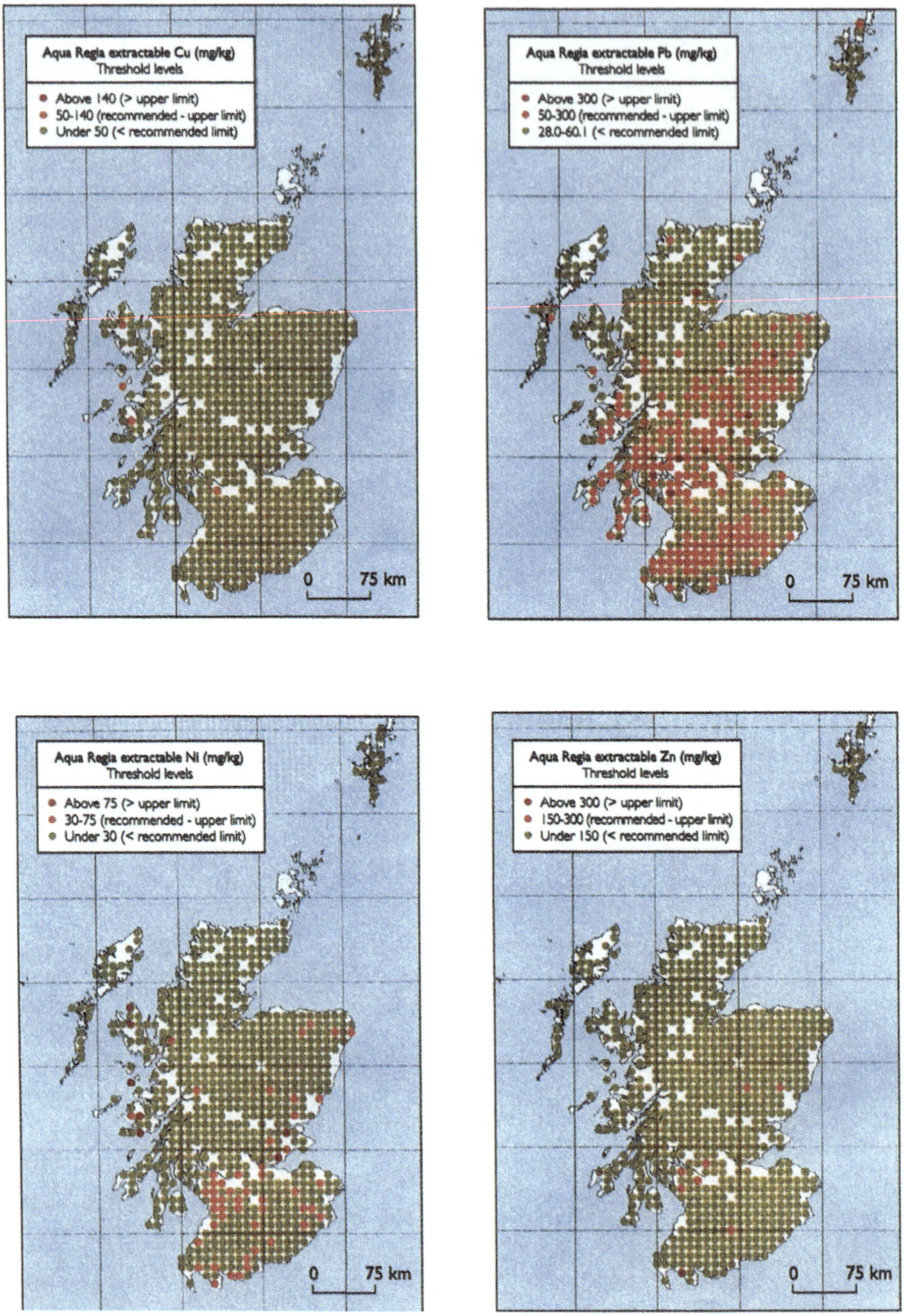

Figure 4 from Paterson et al., p. 200.

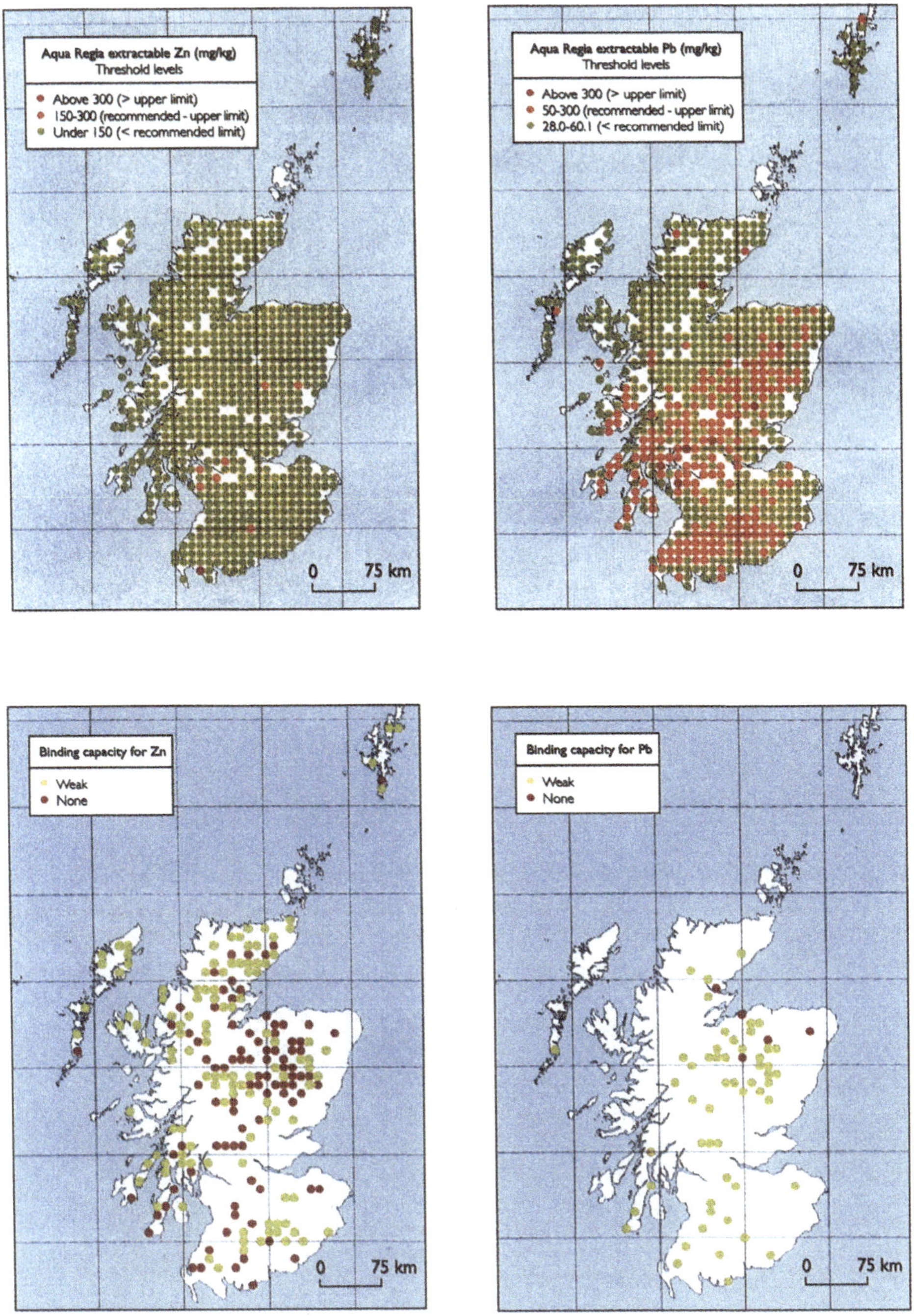

Figure 5 from Paterson et al., p. 203.

GPSR Compliance

The European Union's (EU) General Product Safety Regulation (GPSR) is a set of rules that requires consumer products to be safe and our obligations to ensure this.

If you have any concerns about our products, you can contact us on ProductSafety@springernature.com

In case Publisher is established outside the EU, the EU authorized representative is:

Springer Nature Customer Service Center GmbH
Europaplatz 3
69115 Heidelberg, Germany

Batch number: 10816147

Printed by Printforce, the Netherlands